Achieving Sustainability in Construction

Proceedings of the International Conference
held at the University of Dundee, Scotland, UK
on 5-6 July 2005

Edited by

Ravindra K. Dhir
Director, Concrete Technology Unit
University of Dundee

Tom D. Dyer
Lecturer, Concrete Technology Unit
University of Dundee

and

Moray D. Newlands
CPD/Consultancy Manager, Concrete Technology Unit
University of Dundee

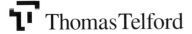 ThomasTelford

Published by Thomas Telford Publishing, Thomas Telford Ltd, 1 Heron Quay, London E14 4JD.
www.thomastelford.com

Distributors for Thomas Telford books are
USA: ASCE Press, 1801 Alexander Bell Drive, Reston, VA 20191-4400, USA
Japan: Maruzen Co. Ltd, Book Department, 3–10 Nihonbashi 2-chome, Chuo-ku, Tokyo 103
Australia: DA Books and Journals, 648 Whitehorse Road, Mitcham 3132, Victoria

First published 2005

The full list of titles from the 2005 International Congress 'Global construction: ultimate concrete opportunities' and available from Thomas Telford is as follows

- Cement combinations for durable concrete. ISBN: 0 7277 3401 6
- Concrete for transportation infrastructure. ISBN: 0 7277 3402 4
- Application of codes, design and regulations. ISBN: 0 7277 3403 2
- Achieving sustainability in construction. ISBN: 0 7277 3404 0
- Repair and renovation of concrete structures. ISBN: 0 7277 3405 9
- Use of foamed concrete in construction. ISBN: 0 7277 3406 7
- Admixtures – enhancing concrete performance. ISBN: 0 7277 3407 5
- Applications of nanotechnology in concrete design. ISBN: 0 7277 3408 3
- Role of concrete in nuclear facilities. ISBN: 0 7277 3409 1
- Young researchers' forum. ISBN: 0 7277 3410 5

For the complete set of 10 volumes, ISBN: 0 7277 3387 7

A catalogue record for this book is available from the British Library

ISBN: 0 7277 3404 0

Printed and bound in Great Britain by MPG Books, Bodmin, Cornwall

PREFACE

Concrete is at something of a crossroads: there are many opportunities and some threats. For these opportunities to transpose into beneficial practice, engineers, material scientists, architects, manufacturers and suppliers must focus on the changes that are required to champion concrete and maintain its dominance within the global construction industry.

The Concrete Technology Unit (CTU) of the University of Dundee organised this Congress to address these changes, continuing its established series of events, namely, Challenges of Concrete Construction in 2002, Creating with Concrete in 1999, Concrete in the Service of Mankind in 1996, Economic and Durable Concrete Construction Through Excellence in 1993 and Protection of Concrete in 1990.

The event was organised in collaboration with three of the world's most recognised institutions: the Institution of Civil Engineers, the American Concrete Institute and the Japan Society of Civil Engineers. Under the theme of Global Construction: Ultimate Concrete Opportunities, the Congress consisted of ten Events: *(i) Cement Combinations for Durable Concrete, (ii) Concrete for Transportation Infrastructure, (iii) Application of Codes, Design and Regulations, (iv) Achieving Sustainability in Construction, (v) Repair and Renovation of Concrete Structures, (vi) Use of Foamed Concrete in Construction, (vii) Admixtures – Enhancing Concrete Performance, (viii) Applications of Nanotechnology in Concrete Design, (ix) Role of Concrete in Nuclear Facilities, (x) Young Researchers Forum.* In all, a total of 420 papers were presented from 68 countries.

The Opening Addresses were given by Sir Alan Langlands, Principal and Vice-Chancellor of the University of Dundee, Mr John Letford, Lord Provost, City of Dundee, Mr Colin Clinton, President of the Institution of Civil Engineers, and senior representatives of the Japan Society of Civil Engineers and the American Concrete Institute. The Congress was officially opened by Mr Peter Goring, President of the Concrete Society, UK. The ten Event Opening Papers were presented by Professor R K Dhir, University of Dundee, UK, Mr E S Larsen, COWI A/S, Denmark, Dr W G Corley, Construction Technologies Laboratory Group, USA, Dr M Glavind, Danish Technological Institute, Denmark, Professor P Robery, Halcrow Group Ltd, UK, Mr D Aldridge, Propump Engineering Ltd, UK, Dr M Corradi, Degussa Construction Chemicals Division, Italy, Professor S P Shah, Northwestern University, USA, Professor C Andrade, Institute of Construction Science, Spain and Professor P C Hewlett, British Board of Agrement, UK. The Closing Papers were given by Professor K Scrivener, Swiss Federal Institute of Technology (EPFL), Switzerland, Mr N Hussain, Ove Arup & Partners, Hong Kong, Professor H Gulvanessian, Building Research Establishment, UK, Mrs G M T Janssen, Delft University of Technology, Netherlands, Dr D R Morgan, AMEC Earth & Environmental, Canada, Dr E Kearsley, University of Pretoria, South Africa, Professor M Collepardi, ENCO-Engineering Concrete, Italy and Mr P Doyle, Jacobs Babtie Nuclear, UK.

The support of 50 International Professional Institutions and 32 Sponsoring Organisations was a major contribution to the success of the Congress. An extensive Trade Fair formed an integral part of the event. The work of the Congress was an immense undertaking and all of those involved are gratefully acknowledged, in particular, the members of the Organising Committee for managing the event from start to finish; members of the Scientific and Technical Committees for advising on the selection and reviewing of papers; the Authors and the Chairmen of Technical Sessions for their invaluable contributions to the proceedings.

All of the proceedings have been prepared directly from the camera-ready manuscripts submitted by the authors and editing has been restricted to minor changes, only where it was considered absolutely necessary.

Dundee
July 2005

Ravindra K Dhir
Chairman, Congress Organising Committee

INTRODUCTION

When attempting to understand the environmental impact of a product, it is important to include its entire life-cycle, from the extraction of raw materials until disposal at the end of its useful life. This event attempts to gather together research from around the world which examines elements of the life-cycle of concrete which can impact the environment, in both positive and negative terms. Moreover, its intention is to explore ways in which negative impacts can be reduced and positive impacts increased.

The environmental impact of an activity is commonly measured using methods such as environmental life-cycle analysis, which requires quantification of inputs and outputs. The first two Themes of this event are dedicated entirely to these inputs and outputs. Inputs take the form of energy, raw materials and water. Outputs are pollution, waste, and the structure itself. The second Theme also addresses the issue of what can be done to divert waste away from disposal and towards re-use and recycling, processes which, in turn, reduce the pressures on primary mineral resources.

The ability to quantify impact is useful, but it is pointless unless action is then taken to modify activities to realise environmental benefits and improved sustainability. Thus, Theme Three examines both ways of measuring environmental impact, as well as actions which can be taken.

Cement and concrete possess great potential to improve the quality of the environment in a variety of different ways. These include the use of concrete in a wide range of structures used in environmental engineering. The properties of cement also make it an ideal material for the stabilisation of soil, and often provides the added benefit of aiding in the immobilisation of contaminants. Indeed, cement and related materials are, in many respects, highly suited for the containment of harmful substances, since they can provide both physical encapsulation and a chemical environment that can reduce the mobility of many chemical species. Recently, the use of concrete surfaces as a means of remediating pollution has become a reality, with further advances likely in the near future.

The Proceedings *'Achieving Sustainability in Construction'* dealt with all these subject areas and the issues raised four clearly defined themes: (i) Efficient Use of Energy and Raw Materials, (ii) Pollution, Waste and Recycling, (iii) Minimising Environmental Impact, (iv) Environmental Engineering with Cement and Concrete. Each theme started with a Keynote Paper presented by the foremost exponents in their respective fields. There were a total of 49 papers presented during the International Conference which are compiled into these Proceedings.

Dundee
July 2005

Ravindra K Dhir
Thomas D Dyer
Moray D Newlands

ORGANISING COMMITTEE

Concrete Technology Unit

Professor R K Dhir OBE (Chairman)

Dr M D Newlands (Secretary)

Professor P C Hewlett
British Board of Agrément

Professor T A Harrison
Quarry Products Association

Professor P Chana
British Cement Association

Professor V K Rigopoulou
National Technical University of Athens, Greece

Dr S Y N Chan
Hong Kong Polytechnic University

Dr N Y Ho
L & M Structural Systems, Singapore

Dr M R Jones

Dr M J McCarthy

Dr T D Dyer

Dr K A Paine

Dr J E Halliday

Dr L J Csetenyi

Dr L Zheng

Dr S Caliskan

Dr A McCarthy

Dr A Whyte

Mr M C Tang

Ms E Csetenyi

Ms P I Hynes (Congress Assistant)

Mr S R Scott (Unit Assistant)

SCIENTIFIC AND TECHNICAL COMMITTEE

Dr J Beaudoin, *Principal Research Officer & Group Leader*
National Research Council, Canada

Dr Jean-Marie Chandelle, *Chief Executive*
CEMBUREAU - The European Cement Association, Belgium

Dr Mario Corradi, *Senior Vice President Technology & Development*
Degussa Construction Chemicals Division, Italy

Mr Peter Goring, *Technical Director*
John Doyle Construction, UK

Professor T A Harrison, *BRMCA Consultant*
Quarry Products Association, UK

Professor Charles F Hendriks, *Professor*
Delft University of Technology, Netherlands

Mr Raymund Johnstone, *Principal Engineer, SE Bridges Section*
Scottish Executive, UK

Professor Vasilia Kasselouri-Rigopoulou, *Professor*
National Technical University of Athens, Greece

Professor Roger J Kettle, *Subject Group Convenor*
Aston University, UK

Professor Johann Kollegger, *Professor, Institute for Structural Engineering*
Vienna University of Technology, Austria

Mr Erik K Lauritzen, *Managing Director*
NIRAS DEMEX Consulting Engineers A/S, Denmark

Professor Tarun Naik, *Professor of Structural Eng & Academic Program Director*
University of Wisconsin - Milwaukee, USA

Professor Ioanna Papayianni, *Head of Laboratory, Building Materials*
Aristotle University Thessaloniki, Greece

Professor Jean Pera, *Professor of Civil Engineering*
INSA Lyon (National Institute of Applied Sciences), France

Professor Subba A Reddi, *Deputy Managing Director*
Gammon India Ltd, India

Dr Vlastimil Sruma, *Managing Director*
Czech Concrete Society, Czech Republic

Dr Tongbo Sui, *Director, Research Institute of Cement & New Bldg Materials*
China Building Materials Academy, China

Professor Kyosti Tuutti, *Research Director*
Skanska International Civil Engineering AB, Sweden

Professor Thomas Vogel, *Professor*
Swiss Federal Institute of Technology, Switzerland

Dr Roger P West, *Head of Department*
University of Dublin, Ireland

COLLABORATING INSTITUTIONS

Institution of Civil Engineers, UK

American Concrete Institute

Japan Society of Civil Engineers

SPONSORING ORGANISATIONS WITH EXHIBITION

Aalborg Portland A/S, Denmark

Aggregate Industries

ARUP

Bid Cities Fund

British Board of Agrément

British Cement Association

Building Research Establishment

Bureau Veritas Laboratories Ltd

Castle Cement Limited

CEMBUREAU

Cementitious Slag Makers Association

Danish Technological Institute

Degussa - Construction Chemicals

Dundee City Council

Elkem Materials Ltd

FaberMaunsell

FEBELCEM

Foam Concrete Ltd

Halcrow Group Ltd

Heidelberg Cement

Jacobs Babtie Group

John Doyle Construction

Master Builders Technologies

SPONSORING ORGANISATIONS WITH EXHIBITION
(continued)

PANalytical

Propump Engineering Ltd

Putzmeister Ltd

RMC Readymix

Rugby Cement

STATS Ltd

Tarmac Group

The Concrete Centre

United Kingdom Quality Ash Association

EXHIBITING ORGANISATIONS

Cambridge Ultrasonics Ltd

Celsum Technologies Ltd

CNS Farnell Ltd

Concrete Repairs Ltd

Germann Instruments A/S

Glenammer Engineering Ltd

John Wiley & Sons Ltd

Metrohm UK Ltd

Retsch UK Ltd

Sonatest Ltd

Wexham Developments

Zwick Testing Machines Ltd

SUPPORTING INSTITUTIONS

Asociacion de Ingenieros de Caminos, Canales y Puertos, Spain

Asociacion de Ingenieros del Uruguay, Uruguay

Association of Slovak Scientific & Technological Societies, Slovakia

Associazione Italiana Ingegneria dei Materiali, Italy

Austrian Society of Engineers & Architects, Austria

Bahrain Society of Engineers, Bahrain

Belgian Concrete Society, Belgium

Brazilian Concrete Institute, Brazil

Canadian Society for Civil Engineering, Canada

China Civil Engineering Society, China

Chinese Institute of Engineers, Taiwan

Colegio de Ingenieros y Agrimesores de Puerto Rico, Puerto Rico

Concrete Institute of Australia, Australia

Concrete Society of Southern Africa, South Africa

Consiglio Nazionale degli Ingegneri, Italy

Construction Institute, USA

Czech Concrete Society, Czech Republic

Danish Concrete Association, Denmark

Deutscher Beton-und Bautechnik-Verein EV, Germany

Engineers Australia, Australia

Federation of Scientific & Technical Unions in Bulgaria, Bulgaria

Feberation de l'Industrie du Beton (FIB), France

General Association of Engineers in Romania, Romania

Hong Kong Institution of Engineers, Hong Kong

SUPPORTING INSTITUTIONS

Hungarian Cement Association, Hungary

Indian Concrete Institute, India

Institute of Concrete Technology, UK

Institution of Engineers, India

Institution of Engineers, Bangladesh

Institution of Engineers, Malaysia

Institution of Engineers, Sri Lanka

Institution of Engineers, Tanzania

Institution of Structural Engineers, UK

Instituto Mexicano del Cemento y del Concreto AC, Mexico

Irish Concrete Society, Ireland

Japan Concrete Institute, Japan

Jordan Engineers' Association, Jordan

Korea Concrete Institute, South Korea

Netherlands Concrete Society, Netherlands

New Zealand Concrete Society, New Zealand

Nigerian Society of Engineers, Nigeria

Norwegian Concrete Association, Norway

Singapore Concrete Institute, Singapore

Slovenian Chamber of Engineers, Slovenia

The Concrete Society, UK

Turkish Chamber of Civil Engineers, Turkey

Yugoslav Society for Materials & Structures Testing, Serbia & Montenegro

Zimbabwe Institution of Engineers, Zimbabwe

CONTENTS

THEME 2 POLLUTION, WASTE AND RECYCLING
Keynote Paper

THEME 3 MINIMISING ENVIRONMENTAL IMPACT
Keynote Paper

THEME 4 ENVIRONMENTAL ENGINEERING WITH CEMENT AND CONCRETE
Keynote Paper

OPENING PAPER

SUSTAINABLE CONCRETE STRUCTURES
A WIN-WIN SITUATION FOR INDUSTRY AND SOCIETY

M Glavind

D Mathiesen   C V Nielsen
Danish Technological Institute
Denmark

ABSTRACT. Development of sustainable concrete structures has been going on for a number of years. A lot of different tools have been developed in order to reduce the environmental impact of concrete and concrete structures and to promote the production of "green concrete". These tools and the technologies behind them vary considerably across Europe due to regional /national differences in legislation, market conditions and traditions in the construction industry. The present article gives an overview of the environmental impact from concrete structures and of the current practice and R&D activities related to sustainability issues. This overview points out the most significant environmental impacts in the life cycle of concrete structures and gives examples of life cycle assessments. Well known ways of improving the environmental performance of concrete are presented such as the use of supplementary cementitious materials, optimisation of concrete mix design, recycling of waste products, improvement of the working environment etc. Furthermore, an overview is shown of recent and on-going activities in Europe related to sustainable concrete structures. The conclusion is that there is a need to continue the work because of the challenges the industry is facing, and an idea of future trends in this development is given.

Keywords: Sustainable concrete structures, Recycling, Supplementary cementitious materials, Future challenges, EU directive, Self-compacting concrete, Blended cements, Communication.

Dr M Glavind, is Manager of the Concrete Centre, Danish Technological Institute

D Mathiesen, is a Researcher and Consultant at the Concrete Centre, Danish Technological Institute

Dr C V Nielsen, is a Researcher and Consultant at the Concrete Centre, Danish Technological Institute

INTRODUCTION

During the last century concrete has developed into the most important building material in the world. This is partly due to the fact that concrete is produced from natural materials available in all parts of the globe, and partly due to the fact that concrete is a versatile material giving architectural freedom.

The production of concrete annually amounts to 1.5-3 tonnes per capita in the industrialised world. This makes the concrete industry including all of its suppliers a major player in the building sector. Thus, improving the sustainability of the concrete industry automatically will lead to significant improvements in the building sector as a whole.

Portland cement, the primary constituent of concrete, is produced and used in large quantities, about 175 million tonnes in the EU and 1.75 billion tonnes worldwide [1]. This equals an average of about 500 kg per capita. Since the production of 1 kg cement generates approximately 1 kg CO_2 emission this corresponds to 500 kg CO_2 annually per capita. The total CO_2 emissions per capita are listed on various web sites [2]. Comparing these total CO_2 emissions per capita with the cement consumption figures in Figure 1 it appears that cement production counts for about 2-3 % in Scandinavia up to about 15 % in Spain and Portugal. These figures do not take the CO_2 uptake (carbonation) during the service life and after demolition of a concrete structure into account and therefore the figures in that case would be slightly smaller.

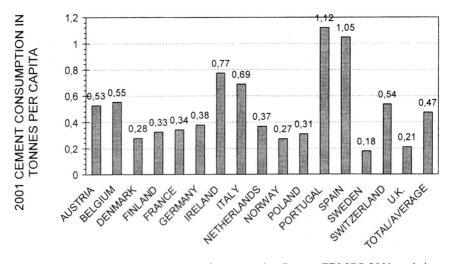

Figure 1 Total annual cement consumption per capita. Source: ERMCO 2001 statistics.

The construction industry as a whole has suffered from an image of being dirty, noisy and environmentally unfriendly, especially in "heavy" concrete construction. The negative image, based on lack of knowledge, needs to be reversed. The amount of information concerning building materials and the environment is large, fragmented and often conflicting. Quite often it can be difficult to distinguish between knowledge and pure marketing.

Development of sustainable concrete structures has been going on for a number of years – especially in countries where the government has a strong environmental profile like the Nordic countries and the Netherlands. A lot of different tools have been developed in order to reduce the environmental impact of concrete and concrete structures and to promote the production of "green concrete". These tools and the technologies behind them vary considerably across Europe due to regional/national differences in legislation, market conditions and traditions in the construction industry.

An important result of the work is the change of attitude in the industry from regarding environmental matters as ideological activities to taking responsibility and working systematically for the environment and the industry itself.

ENVIRONMENTAL IMPACT FROM CONCRETE STRUCTURES

Definition of Sustainable Concrete Structures

In the Nordic network "Concrete for the Environment" one of the main activities was to reach consensus on the definition of a sustainable concrete structure, which can be used as a basis for further work in making concrete even more sustainable [3]. The definition agreed upon by all the Nordic countries reads:

"An environmentally sustainable concrete structure is a structure that is constructed so the total environmental impact during the entire life cycle, incl. use of the structure, is reduced to a minimum. This means that the structure shall be designed and produced in a manner, which is tailor-made for the use, i.e. to the specified lifetime, loads, environmental impact, maintenance strategy, heating need etc – or simply the right concrete for the right application. This shall be achieved by utilising the inherently environmentally beneficial properties of concrete, e.g. the high strength, good durability and the high thermal capacity. Furthermore, the concrete and its constituents shall be extracted and produced in an environmentally sound manner".

Overview of Sustainability Aspects

It is generally accepted that most sustainability aspects of concrete may be considered under one of the following categories [4].
- Natural resources
- Energy consumption/greenhouse effect
- Environmental effects
- Health and safety

The resource consumption in relation to application of concrete is not significant, because all the raw materials are easily accessible in surplus amounts. One distinction from this is the application of stainless steel reinforcement which needs scarce resources, i.e. chromium, nickel and molybdenum. The greenhouse effect from energy consumption is much in focus. A primary source is the production of cement clinker and of steel reinforcement. In addition energy resources are consumed by construction, demolition and recycling and last but not least from the use, operation, and maintenance of buildings and structures.

The environmental effects from the application of concrete are apart from the greenhouse effect relatively badly described. It is possible that these problems are limited, but the following potential problems have been identified:
-   leaching of hydrocarbons from demolished concrete and concrete slurry
-   leaching of heavy metals from concrete and demolished concrete containing residual products with a high amount of heavy metals
-   chemicals in admixtures and repair products

The health effects from the application of concrete are related to the consequences for the indoor climate and to the working environment. Concrete has a bad indoor climate image and this is probably not fair. On the contrary, concrete is likely to be able to improve the indoor climate from its good thermal properties. Working environmental problems stem from noise, vibration, dust and accidents in the construction phase.

Environmental Assessments

When assessing environmental impact of concrete structures it is essential to consider all life cycle phases, i.e. from cradle to grave. In the following, typical examples are shown to illustrate environmental impacts during the life cycle of concrete products.

Materials and production

Since concrete consists of a number of various constituents the environmental impact of concrete production is a complex mechanism partly governed by the individual impacts from each of these constituents, and partly governed by the combined effect of the constituents when they are mixed together.

 The aggregate part of concrete normally accounts for 70-75 % of its volume and therefore the environmental issues of aggregate production strongly influence the concrete production. Furthermore, cement production is associated with large energy consumption and CO_2 emissions. Thus, the sustainability of concrete as a material is strongly influenced by the cement industry and the aggregate industry. However, since concrete is most often reinforced by means of steel bars this material also needs to be included in a total sustainability analysis.

In Figure 2 approximate CO_2 emissions are related to various processes and materials of a prestressed hollow core slab based on Danish experiences. The figures involve the emissions related to production of cement and prestressing steel plus emissions related to transportation, heating of the plant and other sources.

It appears that cement production contributes significantly to the total CO_2 emission (about 55 %). However, if the carbonation of concrete is taken into account (mainly taking place on concrete rubble after demolition) the CO_2 emissions released during calcination may be reclaimed. This contribution theoretically amounts to about 50 % of the CO_2 emissions during cement production, which may be counter balanced giving the negative contribution in Figure 2.

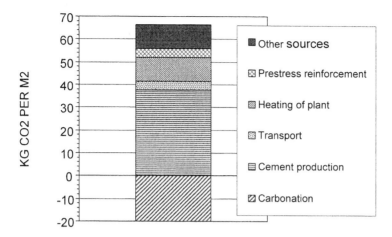

Figure 2 CO_2 emissions from the production of a prestressed hollow core slab [5].

The operation, maintenance, and service phase

Most of the environmental assessments have been focusing on materials and have neglected the operation and service phase, though these life cycle phases can give the largest contribution to the total life cycle environmental impact.

As an example of the importance of the operation and service phase Figure 3 shows the energy consumption of a reinforced concrete office building during its service life [6]. It appears that the contribution from the production of the concrete and its constituents is negligible during the life cycle.

The contribution from heating and power makes up more than 95 % of the total energy consumption. In other words, less than 5 % of the total energy consumed during the life cycle of the building come from the concrete and other building materials used in its construction. This example is of course only valid for cold countries, where heating is needed during wintertime.

Another example of the importance of the operation and service phase is seen in Figure 4 which shows the CO_2 consumption from the life cycle of a concrete bridge inclusive of the contribution from traffic [7]. It appears that the contribution from the traffic is the most dominant part during the bridge lifetime of 74 years with 2/3 of the total CO_2-emission in the lifetime of the bridge.

Figure 5 shows the effect of an alternative structural design and maintenance strategy for a bridge [7]. A significant reduction of CO_2 emissions can be achieved when substituting black steel reinforcement with stainless steel reinforcement, which reduces maintenance activities, and when constructing without asphalt or moisture barrier.

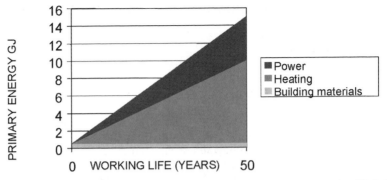

Figure 3 Energy consumption of a reinforced concrete office building in its service life [6].

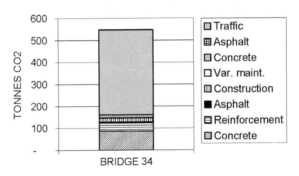

Figure 4 CO_2 emission in the lifetime of a concrete bridge [7].

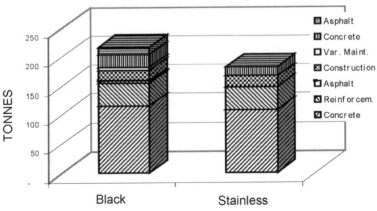

Figure 5 CO_2 emission: concrete bridge. Column 1: Traditional black steel reinforcement & asphalt. Column 2: Stainless steel reinforcement no asphalt and moisture barrier [7].

CURRENT PRACTICE AND STATE - OF - THE ART

In the following an overview is given of current practice and state - of - the art related to sustainable concrete structures.

Cement with Decreased Consumption of Non-Renewable Resources

Almost half of the total CO_2 emission from cement production derives from fossil energy carriers consumed in the process. One of the tools for the cement industry is to use alternative fuels and raw materials, which both results in a reduction of costs and in saving of natural raw materials. In the EU the degree of substitution has reached a level of 12 % in average. Table 1 shows the substitution rate fuels for various EU countries.

Table 1 Substitution rate fuels [1]

COUNTRY	SUBSTITUTION RATE, %
EU	12
Austria	29
Belgium	30
Denmark	4
Finland	3
France	27
Germany	30
Ireland	0
Netherlands	72
Poland	1
Portugal	1
Switzerland	31
United Kingdom	6

Reducing Clinker Content

The most commonly used technologies to reduce clinker content in a concrete are either use of supplementary materials to substitute clinker material or an optimised mix design reducing the total paste content and thereby also the clinker content.

Use of supplementary materials

Minimising the use of Portland cement clinker can be done by blending cement with supplementary cementitious materials. These are pozzolanic or latent pozzolanic materials, which can be ground into the cements or used directly in the concrete. As a result, either the clinker content in concrete is reduced and/or the concrete properties are improved.

The most commonly used supplementary materials are fly ash, silica fume, granulated blast furnace slag and limestone filler. Some of these have been used for a number of years and have been introduced not for environmental purposes but for improving concrete quality or for economic reasons. Other supplementary materials are metakaolin, sewage sludge incineration ash and glass filler. In reference [8] an extensive analysis of the application of supplementary materials is described.

Blending during cement production or at the concrete plant depends on local traditions and level of technology. The utilisation of blended cement across Europe is increasing. In [1] it is stated that in 2000 the market share of CEM I cement in Europe was down to 34.2 %.

The use of by-products and residual products has many advantages with respect to economy, resource consumption and quality of the concrete. In addition, the environment will benefit because CO_2 emissions and other greenhouse gas emissions are reduced and large volumes of waste are upgraded to useful raw materials.

However, it is essential that the solving of environmental problems for example by using residual products from other industries does not introduce any new environmental problems such as e.g. leaching of heavy metals or other harmful substances. Furthermore, it is of course crucial to ensure that the quality of the concrete remains satisfactory. This requires a systematic approach, and a model for the evaluation of new by-products or residual products from an environmental and a technical point of view has been developed at the Centre for Green Concrete in Denmark [9].

Some residual products contain high amounts of heavy metals, and consequently it is not environmentally sound to use these materials in cement or concrete. However, a promising approach is currently being investigated using electrodialytic removal of heavy metals from fly ashes, and thus potentially upgrading residual products for use in cement and concrete production, reference [10].

Optimising mix design

Another way of reducing clinker content in concrete is by optimising the concrete mix design, so that its performance fulfils the specifications with the lowest possible clinker content in the concrete.

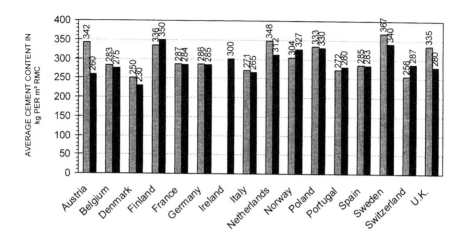

Figure 6 Cement dosages ready-mixed concrete. Column 1: cement consumption divided by the rmc production. Column 2: average cement dosage (ERMCO). Source: ERMCO 2001

Figure 6 shows the amount of cement used to make one m^3 of ready-mixed concrete around Europe. Cement contents are ranging from 250 to 350 kg/m^3. It is possible to lower the cement content in some parts of Europe. One way to do this is by optimising the aggregate composition in order to obtain dense packing of the aggregate particles, thus minimising the need for binder and thereby cement clinker. Several models for this calculation are available, e.g. [11] and [12]. Another way of reducing cement and clinker content in concrete is through a careful use of admixtures.

Recycling of Waste Products

For concrete production the most relevant waste products to consider for re-use may stem from:
- recovered aggregate washed out from fresh concrete
- washing water from saw cutting, cleared from slurry and re-used as mixing water
- construction and demolition waste, i.e. hardened concrete rubble etc.

Working Environment

Improvement of the working environment with SCC

A relatively new way of improving the working environment for the concrete workers during casting is by using self-compacting concrete (SCC) a high-performance concrete that flows into the formwork under its self-weight only, without the need for vibration and compaction. It is generally acknowledged that the use of SCC means an overall improved working environment at the concrete site or production plant, though the improvement has not been systematically quantified.

Working environmental impact from admixtures

Another aspect of the working environment is the enhanced use of chemical substances in order to obtain certain performance improvements of the concrete. These substances may cause problems when exposing skin or inhales to them. Not much effort has been allocated this subject, however, it is generally believed that problems are negligible [13].

Construction Design and Repair and Maintenance Strategies

It is possible to reduce environmental impact from buildings by optimising the construction [9]. An example of such a solution is bridge decks without asphalt and moisture barrier and a more dense and durable concrete. That saves materials and energy both during the construction and when repairing (see Figure 5). Another example is to design a structure that makes it easy to substitute single parts of the construction. These single parts should be parts, which are more exposed. For a road bridge that is e.g. columns.

Repair and maintenance strategies, which reduce environmental impacts, are e.g. use of stainless steel reinforcement as shown in Figure 5. It can also be the use of permanent formwork and a concrete type of lower quality thus reducing the environmental impact or an increased cover, which increases the durability.

Service Life

Concrete buildings can provide excellent energy efficiency and a good indoor comfort if the concrete is utilised in a proper way. Heavy structures are less sensitive to outdoor temperature variations, e.g. on warm summer days the need for cooling will be less than in buildings made of light building materials [14].

Solving Problems with Cement and Concrete

Concrete and cement can be used creatively to solve waste and pollution problems, to achieve energy efficient and comfortable buildings and to increase service life. Concrete can also be used to reduce or eliminate other environmental problems by e.g. stabilising or encapsulating environmental harmful substances.

Examples of this are the use of concrete containers for storing radioactive waste in Russia, and stabilisation of contaminated sediments in Norwegian harbours using cement [15, 16].

RESEARCH & DEVELOPMENT ACTIVITIES

A lot of work is going on within various European industries, associations, institutes, and universities. In table 2 is given an overview of on-going and recent R&D in Europe, which is focused on sustainable concrete structures.

Table 2 An overview of on-going and recent R&D in Europe [8]

NAME OF PROJECT	TYPE OF PROJECT TIME PERIOD	MAIN TOPIC / OBJECTIVES *PUBLICATIONS*	PARTNERS (PROJECT LEADER)
TESCOP	Brite/Euram 3 1997-2000 4-year	Cleaner technologies in the life cycle of concrete products. Develop tools for assessing the environmental impact of concrete products and suggestion for cleaner technologies.	**Danish Technological Institute** Industry and research institutions from Denmark, Greece, Italy, The Netherlands. See Cordis web site for details WEB-LINKS: www.cordis.lu/search/ Search for Tescop
RESIBA (Recycled aggregate for construction)	National (NO) Budget 1 M Euros 1998-2002 3-year	Application of recycled aggregates in unbound use (road construction, drain layers) and as aggregate in concrete or asphalt production. The various applications are documented through demonstration projects. *English summary available from web site. Several reports in Norwegian.*	**Norwegian Building Research Institute**, building owners and contractors. See web site for project participants WEB-LINKS: www.byggforsk.no /prosjekter/resiba
Centre for Green Concrete	National (DK) Budget 3 M Euros 1998-2002 4-year	Production of environmentally friendly (green) concrete. Demonstrate the possibilities for producing and implementing green concrete including experimental documentation of its performance. *Articles and reports downloadable from web site. Project reports in Danish.*	**Danish Technological Institute** Concrete producers and contractors plus research institutions. See web site for project participants WEB-LINKS: www.greenconcrete.dk
Concrete for the environment	Network funded by the Nordic Industrial Fund 2001-2003 3-year	Nordic consensus on sustainability in the concrete industry. Establish common understanding of concrete as a sustainable material and promote its use. *An ACI article is under preparation*	**Danish Technological Institute** See web site for project participants WEB-LINKS: ww.nordicinnovation.net Search for project name
Concrete for the environment	Self-financed network 2004	Forum for exchanging experience and information. *Articles listed on web site.*	See web site for network participants WEB-LINKS: www.concretefortheenvironment.net
Eco-Concrete	Self-financed 2000-2002	To create a life-cycle tool (LCI and LCA) for use in cement and concrete evaluation. EcoConcrete created by INTRON on behalf of the Joint Project Group.	CEMBUREAU BIBM ERMCO EFCA EUROFER UEPG
CO_2 uptake during the concrete life cycle	Funded by the Nordic Industrial Fund 2003-2005 2-year	Document the CO_2 uptake through carbonation and model its effect through a full life cycle. Guidelines for society in how to take the CO_2 uptake into account.	**Danish Technological Institute** Several Nordic participants WEB-LINKS: ww.nordicinnovation.net Search for project name

Table 2   An overview of on-going and recent R&D in Europe [8] continued

Project for concrete products	National (DK) Funded by Danish Environmental Protection Agency 2003-2005	A plan of action for reducing environmental impact from concrete products has been formulated in close co-operation with the concrete industry. Several investigations have been initiated based on that plan: • hydrocarbons in slurry • recycling of crushed concrete waste • thermal properties and drying of concrete	**Danish Technological Institute** Aalborg Portland The council of the Danish concrete industry Support from various industrial partners
ECO-SERVE (European Construction in Service of Society	Brite/Euram 5 2002-2006	The overall objectives of the network are to identify, evaluate and disseminate technologies, which may improve the environmental impact of European construction industry. The means to achieve the objectives are to map technologies, traditions and stakeholders and identify best available technologies within cement production, aggregate production, concrete production and pavement production.	**Dansk Betonteknik,** NCC, intron, Danish Technological Institut, CSTC, Franzefoss, IBRI, VDZ, Norcem, Titan Cement, CTG, Netherlands Energy Research, Dura Vermeer, COWI, Polish Academy of Asciences + approx. 30 associated parternes. WEB-LINKS: www.eco-serve.net

TECHNOLOGICAL FORESIGHT

Although much work has been done already to document and improve the environmental performance of concrete structures, the efforts must continue because of the challenges that the industry is facing. The authors believe that the trends in the future work regarding sustainable concrete structures will be the following,

• A change from material development to optimisation of the concrete structure during its life cycle. When assessing the environmental performance of a concrete structure the total structure in which the concrete is used should be considered. This includes the environmental impact associated with production of the various constituent materials, construction of the structure, use, maintenance, demolition and finally recycling of components or materials.

• A change from a problem solving approach to an approach utilising the environmentally beneficial properties of concrete. Due to the new EU directive concerning energy consumption in buildings, exploitation of the excellent thermal capacity of concrete is expected to receive considerable attention. Concrete structures will help reducing the energy needed for heating or cooling and at the same time ensure a good indoor climate.

• The work related to use of recycled materials and by-products will continue, as it is an important and effective way of reducing the environmental impact of concrete and concrete structures. The increased use of SCC will push this development. However, further investigations are expected to focus on providing necessary documentation ensuring no negative side effects occur and that the concrete quality remains satisfactory.

• The environmental effect from chemicals in admixtures, repair products etc. is expected to be in focus due to the EU directive REACH. This will include the elimination of substances harmful to health and safety, e.g. the reduction of the amount of carbon hydroxide in concrete wastewater and crushed concrete.

In the next generation of European standards, pure technological requirements may be gradually changed with environmental priorities. And in the future, probably only those branches and companies will survive, who can earn their public acceptance from an active use of environmental parameters in their planning and execution of own activities.

CONCLUSIONS

Development of sustainable concrete structures has been going on for a number of years – especially in the countries where the government has a strong environmental profile like the Nordic countries and the Netherlands. This development has primarily focused on materials development like recycling and use of supplementary cementitious materials and from the cement manufacturers' side on the use of alternative raw materials and CO_2 neutral fuels. An important result of the work is the change of attitude in the industry from regarding environmental matters as ideological activities to take responsibility and work systematically for the environment and the industry itself. The industry has recognised that sustainable production and use of concrete can result in economic savings and in improved quality of the concrete.

Due to the challenges facing the industry there is a need to continue this work. The future trends in the work regarding sustainable concrete structures are expected to change from materials development to optimisation of the concrete structure during its life cycle and a change from a problem solving approach to an approach focusing on utilising the environmentally beneficial properties of concrete. Especially the use of thermal capacity to reduce the energy consumption needed for heating and cooling ensuring a good indoor climate will be in focus in the future because of a recent EU directive about energy consumption in buildings. The increased use of self-compacting concrete will push the materials development further towards optimisation of the clinker content, use of supplementary cementitious materials and blended cements.

An additional important challenge for the concrete industry will be to communicate facts on concrete and the environment – also to stakeholders outside the industry.

Both the industry and the environment will benefit from continued activities related to sustainable concrete structures.

REFERENCES

1.   VAN LOO, W, The Production and Use of Cement in a Sustainable Context, Role of Cement Science in Sustainable Development, Proceedings of the International Symposium Celebrating concrete, Dundee 2003, pp 485-494.

2.   http://millenniumindicators.un.org

3.   www.concretefortheenvironment.net

4.   GLAVIND, M., BØDKER, J., MATHIESEN, D. and POMMER, K, Produktområde-projekt vedrørende betonprodukter – Handlingsplan, (in Danish), Danish Environmental Protection Agency, Copenhagen 2004.

5.   GLAVIND, M., DAMTOFT, J.S. and RÖTTIG, S., Cleaner Technology Solutions in the Life Cycle of Concrete Products (TESCOP), in Proceedings of CANMET/ACI International Symposium for Sustainable Development of Cement and Concrete, San Fransisco, September 2001, pp. 313-327.

6.   KUHLMANN, K. and PASCHMANN, H., Beitrag zur ökologischen Positionierung von Zement und Beton, ZKG International, 50(1), pp. 1-18..

7.   TØLLØSE, K., Miljøscreening af betonbro, (in Danish), Centre for Resource Saving Concrete Structures, Danish Technological Institute, www.greenconcrete.dk, 2002.

8.   BASELINE REPORT for the aggregate and concrete industries in Europe, ECO-SERVE Network, Cluster 3 Aggregate and concrete production, June 2004.

9.   HASHOLT, M.T., BERRIG, A. and MATHIESEN, D., Anvisning i grøn beton, (in Danish), Centre for Resource Saving Concrete Structures, www.greenconcrete.dk, 2002.

10. PEDERSEN, A.J., Characterization and electrodialytic treatment of wood combustion fly ash for the removal of cadmium, Biomass & Bio energy 25, pp 446-458, 2003.

11. GLAVIND, M., OLSEN, G.S. and MUNCH-PETERSEN, C., Packing Calculations and Concrete Mix Design, Nordic Concrete Research, Publication no. 13, 1993.

12. LARRARD, F., Concrete Mixture Proportioning – a Scientific Approach, E&FN Spon, London, 1999.

13. DEUTSCHE BAUCHEMIE, Concrete admixtures and the environment, State-of-the-art report, Frankfurt, 1999.

14. ÖBERG, M. Integrated Life Cycle Design, Application to Concrete Multi-Dwelling Buildings, Licentiate Thesis, Report TVBM-3103, Lund University, Division of Building Materials, Lund, Sweden, 2001.

15. LINDERS, H., Trondheim Harbor – containment of polluted sediments with cement, presentation at Nordic seminar Concrete – the sustainable material, Oslo, November 2003, www.concretefortheenvironment.net.

16. GRAN, H.C., Concrete Solutions for containment of nuclear waste, presentation at Nordic seminar Concrete – the sustainable material, Oslo, November 2003, www.concretefortheenvironment.net.

THEME ONE:

EFFICIENT USE OF ENERGY AND RAW MATERIALS

SUSTAINABLE DEVELOPMENT IN CONSTRUCTION - THEORY, FEASIBILITY AND PRACTICE USING RAW MATERIALS AND ENERGY

A Samarin

University of Technology Sydney

Australia

ABSTRACT. The fundamental concepts of sustainable development have been established some time ago, and although their practical applications in different fields of human endeavor do vary considerably, the interdependence of economical, environmental and social aspects of sustainability is never in doubt. However, short term and long term sustainability is affected by a range of factors, some identical, but others dissimilar to a degree which is influenced by the definition of "short" term and "long" term sustainability. In building and construction industries the ultimate depletion of economically accessible non-renewable raw materials and of the sources of non-renewable energy plays a particularly important role in solving the problems of long term sustainability. The paper addresses the solutions of short term sustainability as a prerequisite for the long term sustainability in building and construction industries, and considers the reliability and inherent limitations of the prediction models. It also offers a brief overview of the existing practices for the realization of short term sustainability.

Keywords: Sort term and long term sustainability, Renewable and non-renewable raw materials, Energy efficiency, Chaos theory, Catastrophe theory, Economical requirements.

Dr A Samarin FTSE, Dr. Samarin is a Professor at the Centre for Build Infrastructure Research at the University of Technology, Sydney, Australia, and a private consultant and adviser in the development of Sustainable Energy and Building and Construction industries. He has more than forty years of experience both in academia and in industry and in 1988 he was elected Fellow of the Australian Academy of Technological Sciences and Engineering. He has published widely, received decorations and awards for his work, and has taken a number of patents.

INTRODUCTION

World Bank economist Herman Daly [1] has suggested three simple rules for sustainability:
- For a **renewable resource** the sustainable rate should be no greater that the rate of regeneration. (Thus, for example, the rate of use of timber in construction should not be greater that the growth of new trees planted for this specific use).
- For a **nonrenewable resource**, such as fossil fuel, quarry stone for use as an aggregate in concrete, limestone for Portland cement manufacturing, etc., - the sustainable rate of use should be no greater than the rate at which a renewable resource, used sustainably, can be substituted for it. (eg an oil deposit can be used sustainably if parts of the profits from it were systematically invested in solar, methanol or hydrogen energy sources, so that when the oil is gone, an equivalent stream of renewable energy becomes available).
- For a **pollutant** the sustainable rate of emission should be no greater than the rate at which that pollutant can be recycled, absorbed or rendered harmless to the environment. (For example, sewage can be used as a fuel in manufacturing bricks, in which another waste stream – fly ash from power stations is used as a raw ingredient).

However, the concepts of long term sustainability can vary considerably, depending on the type of industry to which they are applied, and also depending on the end user or on the authority, which implements the rules. At a recent international conference on sustainability, I asked three eminent speakers to give me their perception of the short term and the long term sustainability. For an economist (who is also a politician) short term meant "till the next election", and long term was twenty years (the limit, in his opinion, of the reliability of market movements predictions, using the existing mathematical models). For a general manager of the research and development division of a major motorcar manufacturing company short term was ten years (R&D of a new model), and long term was twenty five years (development of the new types of vehicles – i.e. hybrid or hydrogen fuel types). For the director of the worldwide town planning authority short term was twenty years, and long term a hundred. My personal view of short term is the span of two or three generations, and long term should probably extend to a period of the existence of a nation (such as USSR or Yugoslavia, both of which are no more).

It is also important to emphasize that sustainability in building and construction can not be achieved without a strong economical base, and without the proper concern and preventive measures against the potential negative environmental and social impacts, which these industries may generate. Of the economic, environmental and social considerations, the economy seems to be the most important prerequisite, and those nations which failed to create sufficient wealth, are incapable of satisfying either social or environmental needs and the expectations of the larger part of their population.

FEASIBILITY OF SUSTAINABLE DEVELOPMENT

Problems and Solutions

The original work on sustainability is usually attributed to the treatise "An Essay on the Principle of Population", first published in 1798, and then revised in 1803, by Thomas Robert Malthus (1766 – 1834). The two fundamental propositions in his study were: -when unchecked, population increases in a geometric ratio, and the means of subsistence could not possibly be made to increase faster than in an arithmetic ratio.

Thus, when the population doubling will reach a particular period (Malthus estimated this to be 25 years), the human civilization, as we know it, will collapse. Malthus argued that sustainability can only be achieved by restraining the population growth. As checks on population he suggested war, famine and disease, but later added "moral restraint" as well. The model was strictly speaking applicable only to the U.K., and the opening of the new colonies in North America, Asia, Australia and Africa rendered the concept invalid. However, the basic propositions about the rate of population growth and the rate of food production suggested by Malthus seem to be valid to this day. In 1790 the world population was 900 million, in 1900 it was 1.6 billion, in 1950 it became 2.5 billion, in 2000 it reached 6 billion, and the prediction for the year 2050 is approximately 9 billion (by a billion I mean one thousand million). The sustainability of food production is in question, in spite of the new technologies, as the land recourses are rapidly depleted. In many parts of the world fresh water is already in short supply. For example, in North China plains thirty cubic kilometers more water is being pumped to the surface by farmers each year than is replaced by the rain. In Tamil Nadu ninety five percent of wells owned by small farmers have completely dried up. As a result the famine in many parts of Africa and Asia became a cruel reality.

Some two hundred years later, the most commonly quoted definition of sustainability was formulated at the 1987 UN Conference "Our Common Future", published by the Oxford University Press. It stated that the humanity has the ability to make development sustainable, by ensuring that it "meets the needs of the present without compromising the ability of future generations to meet their own needs". This optimistic view of the future, however, was not universal. A forecast of the potential predicament for the mankind, using a "Mathematical World Model", developed in the early 1970-s by a team at the Massachusetts Institute of Technology under the direction of Professor Denis Meadows, resulted in the Report for the "Club of Rome" entitled "The Limits to Growth" [2]. It predicted the world where industrial production has sunk to zero, where population has suffered a catastrophic decline, and where the air, sea and land are polluted beyond redemption. The most alarming part of the prediction was that this collapse will not come gradually, but with awesome suddenness, and with no way of stopping the calamity. The sequel to this book was "Beyond the Limits" [3] published by the Club of Rome some twenty years later. And in spite of the fact, according to this study, that the exploitation of many world resources already exceeded the capacity for regeneration, it stated, that the catastrophe was not inevitable. To avoid this decline, it suggested, two changes are necessary. The first is a comprehensive revision of policies and practices that perpetuate growth in material consumption and population. (This does sound a little bit like Malthus). The second is a rapid, drastic increase in the efficiency with which materials and energy are used. And this, it seems, is where the success or otherwise of the Sustainable Developments is founded.

Just how Reliable are Predictions of Mathematical Models?

As recently as 1966 an article appeared in "Time" magazine, which read: "By 2000 machines will be producing so much, that everyone in the U.S. will, in effect, be independently wealthy... How to use leisure meaningfully will be a major problem." According to one estimate only ten percent of the population will be working and the rest will have to be paid to remain idle. According to another prediction a pleasure-oriented society will be full of "wholesome degeneracy". Arguably, only the last of the three came close to the truth.

In an article by Sir Ian McLennan, Chairman and Director of BHP Group of Companies (one of the industrial giants in mining and energy use), which appeared in the book: "Australia 2025 [4] written in 1975, Sir Ian made a prediction that the world crude oil reserve will be exhausted by the year 2005 if used at the 1975 rate of consumption. There was, of course, a significant increase in the world wide use of oil since 1975. In spite of the higher demand, in the year 1988 the known world reserves of crude oil were 922.1 billion barrels and the estimated undiscovered reserves were between 275 and 945 billion barrels. In 2002 known USA reserve was 22 billion barrels (exhausted in 9.8 years at the 2002 rate) and that of Iraq was 112 billion barrels (exhausted in 146 years at the 2002 rate of use).

With the rapid increase in the processing capabilities, in the memory capacity and in the sophistication of computer hardware and software, mathematical models became a routine tool in every field of human enterprise. But it is nonetheless apparent, that the reliability of long term predictions, using these models, is often in doubt [5]. There appear to be two main reasons for this unreliability. The first one is the so called Chaos theory.

In 1963 Professor Edward Lorenz of the Massachusetts Institute of Technology developed a meteorological mathematical model. Lorenz had a Royal McBee LGP-300 computer – a rare tool for solving equations in the early 1960-s. His intention was to predict the microclimate behavior over a relatively long period of time. In his original input, the variables were stored to the six decimal places. After the first print-out of the weather forecast, Lorenz decided to round off the variables to three decimal places – much more realistic practical values. The computer was slow, and Lorenz returned to compare the two print-outs some time later. He expected only a slight variation in the results, and initially the two curves coincided, but after some time they widely diverged. At first Lorenz suspected a malfunction, but then he realized that his model reflected chaotic behavior of the non-linear dynamic systems. His results were published in the Journal of Atmospheric Sciences [6]. Many natural processes, including physical, chemical, economical and social are mathematically described as the non-linear dynamic systems. At relatively long periods of time the state of these systems becomes extremely sensitive to even very minute variations in their initial state. Their behavior becomes chaotic and can be predicted with certainty only within a wide range of limiting conditions.

The second reason for the uncertainty in long term forecasts is the Catastrophe theory. In the late 1960-s French mathematician René Thom discovered that the abrupt changes in nature, or catastrophes, as he called them, follow orderly patterns, and can be categorized in terms of seven topological figures [7]. Thom named them the fold, the cusp, the swallowtail and the butterfly (with three varieties of swallowtail and two of butterfly) – seven archetypal catastrophes in all, with the names evocative of their shapes. The simplest of these is the fold catastrophe – its control surface is a single curve, which slides downwards to the brink of a vertical cliff. Thom has modestly described his theory as "a very general form of 'philosophical' biology", and made it clear, that it did not permit quantitative predictions. However, his method of anticipating the likely dramatic changes in such phenomena as the stability of ships, load-bearing capacity of bridges, economical calamities on the stock exchange, civil unrest, wars, and sudden climatic changes was widely and successfully applied by scientific community. The results of the negative effects on sustainability became particularly evident, when both chaos and catastrophe theories are taken in consideration [8].

WE MUST PRACTICE WHAT WE PREACH

Prescriptions for Sustainability

One of the recommendations of the Club of Rome implied that in order to achieve sustainable development governments must undertake a "comprehensive revision" of the existing policies, which would then enforce a significant reduction in material consumption and prevent very rapid population growth. This seems to be an extremely tall order to fulfill, as its basic requirements are inherently abhorrent to the human nature.

Samuel Butler (1835 – 1902), who made his fortune in New Zealand as a sheep rancher, subsequently to become a well known English author, painter and composer, once made an observation that "all progress is based upon an innate desire on part of every organism to live beyond its means". In a democratic society the essential requirement of each politician is to be re-elected and hence, at the least, he or she must promise to satisfy the inspirations of the electorate, no matter how unreasonable these expectations may be. Only those politicians who are sure of their appointments, regardless of the populations' desires, i.e. those, who established themselves in the totalitarian societies, or possibly in the House of Lords in the U.K., can afford to be truthful with the electorate.

Sir Boyle Roche (1743 – 1804), a member for Tralee in the Irish Parliament was audacious enough to declare: "Why should we put ourselves of our way to do anything for posterity? For what has posterity ever done for us?" Most governments actually encourage the population growth of their nations, as it ensures stronger military and possibly even economic influences of these countries in the world. One exception from this rule is China, where the population growth already exceeded the land capacity for its sustainability. Thus, strict limitations on birthrates were imposed. It should be recognized however, that if governments succeed in scaring larger part of population by the inevitability of devastating effects of global warming on one hand, and the potential disasters of the nuclear energy generation on the other, the restrictions on the desire of every person to live "beyond her or his means" are generally tolerated with less resentment.

Many people experience morbid fear as soon as they hear the word "nuclear". In the year 2001 I was appointed by the Australian Government to present a report on the best possible design of a radiation shield for the new Nuclear Research Reactor at Lucas Heights near Sydney. The project was commissioned in collaboration with the commercial arm of the University of Wollongong. However, the insurance company which provided cover for all the University contract work categorically refused to insure this particular project, as soon as they discovered the frightening word "nuclear". And yet, construction of the new research reactor was approved by the Australian Government, and the radiation shield was of course intended to contain, and not to generate radiation. There seems to be a lot of misconceptions about radiation in general, a lack of knowledge about some of the beneficial effects of radiation use in medicine and in industry, and often a complete ignorance about the existence of natural background radiation, to which all humanity is constantly exposed without any apparent ill effects [9].

If we compare two major disasters in chemical and nuclear industries, those of Bhopal (December 2, 1984), and Chernobyl (April 25, 1986), the public reaction to the latter seems completely out of proportion. Bhopal initially claimed at least 2,500 lives, with the subsequent ill effects on another 50,000. As a result of Chernobyl accident 32 people died containing the reactor and some, possibly hundreds, but certainly not thousands were subsequently badly affected. Neither accident should have happened, if proper industrial rules and practices were imposed in each case. There is no wide spread resentment to chemical industries, as they benefit every aspect of our lives, but nuclear – that is a different story. The perception exists, that many other, safer sources of energy can and should be used. And yet, if the onset of runaway conditions will bring a sudden, abrupt and extreme global warming (as formulated by René Thom), the way to prevent this calamity may be the use of nuclear energy in place of fossil fuels, as socially and politically unpalatable this may appear.

One paradox of the global warming is a strong possibility that the Gulf Stream current, diluted by fresh water from the melting polar cups, will be reduced in its intensity, or will even cease to flow completely. In the latter case climate in the large parts of Northern Europe will become similar to that of Siberian tundra. The consequences of such event are even hard to contemplate.

Specifications and Legal Requirements of Sustainability.

Politicians often present facts with a significant bias, in order to achieve their own objectives. As an example, the Chief UN weapons inspector Hans Blix remarked on the interpretation of his report concerning the likelihood of the existence of weapons of mass destruction in Iraq thus: - "They were putting exclamations marks, in place of the question marks". Having convinced larger and the more gullible part of population in the existence of both real and imaginary dangers of nuclear energy, environmental pollution and global warming, some restrains and limitations on material and energy use were gradually enforced by the authorities in different countries.

In Australia, in the State of New South Wales, restrictions on the use of water and energy in newly constructed houses have been introduced as from the 1st of July 2004. Development of each new residential dwelling now requires BASIX Certificate, which ensures, that each new home will be designed and built to use 40% less mains supply of water and produce 25% less greenhouse gas emissions, than was previously the norm. The reduction in the greenhouse gas requirement will increase to 40% as from July 2006. From the 1st of April 2001 Federal Government in Australia introduced mandatory renewable energy targets. The underpinning legislation, the Renewable Energy (Electricity) Act 2000, the Renewable Energy (Electricity) Charge Act 2000, and the Renewable Energy (Electricity) Regulation 2001 require the generation of an additional 9,500 GWh per year renewable electricity by 2010.

One of the potential contributors to this requirement may be the EnviroMission Solar Tower project in the State of Victoria. The design is for a one kilometer tall concrete stack with the base of about seven kilometers across. If built, it will become the tallest man-made structure in the world. Air heated by the sun will rise up the tower where 32 turbines will generate approximately 650 GWh of electricity per year [10]. The concept of solar tower was developed by Jörg Schlaich and in 1988 a 195 meter high "prototype" was built in Spain, near the town of Manzanares, some 150 kilometers south of Madrid. It is expected, that one kilometer tall structure will be five times more efficient than the Manzanares plant.

Although the legal requirement for sustainable development are gradually being introduced in many parts of the world, there are many areas where similar, but non-mandatory practices are becoming more and more evident, due to the environmental, social, and probably most importantly economic considerations, as the cost of non-renewable materials and energy increases. In the development of modern architectural concepts some of the old established principles of designing buildings, which were energy efficient and environmentally friendly, have been largely neglected in the past. These fundamental values became of secondary importance to the demands of open living spaces filled with technical and electronic devices, which modern architecture intended to offer. But now, as the cost of air conditioning and pollution became of real concern, some of those old concepts have been revived and many new ideas, mostly as modifications and improvements of the old, have been developed [11]. Building materials in general and concrete, due to its wide spread use, in particular play an important role in the design of energy efficient, environmentally non-intrusive buildings [12].

Efficient use of Building Materials

There are numerous practices which can be used and, admittedly, many of which have already been implemented in an attempt to ensure sustainability. In building and construction industries the following fundamental principles of sustainable development were suggested, refer Samarin [13]: -
1. The use of industrial wastes as raw materials for structural and building components and products,
2. The recycling of building and structural materials and products,
3. The development of structural and building components and products which require minimal amounts of energy in the process of their manufacturing,
4. The use of renewable raw materials,
5. The development of energy efficient commercial and residential buildings, and,
6. The development of structural and building materials and components which are extremely durable in a particular environment, for which they are designed.

Examples of different waste streams and recycled products, used as raw materials in concrete manufacturing were also provided in some detail, refer Samarin [14]. These include unprocessed waste, such as blast furnace slag aggregate, crushed brick, crushed glass (with special provisions to avert alkali-aggregate reaction), expanded polystyrene granules (when special surface-acting admixtures are used), cork granules, sawdust (again with the use of special admixtures), shredded rubber, bagasse, wood ash, china clay waste, slate processing waste and even paper waste. It is, however, important to remember, that the design life of a structure or a building and the durability requirements in a specific environment will determine the suitability or otherwise of the above practices.

Concrete aggregates can be manufactured from different waste streams, including fly ash, different types of slag and from the blends of household and industrial wastes [14, 15]. In many of these applications there is additional benefit of reduction in the green house gas emissions [16, 17, 18]. Many new types of hydraulic cements manufactured with unconventional raw materials can significantly improve durability of concrete by comparison with plain Portland cement types and thus may be used not only to extend the life-cycle of commercial and residential buildings, but also of engineering structures in an aggressive environments [19]. Yet another useful application of some of these cements is encapsulation of toxic and hazardous wastes [21, 22, 23, 24]. Using special construction methods, even low level radioactive wastes can be physically encapsulated, chemically immobilized and removed from the direct exposure to humans [25].

CONCLUSIONS

As a pre-requisite for long term sustainability, short term sustainable development requires inter alia, very swift and radical increase in the efficiency with which raw materials are used in concrete. Many so-called marginal aggregates, re-cycled concrete aggregates or screened and separated building rubble, unprocessed and processed waste streams can be used in concrete to achieve one or several of the following:

- replacement of non-renewable materials with re-cycled materials or waste,
- reduction in energy requirements in concrete manufacturing,
- improvement of concrete durability in a particular environment,
- reduction in the cost of construction,
- improvement in the thermal efficiency of concrete (for example, conductivity of ordinary concrete is of the order of 1.1 kcal/ m h deg C, and that of expanded polystyrene concrete is approximately 0.05),
- utilization of local unconventional raw materials (for example, bagasse) in a very low cost concrete, designed for a short life-cycle construction (such as temporary buildings or structures), when long-term durability is not required.

It is however very important not to rely on the information about the performance of these resources by simply classifying them by their generic names. Materials from different parts of the world, or even from different regions of the same country or state, which are given identical common names can have dramatically dissimilar properties.

REFERENCES

1. HERMAN DALY, Toward Some Operational Principles of Sustainable Development, Ecological Economics 2, 1990, pp. 1- 6.

2. MEADOWS, D H, MEADOWS, D L, RANDERS, J, BEHRENS, III W W, The Limits to Growth, a report for the Club of Rome's Project on the Predicament of Mankind, A Potomac Associates Book, Earth Island Limited, London, 1972.

3. MEADOWS, D H, MEADOWS, D L, RANDERS, J, Beyond the Limits – Global Collapse or a Sustainable Future, Earthscan Publications Limited, London, 1992.

4. AUSTRALIA 2025, Electrolux Pty Ltd., Pre Press, Melbourne, Australia, 1975, p. 22.

5. SAMARIN, A, Sustainable Development – Can Mathematical Modeling Assist Industrial Progress and Avert Overpopulation, Energy and Environmental Crisis? Invited keynote paper, Proceedings of the Second Biennial Australian Engineering Mathematics Conference, 15 – 17 July 1996, Published by the Institution of Engineers Australia, pp. 637 – 655.

6. LORENZ, E, Deterministic Nonperiodic Flow, Journal of the Atmospheric Sciences, No.20, 1963, pp. 130 – 141.

7. THOM, R, Structural Stability and Morphogenesis: an Outline of a General Theory of Modes, Reading, Benjamin, 1975.

8. SAMARIN, A, Anthropogenic Paradox – Will Homo-Sapiens Self Destruct? Australian Academy of Technological Sciences and Engineering Journal Focus No. 110, January/February 2000, pp. 2 – 8, (also available on the ATSE website: www.atse.org.au)

9. SAMARIN, A, Absorption and Biological Effects of Ionizing Radiation, Australian Academy of Technological Sciences and Engineering, Occasional Paper No. 4, ATSE website: www.atse.org.au

10. NOWAK, R, Power Tower, New Scientist, 31st July 2004, pp. 42 – 45, (Also on the website: www.newscientis.com)

11. McDONOUGH, W, and BRAUNGART, M, Cradle to Cradle: Remaking the Way we Make Things, North Point Press, New York, 2002.

12. ANDERSON, J, SHIERS, D, E, & SINCLAIR, M, The Green Guide to Specification: an Environmental Profiling System for Building Materials and Components, Blackwell Science, Oxford, 2002.

13. SAMARIN, A, Sustainable Development in Building and Construction Engineering, invited paper, Proceedings of the 5-th East Asian Structural Engineering Conference, Gold Coast, Queensland, Australia, Session II-2-E, Criffith University Publication, July 1995.

14. SAMARIN, A, Wastes in Concrete: Converting Liabilities into Assets, invited leader paper, Proceedings of the International Congress, Creating with Concrete, University of Dundee, Scotland, UK, edit. R.K. Dhir and P.C. Hewlett, Thomas Telford, London, 1999, Vol. I, pp. 131 – 151.

15. SAMARIN, A, Total Fly Ash Management – From Concept to Commercial Reality, invited paper, The Australian Coal Review Journal, November 1997, issue No.4, pp. 34 – 37.

16. SAMARIN, A, Utilization of Wastes from Coal-Fired Power Plants – an Important Factor in the Abatement of Greenhouse Gas Emissions, Australian Academy of Technological Sciences and Engineering Journal Focus, No 106, March/April 1999, pp 6 - 9.

17. SAMARIN, A, Hydraulic Cements – New Types and Raw Materials and Radically New Manufacturing Methods, invited leader paper, Proceedings of International Congress: Concrete in the Service of Mankind, University of Dundee, Scotland, UK, edit. R.K. Dhir and M.J. McCarthy, Volume: Appropriate Concrete Technology, E&FN Spon, London, 1996, pp. 265 – 279.

18. DURIE, R,A, SAMARIN, A, Hydraulic Cements of the Future – their Potential Effects on the Abatement of Greenhouse Gas Emissions, Proceedings of the Fifth International Conference on Greenhouse Gas Control Technologies (or GHGT-5), Cairns, Queensland, Australia, CSIRO Publication, 2000, pp. 1248 – 1252.

19. SAMARIN, A, New Cements and Concretes, invited paper, Proceedings of the Symposium of the Australian Academy of Science on Application of New Materials, April 1989, Published by the Academy, pp. 16 – 23.

20. SAMARIN, A, Progressi Nella Technologia del Calcestruzzo: Opportunita per i Prossimi Dieci Anni, L'Industria Italiana del Cemento, No.667, Giugno 1992, pp. 413–418.

21. SAMARIN, A, Encapsulation of Solid Wastes from Industrial By-products, invited keynote paper, Proceedings of the International Conference on Environmental Management, Geo-water and Engineering Aspects, University of Wollongong, Australia, A.A. Balkema, Rotterdam, 1993, pp. 63 – 78.

22. SAMARIN, A, Encapsulation of Wastes in Concrete, Waste Management and Environmental Journal, Volume 6, No.6, April 1996, MPB Waste Management, Victoria, Australia, p. 39.

23. SAMARIN, A, Encapsulation of Hazardous Heavy Metal Wastes in High Performance Concrete, Australian Academy of Technological Sciences and Engineering Journal Focus, No.87, May/ June 1995, pp. 11 - 14.

24. SAMARIN, A, Theory and Practice of Durable Concrete for Encapsulation of Hazardous Wastes, invited keynote paper, Proceedings of the Fourth CANMET/ACI International Congress on Durability of Concrete, Sydney, Australia, 1997, Volume III, pp 833 – 855.

25. SAMARIN, A, Two New Concepts in Sustainable Development of Transportation and Pavement Systems, Australian Academy of Technological Sciences and Engineering Journal Focus, No.95, January/February 1997, pp. 6 – 11, (see also www.atse.org.au)

PROPERTIES OF FLY ASH FROM REAL-SCALE CO-COMBUSTION EXPERIMENTS

A J Saraber

Delft University of Technology

J W van den Berg

Vliegasunie

The Netherlands

ABSTRACT. In most Dutch power stations secondary fuels are co-fired to reduce the emission of greenhouse gases. As the percentages of co-combustion on Dutch power stations increase, more fundamental knowledge of the properties and performance of the generated fly ashes has to be gained. Therefore fly ashes from real-scale co-combustion experiments were investigated. The fly ashes from co-combustion have a chemical composition that is, in general, comparable to the average Dutch fly ashes, with the exception of Ca and P. These components are enriched when sewage sludge, meat and bone meal and chicken manure are co-fired. The most important effect on the mineralogical properties is an increase in the acid soluble fraction, due to the presence of higher percentages of Ca and P components. Only crystalline components that are usually present in Dutch fly ashes were identified. The composition of the glassy phase is comparable to that of fly ashes from 100% coal and to that of natural pozzolans. The amount of reactive silica is within the limit specified in CUR recommendation 83 ($\geq$25% m/m). The physical properties are also comparable to those of average Dutch fly ashes. The performance of fly ashes when used in cement paste, mortar and concrete was not influenced by co-combustion. This is in accordance with the results of the characterisation.

Keywords: Fly ash, Co-combustion, Pozzolanity, Co-firing, Concrete

Ing Angelo J Sarabèr MSc. is a researcher/consultant at Delft University of Technology, Faculty of Civil Engineering and Geoscience, Delft, the Netherlands.

Jan W van den Berg is a quality manager at Vliegasunie, Nieuwegein, the Netherlands.

INTRODUCTION

Since the industrial revolution, energy from fossil sources has played a key role in the technical and economic development of society. However, energy production inevitably comes with CO_2 emission. A major source of CO_2 emission in the Netherlands is power production using coal. The Dutch power companies and the government have agreed to reduce these emissions by signing the covenant "Coal-fired power plants and CO_2 reduction" [1] The power companies have committed themselves to reducing emissions by 3.2 Mton CO_2 on average in the budget period 2008 - 2012 by replacing coal with biomass. The biomass input will correspond to 503 MW_e (electric) of the total capacity of 3875 MW_e. This capacity is realised by five power stations each equipped with one or two pulverished fuel boilers and one power station with a gasification unit.

As dumping is no option in the Netherlands it is important, from an environmental point of view, that the quality of fly ash (PFA) is not diminished by co-combustion, as this would hinder utilisation. Therefore, at the request of the Dutch power companies, KEMA started to investigate the effects of co-combustion on the quality of PFA. In the period 1995 - 2000, co-combustion experiments were carried out and co-combustion was also implemented at Dutch power stations. The experiences with the quality of the fly ash from co-combustion (PFAC) were positive [2]. However, the co-combustion percentage was limited to about 10% m/m on an ash basis (10% of the ash originates from secondary fuels and 90% from coal). In 2000 it became clear that in the near future this limit would be exceeded. The consequence of this development was that more fundamental knowledge had to be gained about the basic characteristics and performance properties of PFAC containing far more than 10% ash from secondary fuels. Therefore, it was decided to start a research programme to develop this knowledge by extensive research of fly ashes from co-combustion experiments. This programme was financed by the Dutch power companies and Vliegasunie. The first step was to investigate fly ashes from pilot-scale experiments. From this research it has become clear that depending on the nature (mineralogy) of the fuel high co-combustion percentages are possible (> 15% ash base) without significantly influencing the technical quality of the fly ash [3]. The second step was to investigate fly ashes from real-scale experiments. The first results are presented in this paper. The aspects that were investigated concerns basic characteristics and properties of cement paste and mortar.

FLY ASH SAMPLES FROM CO-COMBUSTION EXPERIMENTS

PFACs from several real-scale co-combustion experiments were investigated as part of the ongoing research into the effects of co-combustion. Several samples of PFAC from Dutch power stations were collected. Samples were taken during the co-combustion experiments. In Table 1 an overview of these PFACs is given. No further information is given regarding the power stations for confidentiality reasons. The intention was to collect samples with as high a co-combustion percentage as possible and from a wide spectrum of fuels. Meat and bone meal is waste material from the destruction of cadavers. The inorganic fraction of this fuel mainly consists of calcium-phosphor components (bio-apatite). Paper sludge mainly consists of calcium carbonate and kaolinite as the inorganic fraction. Biomass pellets are a commercial product that consists of sewage sludge, paper sludge, wood and RDF. The inorganic fraction of this product consists of several minerals, including quartz, phosphates and clay minerals.

Table 1 Overview of PFACs from real-scale experiments [4]

SAMPLE CODE	COAL	CO-FIRED FUEL	% CO-COMBUSTION	
			m/m coal base	m/m ash base
LVM14	Blend	Meat and bone meal	10.2	
		Biomass pellets	4.2	
		Chicken manure	1.0	
		Soot paste	0.3	
		Sewage sludge	0.1	
		Total	15.8	17.4
LVM12	Blend	Meat and bone meal	6.4	
		Biomass pellets	8.0	
		Chicken manure	1.0	
		Total	15.4	27.7
KSB	Blend	Paper sludge 1	4.4	
		Paper sludge 2	1.5	
		Wood	2.9	
		Total	8.8	24.1
KSBR	Blend	None	0.0	0.0
GREMA		Wood	33	1.0

EXPERIMENTAL DATA

Several methods were used to characterise the fly ashes. A survey is presented in table 2. The mineralogical composition was determined by a special developed characterisation method. The characterisation method consists of a combination of existing analysis methods, intended to provide systematic insight into the mineralogical properties of fly ash. A key feature of the analysis is the cascade approach, which involves removing in turn the fraction that is soluble in acid and the fraction that is soluble in potassium hydroxide (comparable to the determination of the amount of reactive SiO_2), and analysing the eluates each separately. The residue from these steps is analysed chemically and mineralogically (by X-ray diffraction).

Table 2 Methods used to characterise fly ash

PROPERTY	METHOD
Cl	EN 450 [5]
Free CaO	EN 450
Grain size distribution	Laser diffraction and wet sieving 45 um
Loss of ignition	EN 450
Macro-components	X-ray fluorescence
Mineralogical composition	See text
Morphology	Scanning ElectronMicroscopy (SEM)
Reactive silicon oxide	EN 196-2 [6]
Sulphate	EN 450
Volume weight	EN 450

If the LOI of a sample of PFA(C) was higher than the limit of 5.0% m/m, the LOI was reduced to 5.0% m/m before it was used in paste or mortar to investigate the performance.

The procedure was as follows: part of the sample of PFA(C) was smouldered at circa 500 °C for four hours;this part of the sample was mixed with non-smouldered PFA(C) in a ratio sufficient to give a final LOI of just 5.0%. LVM14 was treated in this way. The original LOI was 7.3%. After pre-treatment, the LOI was 4.6%.

The investigated properties of paste and mortar concern setting time, soundness, workability (flow) and activity index (according to EN 450).

RESULTS

Physical properties

The average grain size diameter of the PFACs varies between 17 and 23 µm. This is within the range (21 ± 3 µm) of the Dutch fly ashes [1] . The percentages passing the 45 µm sieves (66-83%) meet the requirements of EN 450 (60%). The morphology of PFAC particles was investigated by SEM. It was concluded that the morphology of PFAC was not different from PFA.

Chemical properties

The amount of $SiO_2 + Al_2O_3 + Fe_2O_3$ is about 80 - 85% m/m (see Table 3). This is more than the minimum required under CUR recommendation 83 and ASTM C618-01 (class F) [7, 8]. The concentration of P_2O_5, CaO of LVM14 and LVM12 is outside the range of Dutch PFA. This is caused by the co-combustion of meat and bone meal and biomass pellets. The high amount of CaO in KSB is due to the co-combustion of paper sludge. Hence, the coal used was relatively poor in CaO.

In a single case, the amount of free lime was higher than 1.0% m/m, but the expansion of the LeChatelier test was lower than the limit of 10 mm. The amounts of Cl and SO_3 meet the requirements of EN 450. The Na_2O equivalent and the amount of CaO total are lower than the maximum required under CUR recommendation 83 (5.0 and 10.0% m/m respectively). The amount of reactive silicon oxide meets the requirement of 25% m/m.

Mineralogical properties

The acid soluble fraction varies between 11 and 24% m/m. This is significantly higher than that of reference PFA (2.4 - 10.6 % m/m). Substantial proportions of the S, P and Ca present are dissolved. It is clear that the percentage of components that is soluble in acid increases when secondary fuels are co-fired.

This is mainly caused by Ca and/or P. However, no calcium and/or phosphor-containing phases were detected by X-ray diffraction (XRD).

Table 3 Chemical composition (macro components) of PFA(C) (n = 1)

	REQ.	REF.	GREMA	LVM14	LVM12	KSB	KSBR
Al_2O_3	-	27.3 ± 2.7	28.12	27.09	29.37	22.90	24.90
Fe_2O_3	< 10.0	7.7 ± 1.9	7.67	4.82	4.46	7.96	9.45
CaO	< 10.0	5.1 ± 1.6	6.90	9.88	9.48	7.96	2.32
Free CaO	< 1.0	0.4 ± 0.3	1.19	1.02	0.37	< 0.05	-
MgO	-	1.7 ± 0.4	1.13	1.36	1.18	1.74	1.49
Na_2O	-	0.5 ± 0.2	-	0.86	0.81	0.34	0.35
K_2O	-	1.5 ± 0.5	-	1.75	1.63	2.28	2.11
TiO_2	-	1.4 ± 0.4	-	1.39	1.42	0.96	1.13
P_2O_5	< 3.0	0.8 ± 0.3	1.18	4.32	3.74	0.83	0.60
SiO_2	-	52.8 ± 2.7	53.04	50.88	47.65	54.90	57.20
Cl	< 0.1	0.01	< 0.01	0.01	0.03	# < 0.08	# < 0.08
SO_3	< 3.0	0.5 ± 0.1	0.89	0.35	0.26	0.44	0.33
Soluble P_2O_5	-	-	< 0.0015	-	0.0053	0.0021	0.0032
LOI	< 5.0	5.3 ± 2.3	3.62	4.62	5.87	4.00	5.30
Acid soluble fraction*	-	2.4-10.6	9.6	24.3	23.9	18.6	9.4
Reactive SiO_2	> 25	-	-	33.6	33.5	-	-
Glass content	-	34.4-56.1	46.1	(32.9)	40.4	42.9	34.4
Al/Si-ratio	-	0.52 ± 0.07	0,53	0.53	0.62	0.42	0.44
Na_2O-equivalent	< 5.0	1.5 ± 0.5	1.38	2.02	1.88	1.80	1.74
$\sum Al_2O_3 + Fe_2O_3 + SiO_2$	> 70.0	87.8	88.9	82.79	81.48	85.76	84.42

Req. = requirement   Ref.=range of dutch fly ashes [1]   * exclusive free lime

Under the Dutch standard NEN 5435 [9], the glass content of PFA is regarded as equal to the fraction that is soluble in potassium hydroxide. The glass content of the PFAC is lower than the reference values. The glass fraction consists mainly of SiO_2 and Al_2O_3 (minimum 83%). This means that, despite co-combustion, the glass phase is still a silica-alumina glass, as in PFA from 100% coal (see Figure 1).

The ratio of network modifiers (NM) to network formers (NWF) and intermediaries (I) increases as a result of co-combustion. This means that theoretically the glass is less resistant to hydroxyl attack in the cement environment as the polymerisation rate of the glass becomes lower. This is basically positive for the pozzolanic behaviour of the PFAC. There is some discrepancy between the glass content of LVM14 and its amount of reactive SiO_2 regarding the Al/Si ratio of the glass fraction.

X-ray diffraction analyses of the ashes show that the same minerals are present as are normally found in PFA. Quartz, mullite, rutile, magnetite and hematite were identified. No other minerals could be identified.

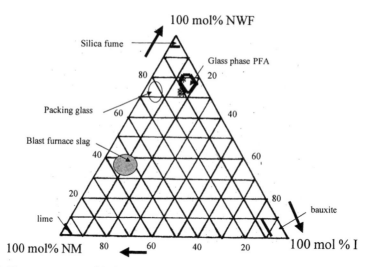

Figure 1   Ternary system with the location of PFAC (*),PFA from 100% coal, lime, blast furnace slag, bauxite, silica fume and packing glass

The results of the cement paste and mortar experiments are presented in Table 5. The activity indices at 28 days vary between 80 and 85%. The activity indices increase up to 91 - 98% after 91 days. These values meet the EN 450 requirements (minimums of 75% and 85% after 28 and 91 days, respectively). The activity indices of the PFACs are very similar to the index of the Dutch PFA.

Table 4 Soundness, setting behaviour, development of compressive strength and activity index of PFACs

PROPERTY	REQ	Ref. [11]	n	GREMA	LVM14	LVM12	KSB	KSBR
Activity index								
28d [%]	75	83.9 ± 3.3	6	85	80	82	82	82
91d [%]	85	98.3 ± 4.3	6	91	94	98	96	95
Setting time								
Initial set [hr : min]	*	3:42 ± 16	1	-	3:39	3:50	3:03	3:07
Final set [hr : min]	-	5:27 ± 32	1	-	4:58	5:40	4:35	4:21
Intial set of paste			3	-	15:10	15:20	15:20	15:26
with retarding agent								
Flow [mm]	-	180 ± 8	2	-	170	171	178	189
Soundness [mm]	<10	1.3 ± 0.7	2	2.0	1.0	2.0	1	0.5

* initial set of cement paste with PFAC may be retarded by up to two hours in comparison with pastes made with pure cement (CUR-recommendation 83)

The initial setting of all cement fly ash pastes is within the range of Dutch PFA. The initial setting of the paste with KSB (3:03 hr) is somewhat faster than the average for Dutch PFA, but is practically equal to that for the reference fly ash KSBR (3:07 hr). The initial setting was retarded up to 15 hours when a retarding agent (Cugla MMV; concentration 25%) was added. The differences between cement pastes with PFAC and the cement paste with KSBR were negligible. The effect of the retarding admixture is in line with the results for other reference fly ashes [2]. The expansion of all samples is relatively small (≤ 2 mm) and meets the requirement of the LeChatelier test (maximum 10 mm expansion).

EVALUATION

The chemical composition of the investigated PFACs from real-scale experiments was significantly influenced by co-combustion. The total CaO concentration increased, with or without an increase of P_2O_5. The chemical composition met the requirements of EN 450 and CUR recommendation 83. These requirements consider Cl, SO_3, Na_2O-equivalent, the sum of $Al_2O_3+SiO_2+ Fe_2O_3$, total CaO, free CaO and reactive silicon acid. Despite the increase of the total CaO content, the free CaO content did not increase, or increased to a limited extent only. PFAC LVM14 had more than 1.0% m/m free CaO, but met the soundness requirements of the LeChatelier test. The amount of components that were soluble in acid increased as a consequence of co-combustion. This was mainly caused by the increase in Ca and P. If the solubility was tested in neutral conditions, hardly any P was dissolved.

The amount of components that were soluble in potassium hydroxide varied as a consequence of co-combustion and/or the coal used. This glassy phase is important for the pozzolanic behaviour of PFA(C). The ratio of network modifiers to network formers is an indication of the resistance of the glassy phase to chemical attack (network theory of Zachariasen). As the dissolution of the glass phase is the first step in the pozzolanic reaction mechanism, it is expected that the shift in this ratio will have no negative influence on pozzolanic behaviour and may even have a positive effect.

The important conclusion from the assessment of the glass phase is that, despite co-combustion, the glassy phase is a silica-alumina glass, as in PFA from 100% coal (class F) and natural pozzolans (such as trass). Figure 1 shows the position of the glassy phase in the ternair system of network formers (NWF), intermediates (I) and network modifiers (NM).

Microscopic observation of the well-known spherical particles indicated that the morphology was not significantly influenced by co-combustion. The grain size, expressed as the D50-value, was within the range of Dutch PFA.

In summary, it may be said that although the chemical composition is influenced by co-combustion, the requirements of EN 450 and CUR recommendation 83 are still met. The glass phase of the investigated PFACs is a silica-alumina glass, as found in PFA from 100% coal. There were no 'PFA-strange' components found in the PFAC. The fraction that was soluble in acid increased due to the increased concentration of Ca and P. Because of the basic character of mortar and concrete, this has no practical significance (limestone is widely used in concrete, but is also soluble in acid). The phosphor present is barely soluble in a neutral or basic environment, as indicated by the solubility experiments and the normal setting behaviour and initial strength development.

CONCLUSIONS

The following conclusions were drawn:

- the coal fly ashes from co-combustion have a chemical composition that is, in general, comparable to the average Dutch fly ashes, with the exception of Ca and P. These components are enriched when sewage sludge, meat and bone meal and chicken manure are co-fired;

- the most important effect on the mineralogical properties is an increase in the acid soluble fraction, due to the presence of higher percentages of Ca and P components. Only crystalline components that are usually present in Dutch fly ashes were identified. The composition of the glassy phase is comparable to that of fly ashes from 100% coal and to that of natural pozzolans. The amount of reactive silica is within the limit specified in CUR recommendation 83;

- the physical properties are comparable to those of average Dutch fly ashes;

- the performance of fly ashes when used in cement paste and mortar was not influenced by co-combustion. This is in accordance with the results of the characterisation.

REFERENCES

1. RIJKSOVERHEID EN DE ELECTRICITEISPRODUCTIESECTOR. Convenant kolencentrales en CO_2-reductie. 2002 (in Dutch).

2. BERG, J.W. VAN DEN ET AL (Vliegasunie), Fly ash obtained from co-combustion. State of the art on the situation in Europe"; Proc. ACAA symposium, USA, 2003.

3. SARABÈR, A.J. AND BERG, J.W. VAN DEN. Influence of co-combustion on the quality of fly ash. Eighth CANMET/ACI international conference on Flyash, silica fume, slad and natural pozzolans in concrete, Las Vegas, USA, May 23-29. 2004

4. VLIEGASUNIE. Written and oral information co-combustion experiments. 2002

5. NEN. EN 196-2. Methods of testing cement. Part 2 chemical analysis of cement. NEN. EN 450. Fly ash in concrete – definition, demands and quality control. 1995.

6. VLIEGASUNIE, Information on quality control fly ash, 2001.

7. CUR, CUR recommendation 83. Toepassing van poederkoolvliegas in mortel en beton, Gouda. 2001 (in Dutch).

8. ASTM. ASTM C618-01. Standard specification for coal fly ash and raw or calcined natural pozzolans for use as a mineral admixture in concrete. 2001.

9. NEN. NEN 5435. Verbrandingsproducten van vaste brandstoffen – Bepaling van het glasgehalte van vliegas als de in kaliumhydroxide oplosbare fractie. 1993 (in Dutch).

WASTE GENERATION ISSUES IN PETROLEUM PRODUCING COMMUNITIES IN NIGERIA

S Dalton

Ambrose Alli University

Nigeria

ABSTRACT. The increasing global concern on the environment demands that waste should be properly managed in order to minimize and possibly eliminate their potential harm to public health and the environment. This paper looks into the various classes of waste and their sources in petroleum producing communities in Nigeria and also possible treatment and method of disposing them are also discussed. From Research, it was discovered that the main sources of wastes in petroleum producing communities in Nigeria are mainly from Petroleum Industry (Upstream Operation). These include drilling mud/fluids and the drilling cuttings which come as a result of well drilling processes. Others are the tank bottom sludge, which come as a result of crude oil sedimentation in the storage tank, the oily water phase, accidental oil spillage, well blow-out, leakage of the pipelines and storage tanks. Also, gaseous wastes are produced as a result of gas flaring, stock flare and the result from the combustion of heavy vehicles during the exploration process. This paper intends to come up with the consequences of the waste generated in the oil producing communities of Nigeria and the ways which are adopted by the oil companies in managing those wastes.

Keywords: Waste, Upstream, Exploration, Environment, Petroleum, Oil, Communities.

O S Dalton, is a final year student of Mechanical Engineering Department of Ambrose Alli University Ekpoma Nigeria.

INTRODUCTION

The discovery of crude oil (Petroleum) in the year 1956 in a town called Oloibiri in Niger-Delta (South-South geo-political zone) of Nigeria brought smile to the faces of many Nigerians especially those living in the region. Though many saw it as an economic instrument, little did they know that it also has its own dangerous implications. It was not until the early 70's that the people started experiencing the by-product of the post war abundance and technology in which there was this uncontrolled pollution of the air, land and water. The atmosphere became polluted by intense Industrial activities; land and water were also not spared.

In this paper, it is the wish of the author to focus on the main source of waste in these communities having proven reserves of crude oil and natural gas of over 27 billion barrels and 120 trillion standard cubic feet. From general observations, it was discovered that the main source of waste in oil producing communities in Nigeria is mainly from Petroleum Industry (Upstream Operation). This consists mainly of solid, or semi solid materials, liquid and gaseous substances resulting from activities and processes necessary for the successful operation of the petroleum Industry. In some ways, the dealings of the industry got man and his environment affected with the associated hazards.

In recent years, due to the consequences of these wastes generated in the Host communities, the Federal Environmental Protection Agency (FEPA) through the Department of Petroleum Resources (DPR) called for a better management of these wastes or a more scrupulous method that will not endanger the biotic and Abiotic environment. It was stated that proper management of wastes must be matched or accompanied by a proper analytical process to ensure that, what is being disposed off as waste does not turn out to be hazardous. This is to say that, all effluents solid, liquid and gaseous wastes must be thoroughly made harmless before being disposed off. In order to effectively evaluate the discharge of these wastes into the environment of the host communities in Nigeria and for the sake of this paper, the oil industry is conveniently divided into three stages of operation, namely:

- Exploration
- Production and
- Transportation

DEFINITION OF SOME TERMS

Before going into detail, let us define some terms as used in this paper.

Waste: Is any substance, solid, liquid and gas that remains as a residue or an incidental by-product of the processing of a substance and for which no use can be found by the organisms or systems that produce it.

Upstream Operation: This covers all the activities related to exploration, discovering and extraction of petroleum and gas and their treatment, transportation and delivery to designated export terminals or otherwise to the processing plants e.g. Refineries. Also included are the auxiliary services related to petroleum and gas production, transportation and termination.

Petroleum Producing Communities: These are the region where crude oil is discovered. In other words, they are the Host communities of the various petroleum Industries in Nigeria.

WASTE GENERATION

Generally, Upstream Operation in petroleum Industry generates wastes which could be solid, liquid and gaseous.

Solid Waste

The main sources of solid wastes in Upstream Operation are the drilling cuttings, Tank bottom sludge and contaminated soils which are the combination of both sludge and other impurities which may tend to stay in the tank bottom where they are deposited.

Liquid Waste

The main source of Liquid waste is the expressed liquid when oil are squeezed by machines (busters), Drilling fluids (mud), deck drainage, well treatment and oil spillage. On production process, oil is pumped up by buster machines and collected into a storing tank (Flow station). This petroleum is contaminated with sand and other impurities. It is then allowed to settle down; from here tank bottom sludge are formed and only water is separated.

Gaseous Waste

This is the atmospheric emission from the rigs which consist mainly of exhausts from diesel engines supplying power to meet rig requirement. These emissions may have small amount of Sulphur dioxide. (dependent upon fuel Sulphur content) and exhaust smoke (heavy hydrocarbons). An unexpected over pressure formation encountered during drilling (i.e. formation pressure above normal hydrostatic pressure gradient of 0.465psi/ft) may result in a blow-out or gas discharge. If the mud used does not provide adequate hydrostatic head balance over the reservoir pressure, transient emission of light hydrocarbons and possible hydrogen sulphide may occur. Apart from those major wastes which are deliberately discharged and planned, several factors during production operation and processing sometimes result in hazardous unplanned oil discharges. These are what are referred to as accidental discharges which are principally due to system leakage or mechanical failure:- The "system" here refers to the network of gas pipelines, compressors, vessels, the control loops e.t.c which are being operated under high pressure for petroleum transportation and distribution between the source (well) and the flow stations. Nigeria has a pipeline network of over 5,000km. When this happens, it will result to uncontrolled escape of oil and gas at high pressure into the land and atmosphere respectively. The land and ambient air get polluted

RATE OF GENERATION

Table 1 Rate of major discharges from offshore oil and gas operation

DISCHARGE CATEGORY	EXPLORATION/APRISAL	DEVELOPMENT
(i) Drilling Fluid (total addictive to water based) system and volumes discharge	(i) Well depths less than 10,000ft (3050m) 300 600 tons/well 240-489cm^3	
	(ii) Well depth greater than 10,000ft (3050m) 500-850tons/well 400-680m^3/well	250-500tons/well 200-400m^3/well
(2) Drilling cuttings	100-1600tons/well	900-1400tons/well

For Gas flaring it is estimated at 2 billion standard cubic feet per day out of total daily production of 2.5 billion standard cubic feet.

CLASSIFICATION OF WASTES

Broadly speaking, wastes can be classified as:

- Corrosive (types)
- Toxic (types)
- Flammable (types)

CONSEQUENCES OF WASTE GENERATION

Hazardous waste generation from petroleum industry and formation water causes considerable environmental pollution. In addition to hydrocarbons, industrial waste may also contain phenols, ammonia, sulphides, cyanide and metals all of which may constitutes health hazards. As a result of these, it would be worthwhile to review the effects of wastes generated from upstream operation on Agricultural land, aquatic life and ecosystem in General.

Consequences of Waste on Land

Effect of wastes on land can be studied from the agricultural stand-point and be looked at from two angles.
(i)   its effects on the soil and
(ii)   its effects on the crop plants.

Consequences of Waste on Sea and / or Rivers.

The effect of petroleum on aquatic organisms, mangrove and its community are very diverse and complex. But a careful consideration of the biological and ecological effects of petroleum hydrocarbons is important for the prevention and control.
Pollution of water means interfering with various uses of water: Recreational, drinking, navigation, irrigation, sea food, farming etc. Sea foods are exterminated, marine life may be destroyed. On the surface of water, petroleum may limit oxygen exchange, entangle and kill surface organisms and cut the gills of fishes. However, there are only a few report of oil pollution causing severe biological damage.

Consequences of Wastes in Air

Climatic condition, human and machine activities often combined to highlight the effect of air pollution on human health and wellbeing. We are all aware that polluted air makes the eyes water, bounds the nose and stifles the lungs. Fumes of sulphur and nitrogen oxide, sulphuric acid and photo-chemical oxidants irritate the respiratory system causing coughing, chest discomfort and impaired breathing. Carbon monoxide interferes with the ability of the red blood cells (R B C_S) to carry oxygen. Heat and nerves tissue are particularly susceptible to oxygen deficiency. Carbon monoxide pollution can seriously impair coronary and central nervous system function.

Economic Consequences of Waste Generation

Apart from the consequences of waste generation already highlighted. Waste generation also has economic consequences such as wastage of large Acres of land and gas flaring.

Wastage of large acres of land: Solid wastes occupy space on land. Many acres are wasted by dumping of solid wastes and landfill process. Other factors which contribute to large space occupied are the development process of the petroleum exploration and exploitation. This involves:

(a) Implementation of oil concession
(b) Shooting of seismic lines
(c) Digging of oil well

WASTE MANAGEMENT/TREATMENT APPLIED BY VARIOUS PETROLEUM INDUSTRIES IN NIGERIA.

Waste minimization goals are both necessary and desirable but most oil industries in Nigeria still create waste product that will ultimately need to under go treatment to destroy the waste or render them harmless to the environment. This is due to improper or inefficient waste management/treatment adopted by some petroleum companies. Some companies adopt proper techniques used to manage hazardous wastes which have broad acceptance by the government, industry and public alike. These methods include:

(i)   Chemical waste treatment
(ii)   Biological waste treatment.
(iii)   Physical waste treatment.

Chemical Treatment

Chemical treatment involves the use of reaction to transform hazardous waste streams into less hazardous substances. Chemical treatment can be useful in promoting resources recovery of hazardous substances. In that, it can be employed to produce useful by- products and residual effluents that are environmentally acceptable. Chemical treatment is a far better method of waste management than the traditional method of disposal at a landfill. Majority of the petroleum industries in Nigeria do not adopt this method fully, due to high cost and restrictive regulation.

There are many different forms of chemical treatment used in this method of hazardous waste management. These are:

* Solubility.
* Neutralization.
* Precipitation
* Disinfection
* Coagulation and Flocculation

Biological Waste Treatment

Organic wastes can be stabilized by biological treatment process through the metabolic activities of heterotrophic micro-organisms. These micro-organisms convert organic to end product such as carbondioxide, water and methane gas. The biological treatment process Adopted by some petroleum industries in Nigeria are:

- Activated sludge process
- Lagoons
- Biological filters, Trackling Filters, and oxidation (cooling) towers.

Physical Waste Treatment.

Wastes are also treated by physical methods. These methods are sedimentation, filtration, flotation, reverse osmosis and submerged combustion. Others are landfill, incineration, and Gravity separation, Stripping, Adsorption and Extraction.

CONTRIBUTIONS/COMPENSATION BY THE VARIOUS OIL COMPANIES TO THE DEVELOPMENT OF THE REGION

Though there are a lot of community developmental projects embarked upon by the various oil companies such as building of community Health care centre, provision of portable water, awarding of scholarship to deserving students from the community, constructions of roads, sponsoring of sporting events e.t.c Still, this not enough to applause the oil companies and for the fact that there is still a great deal of under development and abject poverty in the region signified that there is much to be done in the region.

RECOMMENDATION/CONCLUSION

In order to protect and preserve our environment from pollution caused by petroleum-related operations, each petroleum producing company should enhance their existing oil spill contingency plan and it is necessary to continuously update and extend the baseline data using any further information that the petroleum may collect from their areas of operation. Co-operation amongst the industries, national and local authorities is a further step towards obtaining significant result. The computerized oil spill record system for its wild applicability range, is an advantageous tool that would give a practical answer to waste problems.

Then, for the strategy of waste management, waste should be divided into three broad categories in order to achieve better results:- these are domestic, industrial and hazardous wastes. The first two categories of wastes are non-hazardous. They are normally encountered in municipal waste management. Hazardous wastes (some times called special or chemical wastes) require specific treatment and disposal methods. For hazardous it will be required that, the generators of the waste should provide facilities for treatment/disposal or alternatively enter into contract with specialized firms to provide such services. The contract for disposal of hazardous waste will not however relieve the generator of his "cradle to grave" responsibility and he must ensure that the wastes are disposed off safely with adequate records and proofs. Subsequently, the modern waste management practice should be adopted which are the principles of Reduce, Reuse, Recycle and Recovery.

More so, the various petroleum industries should ensure that they invest in researches and development, finding design to improve on the already existing environmental control devices and process. Furthermore, effective measures and stern decision should always be taken in advent of obvious problems, while awaiting full proof of the cause and effect.

Educating the people to understand the basic consequences (effects) of wastes found in their environments, whether they are the ones producing them or are being produced by the neighbouring industries. This can obviate the individual from being exposed to the pollutants and thus, the public can carefully handle their own produced wastes.

REFERENCES

1. N.N.PC (department of petroleum Resources), Environmental Guidelines, 1990

2. FEDERAL ENVIRONMENTAL PROTECTION AGENCY DECREE, A publication of (FEPA), 1988

3. LUND, Industrial Pollution Control Handbook, pp7-38, McGraw-Hill Company

4. OGBA JAMILU LIKITA, Oil Companies and pollution.   pp.37, The Monitor (4th Quarter 1991) An in-house newsletter of (NPDC)

5. CHEREMISINOFF and YOUNG, Pollution Engineering Practice handbook

6. WARREN E. ISMAN, GENE P. CALSON, Hazardous materials

7. NIGERIAN SOCIETY OF ENGINEERS, Warri Branch, Proceedings of Seminar on Municipal Waste management, 2001

8. LAMID ALIU YAMAH, Waste management (Chemical treatment) and Remediation of tank bottom sludge, oily and contaminated soils. Paper presentation

9. FEDERAL REPUBLIC OF NIGERIA, Obasanjo's Economic Direction, 1999-2003

10. THE GUARDIAN NEWSPAPER, Oil companies and NDDC, March 16th, 2004

11. HARDAM SINGH AZARD, Industrial waste management handbook

12. RON BAKER, A primer of offshore operations Second edition Pp 32-47, 1985

13. LAWRENCE PRINGLE, Recycling resources pp14-68 Macmillan publishing company Inc. New York Collier Macmillan Publishers London

14. DAPO OGUNTOYINBO, Municipal waste management- an overview, Pp 6-8, waste management department S.P.D.C. waste Division

15. ALBERT PARKER, Industrial Air pollution Handbook Pp578-596, Mc-Graw-Hill Book Company (UK) Limited

STUDY OF EARLY HYDRATION OF CEMENT PASTES CONTAINING ALTERNATIVE CALCIUM SULFATE BEARING MATERIALS

G Tzouvalas   **A Papageorgiou**   **S Tsimas**

National Technical University of Athens

D Papageorgio

Titan Cement Corporation

Greece

ABSTRACT. The majority of cement plants mainly uses natural gypsum ($CaSO_4 \cdot 2H_2O$) to regulate cement setting. The partial or total replacement of gypsum by materials, which contain calcium sulfate, has been instigated either by the increasing availability of low-cost by-products containing calcium sulfate or by the possibility that quarries would mine rock that is a mixture of gypsum and anhydrite. This paper is part of an extent research program aiming to investigate alternative calcium sulfate bearing materials, such as anhydrite and FGD gypsum. These materials have been examined in order to study the influence of their addition in cement physicomechanical properties (compressive strength, setting time). Hydration of cement pastes containing these CSBM was interrupted at specific time intervals within 5 min to 12 hours, and cement-setting reactions were characterized with the use of DTA, isothermal calorimetry and ESEM. In the same time SO_3^{2-}, Ca^{2+} and pH of the pore solution were also evaluated. It is concluded that, compared with natural gypsum, FGD gypsum prolongs the dormant period of hydration, while anhydrite accelerates it. Moreover, at a specific hydration time, cements with FGD gypsum seem to form more ettringite than natural gypsum and anhydrite does.

Keywords: Gypsum, Anhydrite, FGD, Setting, Ettringite, Induction / dormant period.

George Tzouvalas is a PhD Chemical Engineer and researcher in the National Technical University (NTU) of Athens. His major interests are in aluminosilicate technology and he is currently undertaking research into the evaluation of alternative setting retarders in the production of cement.

Aristotelis Papageorgiou is a research PhD Chemist. He has long experience in cement chemistry and technology.

Dimitris Papageorgiou is a PhD Chemical Engineer and researcher in Titan Cement Co. His major interests are in cement chemistry, technology and microscopy.

Stamatis Tsimas is a Chemical Engineer and Professor of Inorganic Materials Techniques at the NTU of Athens. His interests are in chemistry and technology of aluminosilicates, specializing in cement technology and upgrade techniques of industrial minerals and by-products. He has published over 80 papers in related Journals and Congresses.

INTRODUCTION

The most common cement setting retarder is natural gypsum, a dihydrated product of calcium sulfate ($CaSO_4 \cdot 2H_2O$). Gypsum quarries are steadfastly moving into mining rock that is a mixture of gypsum and anhydrite. Partial replacement of the gypsum with anhydrite is possible for the majority of types of clinker [1, 2, 3]. At the same time, due to environmental concerns, it is essential to examine the possible use of by-product gypsum. An increasingly available low-cost by-product is Flue Gas Desulfurisation (FGD) gypsum, the product of the desulphurization process of residual gases with limestone in coal burning power plants. Strong efforts have been made to utilize FGD gypsum from the technical and economic point of view [4, 5, 6, 7, 8]. Recently Tzouvalas et al. [9, 10] attempted to evaluate different calcium sulfate bearing materials (CSBMs) as setting retarders on laboratory and industrially-produced cements with various admixtures with very encouraging results. Compared with natural gypsum, anhydrite reduces setting time while FGD prolongs it.

When cement is mixed with water, a strong exothermic reaction takes place indicating the pre-induction period of hydration, which lasts a few minutes. During this period Ca^{2+}, SO_4^{2-} and Si^- are released in the solution and only a 2 to 10 % of C_3S is hydrated [11, 12]. For the next few hours the reaction rate is reduced to a large extent, although an ion exchange between unhydrated grains of cement and the pore solution goes on. This is attributed to the reaction of C_3A with CSBM and the formation of ettringite. It is assumed that ettringite forms initially on the reacting C_3A surface, a more-or-less impermeable coating that impedes diffusion of the ions needed to form the hydrates that cause setting [1, 3, 13]. According to another theory C_3A dissolves incongruently in the liquid phase, leaving an alumina-rich layer on the surface with adsorbed Ca^{2+}, thus lowering the number of active dissolution sites and the reduction rate. A subsequent adsorption of sulfate ions results in a further reduction of C_3A hydration [11, 12]. This is the induction or dormant period of hydration.

Increase of Ca^{2+} and OH^- in the solution leads to a considerable acceleration of hydration rate (acceleration period), where $Ca(OH)_2$ concentration to the liquid phase reaches to a maximum value and portlandite starts to precipitate. The initiation of this period reflects macroscopically the initial setting time of cement. Hydration products are formed and pores are drastically reduced. Afterwards decreasing hydration rates lead to a thermal relaxation, ion exchange is inhibited and reaction rate is slow down. This is the deceleration period of hydration which is controlled by diffusion and calcium silicate hydrates form denser structures. Strength development slowly starts after the setting, is relatively fast during the first days and slowly continues for months and years [13, 14].

The variable solubility of the calcium sulfate bearing materials induces a range of available sulfate concentration during cement hydration. Previous investigations have demonstrated that each form of $CaSO_4$ has different solubility [3, 15, 16]. This results in different strength of the hardened paste. These authors along with Xiugie [17] believe that absolute solubility for both gypsum and anhydrite is similar, while the dissolution rate of gypsum is faster than that of anhydrite. Kumar [18] has also studied the hydration of cement based on the solubility of the cement components and states that the presence of a strong anion such as SO_4^{2-} in the solution reduces the solubility of less strong anions, such as silicates and aluminates, but tends to accelerate the solubility of calcium anions, thus effecting the setting time.

The present work studies the influence of CSBMs on the early hydration of cement by means of DTA-TG, calorimetry, ESEM for the cement paste and determination of SO_3, Ca^{2+} and pH for the pore solution.

EXPERIMENTAL

Three laboratory produced cements (CEM I) with natural gypsum (G), anhydrite (A) and FGD (F) as setting regulators and 3.5% SO_3 addition, which was found to be the optimum for the setting time and compressive strength performance in previous works [4, 9, 10], were examined. Chemical analysis of clinker and CSBMs according to ASTM C471M-95 [19], is summarized in Table 1.

Table 1 Chemical analysis of calcium sulfate-bearings and clinker

	COMBINED WATER	SO_3	SiO_2	CO_2	CaO	R_2O_3*	MgO	C_3S	C_2S	C_3A	C_4AF
Gypsum	19.30	43.4	0.65	2.51	32.4	0.03	0.92				
Anhydrite	1.81	51.3	27.7	3.44	27.6	0.05	2.14				
FGD	18.05	42.2	0.30	1.58	33.4	0.10	0.10				
Clinker	-	0.71	22.0	-	66.2	8.52	1.92	63.9	14.9	6.7	11.2

* R_2O_3: $Fe_2O_3 + Al_2O_3$

Cements were mixed with water at a w/c ratio 0.5. At specific time intervals (5, 15, 30 min, 1, 2, 3, 6, 12 hours) the pore solution was separated from the paste by vacuum filtration. Sufficient quantities of the pore solution for the chemical determinations were filtrated up to 6 hours. The solids were then immersed successively in acetone and diehylethere solution to stop all hydration reactions and kept to a vacuum device at ambient temperature for 24 hours. For each hydration step, sulfate concentration in the pore solution was measured by precipitation of $BaSO_4$ with a 10 % (w/w) solution of $BaCl_2$ according to ASTM C471M-95 [19]. Calcium concentration was determined by AAS (Atomic Absorption Spectrometry) and pH with an electronic device.

Hydration products of the pastes were characterised with thermogravimetry and differential thermal analysis (DTA-TG). Measurements were performed in a Netzsch STA 409 – DSC/TG in a platinum crucible of 70 µl capacity in a nitrogen atmosphere (50 ml/min) at a heating rate of 10°C/min from ambient temperature to 1000°C.

The three non hydrated cements were also measured in a calorimeter C80 II (SETARAM) under isothermal conditions at 25°C. The calorimeter was consisted of two membrane mixing vessels, one for the specimen (250 mg) and another which was empty (reference vessel). Each membrane mixing vessel had two compartments, separated by a metal membrane. In the upper compartment 0.5 ml H_2O was placed. Mixing was done by breaking the membrane using a metal rod, operated from outside. The evolution of heat flow as a response to exothermal hydration reactions is documented over the first 36 hours [20]. The first peak indicates the pre-induction period. The period where the heat flow decreases and then remains stable is the induction/dormant period, followed by the acceleration one where the heat flow is increased dramatically (second peak).

Microstructural aspects of selected pastes (hydration step: 30 minutes, 3 and 12 hours) were investigated in an environmental scanning electron microscope.

RESULTS AND DISCUSSION

SO₃ concentration (Figure 1) in the pore solution of cement with anhydrite precedes followed by gypsum and finally FGD. Since all cements had the same SO_3 addition (3.5 %), the authors conclude that SO_3 in the solid paste, which is expected to precipitate as ettringite, should follow just the opposite range; that is more ettringite is formed in cement pastes containing FGD than in the pastes containing gypsum and anhydrite. In the same time it is obvious that SO_3 concentration remains almost stable in FGD pore solution much more time (from 30 to 180min of hydration) than in the other CSBMs (gypsum: 60 to 120 min, anhydrite: 30 to 60min). This is a strong indication that the induction/dormant period lasts longer when FGD is used, followed by gypsum and finally anhydrite. The dormant period is followed by a period (acceleration) of reduction of SO_3 concentration in the pore solution, where all sulfate anions are precipitated, CSBM is exhausted and its regulation action to the setting is over.

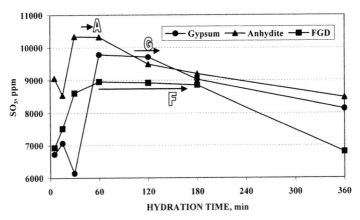

Figure 1 Measured SO₃ concentrations in the pore solution of cements with different CSBMs

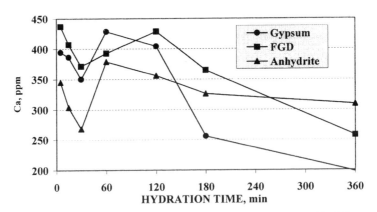

Figure 2 Measured Ca concentrations in the pore solutions of cements with different CSBMs

As far as the calcium concentration is concerned, it is obvious from results shown in Figure 2 that there is a reduction during the first 30 minutes due to the ettringite and some C-A-S-H (hydrogarnets) formation. The increase of Ca concentration in the pore solution reflects the dormant period [13, 14], where no hydration reactions take place. It is also concluded that the duration of dormant period follows the range: FGD>Gypsum>Anhydrite. This fact justifies the delay of setting caused by FGD as well as the acceleration of setting due to anhydrite addition compared with natural gypsum.

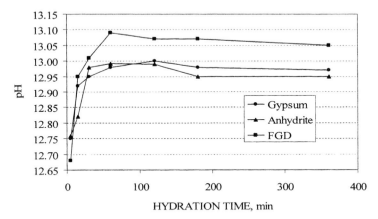

Figure 3 Measured pH values in the pore solutions of cements with different CSBMs

Measurement of $(OH)^-$ concentration (pH) is of great importance, since alkaline solution favours ettringite precipitation. It is also referred that the thickness of ettringite protective layer around the grains of C_3A, which delays the setting, depends on the pH. In Figure 3 the pH of all cements starts from 12.7 at the beginning of the hydration, but changes after the first 30 minutes. The pH of cement with gypsum stabilizes at 12.95 – 13.0 while that of anhydrite cement is slightly less. As far as the FGD is concerned, it shows the higher values (13.05-13.1). This must be attributed to the higher solubility of FGD (0.28 g / 100 g H_2O) compared with anhydrite (0.19 g / 100 g H_2O) and gypsum (0.26 g / 100 g H_2O), which releases more alkalis ions in the pore solution according to the reactions: $K_2O \rightarrow O_2^-+2K^+$, $Na_2O \rightarrow O_2^-+2Na^+$ and $O_2^- + H_2O \Leftrightarrow 2OH^-$. [14] This leads to the precipitation of more ettringite, which forms a thicker layer around C_3A than gypsum and anhydrite do. Halvica et al. [21] and Stein et al. [22] also confirm that in a low $[OH^-]$ solution ettringite is formed at a distance from the surface of the solid phase, and the hydration of calcium silicates is accelerated. Thus, a more uniform and denser structure is formed.

Figures 4, 5 and 6 show the mass loss at specific temperature ranges derived from DTA-TG measurements. Mass loss at 50 – 110°C (Figure 4) is attributed primarily to the dehydration of ettringite and to a small extent of CSH. During the first hour of hydration more ettringite is formed to cement with FGD. This happens because FGD is more soluble than gypsum and anhydrite and thus its addition delays cement hydration. Figures 5 and 6 present the main hydration products, calcium silicate hydrates, calcium aluminate hydrates and portlandite. Cement with anhydrite produces more CSH, CAH and portlandite followed by gypsum and FGD, since it sets faster than those with gypsum and FGD.

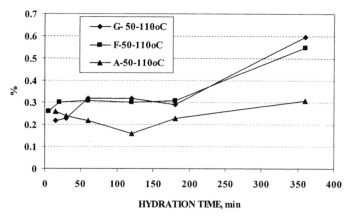

Figure 4 Mass loss at 50-110°C – Dehydration of ettringite and CSH

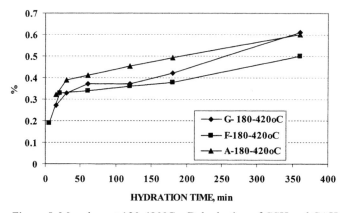

Figure 5 Mass loss at 180-420°C – Dehydration of CSH and CAH

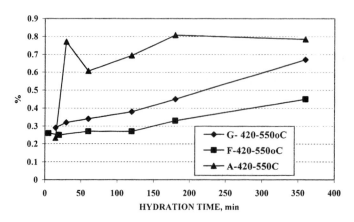

Figure 6 Mass loss at 420-550°C – Dehydration of $Ca(OH)_2$ (portlandite)

During the transformation of fresh paste from a suspension to a solid material, four characteristic periods are generally distinguished. In Figures 7 and 8, where the heat flow evolution of the three cements is presented, the induction/dormant (2nd period), the acceleration (3rd) and the retardation period (4th) are clearly shown. The pre-induction period (1st) cannot be distinguished in these figures since it lasts only a few minutes. The duration of the induction and the acceleration period can be determined from these flows. Results are summarized in Table 2. FGD, which has shown the higher setting times, strongly prolongs the induction period of hydration, while anhydrite reduces it compared with natural gypsum. The acceleration period does not seem to be affected from the different CSBMs. These results are in full accordance with the chemical analysis of the pore solutions, which also confirm the prolongation of the induction period due to the presence of FGD gypsum.

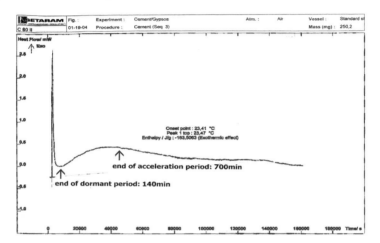

Figure 7 Heat of hydration of cement with gypsum as setting regulator

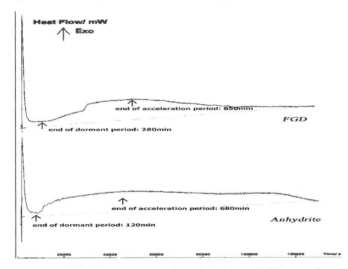

Figure 8 Heat of hydration of cement with anhydrite and FGD as setting regulator

Table 2 Determination of hydration periods for cements with different CSBMs

CSBM	INDUCTION / DORMANT PERIOD, MIN	ACCELERATING PERIOD, MIN
Gypsum	140	560
Anhydrite	120	560
FGD	280	570

Hydration study with ESEM is presented in Figures 9 to 12. After thirty minutes of hydration, the development of AFt crystals is in full progress. Ettringite is grown and needle-formed in the free space at the interface of gypsum and CA grains (Figure 9). AFt crystals have about 10 to 15 μm length.

(a)   (b)

Figure 9 Morphology of cement pastes containing gypsum after 30 minutes of hydration

(a)   (b)

Figure 10 Morphology of cement pastes after 30 minutes of hydration containing: (a) FGD - massive formation of ettringite crystals, (b) anhydrite - formation of monosulfate

In the case of FGD (Figure 10a), it is obvious that the needles of ettringite crystals are thicker than in the case of gypsum forming a more compact structure between cement grains. Less free space is disposable for the growth of hydration products. In this way the hydration of C_3A and C_3S is blocked (dormant period) and retarded compared with the hydration of cement which contains gypsum. Anhydrite cement structure (Figure 10b) is even more compact confirming that the hydration is accelerated compared with gypsum. It is remarkable that the transformation of ettringite to monosulfate (point 1) has already started. Chemical analysis in point 1 with the aid of EDAX (Energy Dispersive Analysis) has shown the following results: 14% SO_3, 55% CaO, 16% Al_2O_3, 8% SiO_2, 6% Fe_2O_3.

After 3 hours of hydration, in the case of gypsum (Figure 11a), ettringite is dramatically reduced as it has been transformed to monosulfate (point 2). CAH hydrates and CSH (points 1, 3, 4) are grown. In the case of FGD (Figure 11b), where the structure is less cohesive, there are still AFt phases (2), as well as AFm phases (1, 3) and formation of CSH (4).

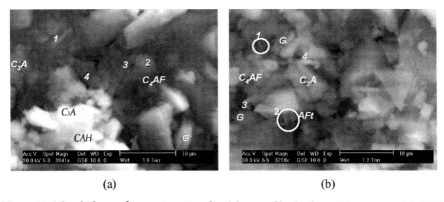

(a)   (b)

Figure 11 Morphology of cement pastes after 3 hours of hydration - (a): gypsum, (b): FGD

After twelve hours of hydrations, both in cement with gypsum (Figure 12a) and FGD (Figure 12b) CSH and CAH are being developing. AFm phases are still shown, which seem to be, in the case where FGD is used, more clearly formed and longer than in gypsum cement. This is due to the retardation of dormant period caused by FGD.

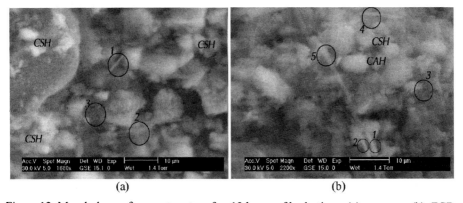

(a)   (b)

Figure 12 Morphology of cement pastes after 12 hours of hydration - (a): gypsum, (b): FGD

CONCLUSIONS

The study of the early hydration of cements with CSBMs with the aid of pore solution analysis (SO_3 and Ca determinations) proves that compared with natural gypsum, FGD increases the induction period of hydration, while anhydrite reduces it. This is also confirmed by the evolution of heat hydration, which was measured with isothermal calorimetry. The acceleration period is not affected by the differentiation of the setting regulator.

The acceleration of hydration that anhydrite causes is proved by the production of more portlandite and other hydration products (C-A-H, C-S-H) measured by DTA/TG analysis. Moreover, from the results of DTA/TG analysis and pH measurements, it is concluded that during the first hours of hydration, more ettringite is produced when FGD is used than in the cases of gypsum and anhydrite. ESEM analysis confirms these results showing that crystals of ettringite produced in FGD cement are longer, thicker and are transformed in monosulfate more slowly than in gypsum and anhydrite cements. This must be attributed to the different solubility of each CSBM and consequently to the different quantities of soluble SO_3 which exist in cement after its mixture with water and react with C_3A retarding its hydration.

ACKNOWLEDGEMENT

The authors wish to thank the Associate Professor Vaimakis T. (Dep. of Industrial Chemistry, University of Ioannina) for his contribution on the measurements of isothermal calorimetry.

REFERENCES

1. LOCHER, F.G., RICHARTZ, W., SPRUNG, S., Setting of cement – Part II: Effect of adding calcium sulphate, ZKG. Vol. 6, 1980, pp 271-277.

2. SINGH, M., GARG, M., Making of anhydrite cement from waste gypsum, Cement and Concrete Research, Vol. 30, 2000, pp 571-577.

3. ZHANG, H., LIN, Z., TONG, D., Influence of the type of calcium sulfate on the strength and hydration of Portland cement under an initial steam-curing condition, Cement and Concrete Research, Vol. 26, 1996, pp 1505-1511.

4. TSIMAS, S., DELAGRAMMATIKAS, G., PAPAGEORGIOU, A., TZOUVALAS, G., Evaluation of industrial by-products for the control of setting time of cements, Cement and Concrete World, Vol. 6, 2001, pp 54-62.

5. HULUSI, M.O., Utilization of citro- and desulphogypsum as set retarders in Portland cement, Cement and Concrete Research, Vol. 30, 2000, pp 1755-1758.

6. ARIOZ, O., TOKAY, M., FGD gypsum as cement retarder, Proceedings of International Symposium on mineral admixtures in cement, 6-9 November 1997, Istanbul, pp 168-175.

7. SINGH, M., Influence of blended gypsum on the properties of Portland cement and Portland slag cement, Cement and Concrete Research, Vol. 30, 2000, pp 1185-1188.

8. HAMM, H., Coping with FGD gypsum, ZKG, Vol. 8, 1994, pp 443-449.

9. TZOUVALAS, G., RANTIS, G., TSIMAS, S., Alternative calcium sulfate bearing materials as cement retarders, Part I: Anhydrite, Cement and Concrete Research, 2004, in Press.

10. TZOUVALAS, G., DERMATAS, N., TSIMAS, S., Alternative calcium sulfate bearing materials as cement retarders, Part II: FGD gypsum, Cement and Concrete Research, 2004, in Press.

11. TAYLOR, H.F.W., Cement Chemistry, Thomas Telford, 2nd edition, London, 1997.

12. LEA, F.M., Chemistry of Cement and Concrete, Arnold, 4th edition, London, 1998.

13. HOLZER, L., WINNEFELD, F., LOTHENBACH, B., ZAMPINI, D., The early cement hydration: A multi-method approach, Proc. of the 11th International Congress on the Chemistry of Cement (ICCC), 2003, Durban, South Africa, pp 236-247.

14. ROTHSTEIN, D., THOMAS, J., CHRISTENSEN, B., JENNINGS, H., Solubility behaviour of Ca⁻, S⁻, Al⁻ and Si⁻ bearing solid phases in Portland cement pore solutions as a function of hydration time, Cement and Concrete Research, Vol. 32, 2002, pp 1-9.

15. XIYU, C., Journal of Chinese Silicates Society, Vol. 15, 1987, pp 152-160.

16. MEIFEI, S., ZHONGYAN, D., KEZHONG, L., Journal of Chinese Silicates Society, Vol. 10, 1982, pp 298-305.

17. XIUJIE, F., HUI, W., Journal of Chinese Silicates Society, Vol. 12, 1984, pp 166-206.

18. KUMAR, S., RAO, K.C.V., Effect of sulphates on the setting time of cement and the strength of concrete, Cement and Concrete Research, Vol. 24, 1994, pp 1237-1244.

19. ASTM C 471M – 95, Standard test methods for chemical analysis of gypsum and gypsum products.

20. TZOUVALAS, G., DERMATAS, N., VAIMAKIS, T., TSIMAS, S., Investigation of the influence of alternative calcium sulfate bearing materials on the early hydration of cement with thermal techniques, Proc. of the 2nd congress of thermal analysis, 2004, Ioannina, Greece, pp 133-140.

21. HALVICA, J., SAHU, S., Mechanism of ettringite and monosulfate formation, Cement and Concrete Research, Vol. 22, 1992, pp 671-677.

22. STEIN, H N , STEVELS, J.M., Influence of silica on the hydration of $3CaO.SiO_2$, Journal of Applied Chemistry, Vol. 14, 1964, pp 338-346.

PECULIARITIES OF SYNTHESIZING ARTIFICIAL ZEOLITE ON A FLY ASH BASIS AND THEIR USAGE FOR MODIFICATION OF SPECIAL DESTINED CONCRETE

E K Pushkarova

O A Gonchar

Kiev National University of Construction and Architecture

Ukraine

ABSTRACT. The given scientific research is dedicated to the problems of modification of various binding materials by artificial zeolites synthesized on the basis of fly ash. The inputting of zeolite phases in composition of a synthetic stone on the basis of inorganic cementitious materials allows to regulate effectively physico-mechanical, exploitational and special properties of obtained composites. The problem of new technology development of deriving artificial zeolites is quite actual, as opens out new possibilities for composites creation with beforehand given properties at the expense of modifications of various cementitious material by similar to zeolite products. The selection of artificial zeolites for further usage as modified additives, which was stipulated by their ability to recrystallize from low-Si zeolitic phases to high-Si zeolitic phases, since the use of modified additives in unstable position remains most effective was proved. According to the results of the experiments inputting of synthesized artificial zeolites for modifications of cementitious materials will ensure the forming of artificial stone in the composition of hydration products on the basis of such cementitious materials as not only low-basic hydrous calcium silicate but also zeolitic phases of analcime. It will promote the improving of phisico-mechanical properties of the obtained artificial stone such as firmness growing up to 25-35%, increasing of corrosion resistance and such working properties as weatherability and frost resistance.

Keywords: Artificial zeolites, Synthetic stone, Modified additives, Corrosion, Frost resistance, Weatherability.

Professor, DrSc E K Pushkarova, is Head of the Special Cements and Concretes Division, Scientific Research Institute of Binders and Materials, Kiev National University of Construction and Architecture. Since 1980 she specializes in development of formulation and technologies of producing the composite materials on alkaline cementitious materials for special use.

Dr O A Gonchar, is a Researcher in Scientific Research Institute of Binders and Materials, Kiev National University of Construction and Architecture. Her research topic is to investigate of peculiarities of synthesizing artificial zeolites on a fly ash basis and their usage for modification of composition binder materials.

INTRODUCTION

Nowadays the deriving of composites with improved physico-engineering and service properties is reached by activating and modifying of already existed building materials which is more effective than developing a new composite systems. The economical and technological advantages of modifying mineral cementitious materials stipulate interest both to natural and artificial zeolites, which frame structure makes possible their usage as modified components. The inputting of zeolite phases in composition of a synthetic stone on the basis of inorganic cementitious materials allows to regulate effectively physico-mechanical, exploitational and special properties of obtained composites. It is known, the addition of natural zeolites in composition of Portland cement and blast-furnace cement in an amount of 10-25 % increases density and slashes infiltration of the derivated cement stone, [1] and results in force increasing [2]. However, the usage of natural zeolites as modified components is not always effective because of their limited number and amount, unstable structure and elemental composition. On the other hand, a high cost price of artificial zeolites, obtained on the basis of chemically pure raw material, makes economically unprofitable their usage for modification of cementitious materials.

Alternative to the above mentioned modified components can be artificial zeolites, obtained with utilizing of waste raw material. It will allow to reduce their cost price and to receive zeolite phases of given composition and structure indispensable for the regulation of artificial stone properties. The problem of new technology development of deriving artificial zeolites is quite actual, as opens out new possibilities for composites creation with beforehand given features at the expense of modifications of various cementitious material by similar to zeolite products. Considering all requirements to raw material, which could be used for synthesizing artificial zeolites, we suggested that it is possible to use fly ash as aluminosilicate component of reactionary mixtures. To motivate the actual possibility of deriving artificial zeolites in Na_2O-Al_2O_3-SiO_2-H_2O system, using waste raw material we researched the reactionary mixtures based on fly ash and alkaline components represented by mixture of liquid glass and NaOH.

RESULTS AND DISCUSSIONS

Proceeding from scientific sources on fly ash usage as one of the component of cementitious material and concrete we can draw a conclusion about it's reactivity which is valued on pozzolanic activity, which is, in turn, defined as a measure of degree and speed factor of the reaction between pozzolan and composites contain Ca^{2+} or $Ca(OH)_2$ and a water, or reaction between pozzolan, water and material that form $Ca(OH)_2$ with presence of water [3]. The two types of fly ashes with difference content of cristallisational component were used for given reserch. Further we will consider the two following types of fly ashes conventionaly named as fly ash (A) and fly ash (B). The pozzolanic of used fly ashes had being researched by method [4] depending on time of sols milling in a ball mill (1, 1.5, 2 hours).

Coefficient of pozzolanic of either fly ash (A) or fly ash (B) grows during it's milling from 1 up to 1.5 hours comparing to not milled ash from 138% up to 145 % for fly ash (B) and 174% for fly ash (A) (see Figure 1). Increasing the mechanical activation was proved useless as it leads to decreasing of pozzolanic coefficient in 2 hours milling. According to obtained results most reactively capable is fly ash (A) after it's mechanical activation during 1,5 hours. It's is confirmed also by the results of XRD-analysis (see Figure 2).

Thus, fly ash (B) is characterized by the content of mullite and silimanite crystal phases that reduce chemical activity of siliceous component of fly ash.

Using as the main component fly ash (B) results in direct synthesizing of low-Si zeoilitic neo-formations such as laumontite, garronite, faujasite and zeolite Zh with basic formula $Na_2O \cdot Al_2O_3 \cdot (2.14)SiO_2 \cdot (4-7)H_2O$, at 100-120°C   heating temperature. The rising of temperature up to 160-180°C leads to gamma expanding of synthesized zeolites structurally similar to analcime, natrolite, erionite, chabasite and 5A zeolite with basic formula $Na_2O \cdot Al_2O_3 \cdot (2-6) SiO_2 \cdot (2-6)H_2O$ (see Figure 3).

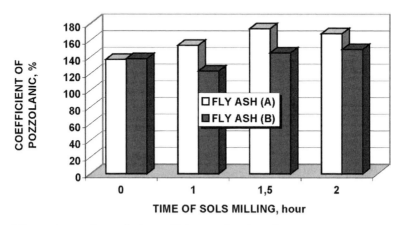

Figure 1 Dependence of pozzolanic coefficient on duration of mechanical activation of fly ash (A) and fly ash (B).

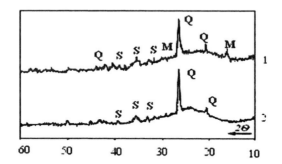

Conventional signs: Q-β-quartz, M-mullite, S-silimanite

Figure 2 XRD patterns of fly ash (A) source (1) and fly ash (B) (2).

Using as a source component fly ash (A) with high content of glassy phase (see Figure 4) at a low heating temperature results in crystallization of zeolitic neo-formations such as natrolite, faujasite and zeolite Zh with basic formula $Na_2O \cdot Al_2O_3 \cdot (2.1-3)SiO_2 \cdot (2-6)H_2O$ but increasing the temperature results in a crystallization of mainly high-Si zeolitic phases such as analcime, laumontite, garronite and zeolite Zh with basic formula $Na_2O \cdot Al_2O_3 \cdot (2.1-6)SiO_2 \cdot (2-6)H_2O$.

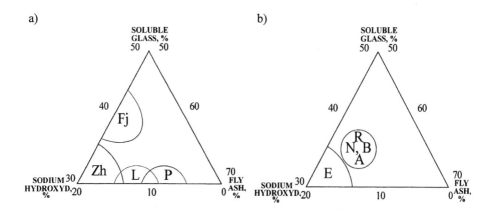

Conventional signs: L - laumontite $Na_2O \cdot Al_2O_3 \cdot 4SiO_2 \cdot 4H_2O$, P - garronite
$Na_2O \cdot Al_2O_3 \cdot 3.6SiO_2 \cdot 4.2H_2O$, Fj- faujasite $Na_2O \cdot Al_2O_3 \cdot 2.4SiO_2 \cdot 7H_2O$, Zh - zeolite Zh
$Na_2O \cdot Al_2O_3 \cdot 2,1SiO_2 \cdot nH_2O$, E - erionite $Na_2O \cdot Al_2O_3 \cdot 6SiO_2 \cdot 6H_2O$, A -5A zeolite
$Na_2O \cdot Al_2O_3 \cdot 2SiO_2 \cdot nH_2O$, N - natrolite $Na_2O \cdot Al_2O_3 \cdot 3SiO_2 \cdot 2H_2O$, R - chabasite
$Na_2O \cdot Al_2O_3 \cdot 4SiO_2 \cdot 6H_2O$, B - analcime $Na_2O \cdot Al_2O_3 \cdot 4SiO_2 \cdot 2H_2O$

Figure 3 Probable zones of artificial zeolite crystallization obtained on the basis of
reactionary mixtures using fly ash (B) under condition of autoclave at:
a)100-120°C; b)160-180°C of isothermal standing.

The generalization of syntheses conditions that can be determined by the content of the
source substances of reacted components and also by treating terms allows to make a
conclusion that the most effective forming of either low-Si zeolitic phases with basis formula
$Na_2O \cdot Al_2O_3 \cdot (2.1-3)SiO_2 \cdot (2-6)H_2O$ or high-Si zeolitic phases with basic formula
$Na_2O \cdot Al_2O_3 \cdot (2-6) \cdot SiO_2 \cdot (2-6)H_2O$ is observed in reactionary mixtures based on fly ash (A).

To determine physico-mechanical influence of artificial zeolites on the cementitious materials
properties, as a modified additive we chose low-Si zeolitic ($SiO_2/Al_2O_3=2,1$) and medium-Si
zeoilite of garronite type ($SiO_2/Al_2O_3=3,6$), synthesized with adding of fly ash (A) by
methods elaborated on prier stages of research [5].

The selection of artificial zeolites for further usage as modified additives was stipulated by
their ability to recrystallize from low-Si zeolitic phases to high-Si zeoilitic phases, since the
use of modified additives in unstable position remains most effective. Beside the fact that
low-Si zeolitic phases are more reactive, they synthesizing at lower temperatures. The
process of transformation from unstable zeolitic phases to stable promotes crystallization of
such hidrational products of concrete stone as low-basic hydrous calcium silicate and zeolitic
neo-formation such as analcime. It was proved by the results of XRD-analysis and electronic
microscopy carried out together with electronical probe analysis.

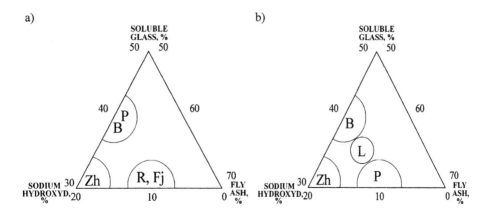

Conventional signs: L - laumontite $Na_2O \cdot Al_2O_3 \cdot 4SiO_2 \cdot 4H_2O$, P - garronite $Na_2O \cdot Al_2O_3 \cdot 3.6SiO_2 \cdot 4.2H_2O$, Fj - faujasite $Na_2O \cdot Al_2O_3 \cdot 2.4SiO_2 \cdot 7H_2O$, Zh - zeolite Zh $Na_2O \cdot Al_2O_3 \cdot 2,1SiO_2 \cdot nH_2O$, R - chabasite $Na_2O \cdot Al_2O_3 \cdot 4SiO_2 \cdot 6H_2O$, B - analcime $Na_2O \cdot Al_2O_3 \cdot 4SiO_2 \cdot 2H_2O$

Figure 4 Probable zones of artificial zeolite crystallization obtained on the basis of reactionary mixtures using fly ash (A) under condition of autoclave at: a)100-120°C; b)160-180°C of isothermal standing.

According to obtained results in reactionary mixtures that contain 50% of fly ash (A) and 30% of soluble glass and 20% of NaOH zeolitic phases such as zeolite Zh ($Na_2O \cdot Al_2O_3 \cdot 2,1SiO_2 \cdot nH_2O$) (d=0.210; 0.256; 0.314; 0.363; 0.628 nm) were synthesized and than after a while were recrystallized to zeolitic phases similar to analcime ($Na_2O \cdot Al_2O_3 \cdot 4SiO_2 \cdot 2H_2O$) (d=0.174; 0.293;0.343; 0.560 nm). (fig.5. 7a). Similar recrystallization is taking place when garronite is synthesized ($Na_2O \cdot Al_2O_3 \cdot 3.6 \cdot SiO_2 \cdot 4.2H_2O$) (d=0.178; 0.198; 0.269; 0.319;0.411; 0.505; 0.713 nm) on the basis of reactionary mixtures that contain 50% of fly ash (A), 40% of soluble glass and 10% of NaOH (see Figure 6, 7b).

According to the results of the electronic microscopy the sizes of crystals of synthesized zeolite Zh do not exceed 2-5 mkm (see Figure 5a), while crystals of garronite are 10-20 mkm. At recrystallization of synthesized artificial zeolitic neo-formations to zeolitic phases of analcime the difference in dimensions of zeolitic phases of analcime was noticed. It is stipulated by the size of source crystals of zeolite Zh and garronite. Thus, in case of recrystallization of zeolite Zh the sizes of crystals of derived analcime phases do not exceed 10-15 mkm. (see Figure 5b). According to the results of electronic microscopy (see Figure 6b) recrystallization of garronite results in deriving of zeolitic phases of Na-K-analcime, with the size of crystals 15-20 mkm. The presence of potassium in zeolitic neo-formation composition is stipulated by chemical composition of fly ash, that was used as aluminosilicate composition for zeolitic phases synthesizing and lower content of alkaline in reactionary mixtures. On their bases garronite was synthesized comparing to mixtures, used for zeolite Zh synthesizing.

a)   b)

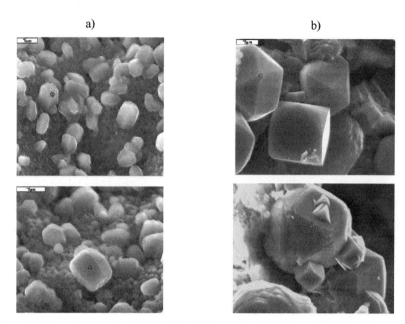

Figure 5 Microphotographs (x500) of cracked samples surface on the basis of alkaline aluminosilicate composition of $3,2Na_2O \cdot Al_2O_3 \cdot 5,6SiO_2 \cdot 13,5H_2O$ structure, that ensures synthesizing of zeoilite Zh after autoclaving and hardening for 28 days (a) and 90 days (b).

However, the character of X-ray photographs made for synthesized zeolitic neo-formations also proved the difference between crystal dimensions. So, the most intensive peaks of garronite testify about largely crystallized neo-formations that in time almost completely recrystallized to phases similar to analcime (see Figure 7b). The intensity of peaks of synthesized zeolite shows that the process of recrystallization to zeolitic phases of analcime happens slower and new formed crystals have smaller dimensions (see Figure 7a). The difference in influence of derived zeolite additives upon the process of artificial stone structurization on the bases of cementitious materials which were modified by artificial zeolites was stipulated by the different sizes of synthesized artificial zeolites like Zh type and garronite and also zeolitic phases similar to analcime which were formed as a result of recrystallization of synthesized artificial zeolites.

CONCLUSIONS

According to the results of the experiments inputting of synthesized artificial zeolites similar to zeolite Zh and garronite for modifications of cementitious materials will ensure the forming of artificial stone in the composition of hydration products on the basis of such cementitious materials as not only low-basic hydrous calcium silicate but also zeoilitic phases of analcime. It will promote the improving of phisico-mechanical properties of the obtained artificial stone such as firmness growing up to 25-35%, increasing of corrosion resistance and such working properties as weatherability and frost resistance.

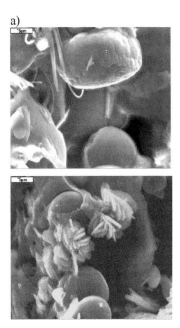

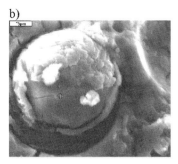

Figure 6 Microphotographs (x500) of cracked samples surface on the basis of alkaline aluminosilicate composition of $1.6Na_2O\cdot Al_2O_3\cdot 5,6SiO_2\cdot 10.1H_2O$ structure, that ensures synthesizing of zeoilite Zh after autoclaving and hardening for 28 days (a) and 90 days (b).

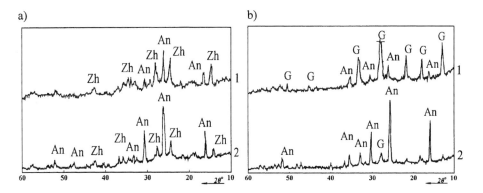

Figure 7 X-ray photographs of the samples of synthesized artificial zeolites (zeolite Zh (a) and garronite (b) had been harded for 28 days (1) and 90 days (2) were prevailed initially in the composition of synthesized artificial zeolites). Conventional signs: Zh - zeolite Zh $(Na_2O\cdot Al_2O_3\cdot 2.1SiO_2\cdot nH_2O)$, G – garronite $(Na_2O\cdot Al_2O_3\cdot 4SiO_2\cdot 2H_2O)$, An – analcime $(Na_2O\cdot Al_2O_3\cdot 4SiO_2\cdot 2H_2O)$.

REFERENCES

1. WAGNER, G R. Physico-chemical process of activating concrete dispersions, K, Naukova dumka, 1980, 200pp.

2. VOLKOVA, S N AND SELYAEV, V P. The properties of concrete compositions filled by zeolites / Proceeding of the 37th International Seminar on Modelling and Optimisation of Composites, Odessa, Astroprint, 1998. p. 174.

3. TAKEMOTO, T AND UCHIKAVA, X. Hydration of Pozzolan Cements / 7th International Congress on the Chemisty of Cement, Paris, 1980. p. 243.

4. SN 306. Protaction of Alkaline Concretes by application of mineral admixtures, Varshava, 1991, 14pp.

5. PUSHKAROVA, K K AND NAZIM, O A. Establishment of parameters the tecnology receiption artificial zeolities on basis the fly ash Tripilskaya DRES and alkaline components with difereht chemical activity / Collection of work "Scientific Bulletin of Constraction", Harkov, No12, 2000. pp. 81-88.

PERFORMANCE OF CONCRETE WITH MINING SAND AS AN ALTERNATIVE NON-CONVENTIONAL MATERIAL

Z Y H Harith

K A Shavarebi

Universiti Teknologi MARA

Malaysia

ABSTRACT. This research paper attempts to analyse an alternative replacement for river sand in concrete with sand from tin mining as a non-conventional material, which is widely available in Malaysia and in particularly in the State of Perak, and will be wasted if not being fully utilised. Three types of mining sand from three locations in the State of Perak, have been chosen. Compressive strength testing using mining sand in place of river sand was carried out. Concrete was designed to have target strengths of 30 and 40 N/mm^2 using mining sand and river sand as control. Compressive strength tests at the 3, 7, 28, 60 and 90 days were carried out. The compressive strength results indicated that concrete using mining sand has slightly higher compressive strength at 28 days for all grades. The permeation properties of concrete in terms of permeability and resistance to carbonation were investigated. The results indicate that concrete from mining sand has comparable with river sand. It can be concluded that mining sand has the same performance as river sand and can be used and marketed as an alternative to river sand. Therefore, excavation of riverbeds in Malaysia can be minimised and rivers can be maintained.

Keywords: Demand on construction materials, Uncontrolled environment, Riverbed aggravation, Non conventional material, Concrete strength, Concrete durability

Zarina Yasmin Hanur Harith is Associate Professor and Head of Department of Building, Faculty of Architecture, Planning and Surveying at Universiti Teknologi MARA. Malaysia. Her research focuses on industrial waste and Building Environment. She is a member of Chartered Institute of Building UK.

Kamran Shavarebi Ali is Senior Lecturer in Building Department of Faculty of Architecture, Planning and Surveying at Universiti Teknologi MARA, Malaysia. His research focuses on usage of Non Conventional Material in Concrete Production. He is a member of Chartered Institute of Building UK.

INTRODUCTION

Urbanisation and fast population expansion are the factors that lead to the growth of construction industry, hence the high demand for construction materials. We have been consuming non-renewable material and energy resources too quickly and inequitably. In the process, we have created uncontrolled environmental pollution. Increasing demand for pollution reduction has led to a change in the characteristics of the standard materials commonly used in concrete. Demand to reduce pollution has also created the need to find a use for waste products instead of disposing of the materials. This practice may be beneficial to the performance of concrete, or may compromise its quality and severely reduce its performance and durability.

As we are in the new millennium, it could be said that the concrete industry has been prudently conservative in replacing material as its main ingredient, especially where changes in technologies, design and methodology are concerned [2].

Using mining sand is a good example of avoiding the cause of environmental problems such as excavation of riverbeds. Rivers are suffering serious bed aggravation and degradation. The consequences of bed degradation lead to unstable riverbanks causing erosion.
The concrete industry is concerned with these problems and is looking for alternatives to reduce usage of customary river sand. This research attempts to analyse alternatives to river sand such as mining sand/tailing sand from tin mining, which is largely available in the State of Perak, Malaysia. The State of Perak is known to have silica sand deposits amounting to 56.1 million metric tonnes on existing mining land. These by-products of large-scale mining operation are an eyesore and will be wasted if it is not been fully utilised and exploited.

WHY MINING SAND?

Silica sand raw materials are common in every part of the country. The most abundant are naturally occurring silica sand and mine tailing sand deposits. The deposits vary in terms of quantity, quality and particle size distribution, and are almost always contaminated with heavy minerals. The occurrences of very high purity raw silica sand/quartz rocks are very limited and found scattered throughout the country, which are mainly unavailable for economic exploitation. As such, most local extractions are carried out on the so-called market potential deposits of natural silica sand or tailing sand, in which the sand is rarely used without subjection to some level of beneficiation.

CHARACTERISTICS OF MINING SAND AS FINE AGGREGATE

There has been very little use reported of the vast quantities of wastes generated by mining. Mining wastes consist essentially of waste rock, which is the coarse material excavated to expose the ore, and mill tailings, which are the residue obtained from the separation of minerals from their ores. Because mining is conducted in geographically remote areas, very little attention has been paid to them. Mining sand is formed from mining waste consisting essentially of waste rock, which is the residue obtained from the separation of minerals from their ores during the process of washing.

River sand and mining sand both have different shapes and textures. River sand is rounded and has smooth texture, but mining sand has an angular shape due to an absence of abrasion processes during its creation, which are associated with physical weathering. This is the most important factor because the interlocking between the particles is good, thereby providing a good bond and would influence the concrete strength considerably.

EXPERIMENTAL PROCEDURES

A number of cubes (150mm x 150mm x 150mm) were cast using river and mining sand as variable to carry out the following:

♦ Experimental programmes based on comparison of the physical properties, which influence the concrete properties of mining sand and river sand as a control. The following tests were considered: which influence the concrete properties, sieve analysis, specific gravity, rate of absorption and silt content.
♦ Experimental programmes to evaluate strength development of concrete using mining sand. The following tests were considered: slump test, compacting factor test and compressive strength test.
♦ Experimental programmes to assess the durability performance of concrete. The following tests were considered: carbonation resistance and gas permeability

RESULTS AND DISCUSSION

Compressive Strength of Concrete with Mining Sand

The workability was measured by the slump test in accordance with BS 1881 Part 102: 1983 [1] and the compacting factor was measured in accordance to BS 1881 Part 103: 1983. The result indicates that no significant effect on slump and compacting factor was observed when using mining sand in comparison with control sand. The results of slump tests are consistent and vary within the tolerable region of 100 ± 25mm, which satisfies the standard specification.

The compressive strength results from cubes using sand from various mining sand sites at 3, 7 and 28 days, indicated that all sand taken from 3 sites, Tapah, Batu Gajah and Tronoh (in state of Perak, Malaysia) the requirement of Grade 30, 35 and 40 N/mm^2 by an age of 28 days. These results are shown in Table 1 and 2.

Table 1 Compressive strength result of Grade 30 N/mm^2 (water cured) concrete

FINE AGGREGATE (SAND)	STRENGTH, N/mm^2				
	Age in Days				
	3	7	28	60	90
River (cont)	12.5	21.9	31.3	34.4	40
Tronoh (mining)	13.1	23.6	32.9	36.1	42.1
Tapah (mining)	13.2	22.4	31.6	34.7	41.5
Batu Gajah (mining)	12.0	21.1	30.2	33.2	40.2

Table 2 Compressive strength result of Grade 40 N/mm² (water cured) concrete

FINE AGGREGATE (SAND)	STRENGTH, N/mm²				
	Age in Days				
	3	7	28	60	90
River (cont)	17.4	29.7	42.5	46.7	53.1
Tronoh (mining)	18.4	31.2	44.0	48.4	55.8
Tapah (mining)	17.9	30.6	43.2	47.5	54.8
Batu Gajah (mining)	17.1	30.3	42.8	47.0	53.9

The results indicate that mining sand from Tapah displays higher strength in early age and also at 28 days compared to control sample. The Batu Gajah mining sand sample displays slightly lower performance at 28 days in comparison to other mining sands due to its finer particle size distribution. All concrete specimens in this study could be classified as high early strength concrete where more than 70% of the 28 day strength is obtained at an age of 7 days.

Carbonation Resistance of Mining Sand Concrete

Since natural atmospheric carbonation is a slow process, an accelerated carbonation chamber was used [3]. Carbonation depth was monitored at 5, 15, 20 and 25 weeks intervals. Prior to placing in the accelerated carbonation tank, all specimens were dried for 14 days after the 28 days water cured process in the laboratory, to ensure that the specimens were in a surface-dry condition. In order to relate the differences between strength development rates and carbonation rates from the use of the different types of sand, the same grades of mix using the designed water cement ratio of 0.52, 0.48 and 0.43 respectively were studies. From the result obtained, it can be concluded that for concrete of Grade 30 N/mm² using the higher water cement ratio of 0.52, the carbonation rate is the highest compared to concrete of Grade 40 N/mm². In all cases, the depth of carbonation increases with time, and decreases as the strength of the concrete increases Figure 1 and 2.

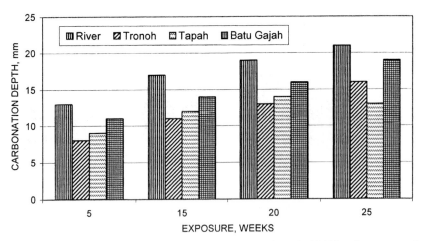

Figure 1 Comparison between the carbonation depths of Grade 30 N/mm² water-cured concrete samples of 4 different types of sand versus duration of exposure

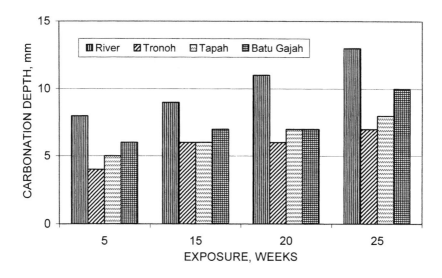

Figure 2 Comparison between the carbonation depths of Grade 40 N/mm² water-cured concrete samples of 4 different types of sand versus duration of exposure

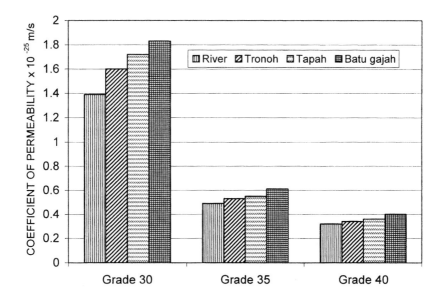

Figure 3 Comparison between the coefficients of permeability of water cured concrete sample of 4 different types of sand with different Grade concrete at age of 28 days

The carbonation depth of the Tronoh sample showed higher resistance to carbonation at all Grades in comparison to the other mining sands and control sand. The carbonation depth for the Batu Gajah mining sand samples was slightly higher compared to the Tronoh and Tapah mining, but lower in comparison to the control sand. However, the rate of carbonation in all specimens shows a similar characteristic. It can be concluded that using mining sand in concrete has no significant effect on the carbonation resistance of concrete as compared to the control mix.

Permeability of Mining Sand Concrete

Gas permeability tests were carried out on samples cured for 7 and 28 days. Figure 3 shows the coefficient of permeability values for all samples at all Grades. It can be seen that the coefficient of permeability decreases with curing duration (from 7 to 28 days). Older samples have lower coefficients of permeability. From these of results, it can be seen that concrete with higher Grades is less permeable and therefore, more durable. The coefficient of permeability in all specimens shows a similar characteristic.

CONCLUSIONS

It can be concluded that theoretically, the mining sand has the same performance and, in some cases, slightly better performance than river sand and can be accepted and used as fine aggregate as an alternative to river sand with no significant effect on mix proportions or workability. Therefore, excavation of riverbeds can be minimised and rivers can be maintained. In view of the contamination of the aggregate with dust and silt, it has been suggested that, in many cases, washing aggregate before use would realise significant benefits. The analysis of the cost of transportation and treatment for two types of sand (mining and river) did not indicated critical differences, and in some cases it was even cheaper to use mining sand in concrete.

REFERENCES

1. BRITISH STANDARD INSTITUTION, BS 1881: Part 116, Method for Determination of Compressive Strength, London, 1983

2. FALADE F., A Comparative Study Of Normal Concrete With Concrete Containing Granite and Laterite Fine Aggregates, Fifth International Conference on Concrete Engineering And Technology, 1997

3. DIAH A.B.M., MAJID T.A., KAMARULZAMAN K.B., KAMRAN S.A., Carbonation of Concrete Exposed To Natural Laboratory Environment, Proceeding of Conference on New Frontiers & Challenges Bangkok, Thailand, 1999

INFLUENCE OF MODIFYING ADMIXTURES ON PROPERTIES OF FOAM GLASS OBTAINED BY USING ASHES RESULTING FROM INCINERATION OF HOUSEHOLD WASTE

V I Gots

K M Germash

Kiev National University of Construction and Architecture

Ukraine

ABSTRACT. The actuality of energy saving problem was noted in this article. The principal advantages of foamglass is compared with other heat insulation materials in this paper and addresses the possibility of using the ashes resulting from incineration of solid household waste in foamglass technology. It was found that the increase of quantity of ashes resulting from incineration of solid household waste in raw mixtures results in the rise of temperature interval of emolliating process and viscosity of mixtures. It is necessary to add the modifying admixtures into raw mixtures for decrease temperature interval of emolliating process and improve the properties of material. The soda Na_2CO_3, the liquid glass and the silicate aggregate (silicate lump) were held as the modifying admixtures. It was established that it is more effective to use the silicate aggregate as modifying admixture in foamglass technology with using the ashes resulting from incineration of solid household waste. The using of ashes resulting from incineration of solid household waste in foamglass technology is efficient technological solution for utilizing of dangerous wastes and environmental protection.

Keywords: Foamglass, Energy Saving problem, Heat insulation materials, Silicate melt, Viscosity of mixtures, Liquid glass, Silicate aggregate.

V I Gots, PhD is Dean of the building-technological faculty of the Kiev National University of Construction and Architecture

Germash Kate is a Post graduate student, Kiev National University of Construction and Architecture.

INTRODUCTION

This paper is about possibility of production foamglass using the ashes resulting from incineration of household waste. It was found it is necessary to add the modifying admixtures into raw mixtures with ashes resulting from incineration of household waste. The aim of this paper is to prove the efficacy of using the liquid-glass and silicate-aggregate as the modifying admixtures in technology of foamglass obtained by using the ashes resulting from incineration of household waste.

Constantly increasing requirements for thermal resistance of external constructions of buildings and actuality of energy saving problem determine the growth of demand for efficient heat insulating materials.[2,6]

Foam glass is one of the most efficient and perspective heat insulating material. Among its principal advantages in comparison with other heat insulation materials are inorganic origin and combination of high heat insulation properties with considerable strength characteristics. The conventional foam glass production technology is power intensive as if provides a stage of express glass melting. Therefore the refusal from this power intensive technological stage and obtaining foam glass directly by foaming raw mixtures is the matter of contemporary direction of scientific research of this problem. As raw mixtures it is possible to use natural and artificial materials such as volcanic glasses, fuel ashes, drosses and others which chemical and mineralogical structure which is similar to structure of traditional raw materials. From this point of view the possibility of using the ashes from the incineration of solid household waste in foam glass production technology is of great interest.

Such a technological solution enlarges and enriches the sources of raw materials for production of foam glass and enables the utilization of injurious wastes such as ashes from the incineration of solid household wastes.(надалі – AI)

THE PRELIMINARY TESTS

Preliminary tests have confirmed the possibility of obtain foam glass using the ashes by incineration of solid household wastes. Broken window glass and AI from Kiev wastes incineration factory "Energy" were used. Limestone and coke were used as the gas-agents. The raw mixtures were prepared by separate grinding of components with further grinding-mixing in ceramic milt up to value of specific surface 4500-4700 cm^2/g.

The investigation of basic building-technical properties of foam glass was conducted on the samples of foamglass. Basic building-technical properties of foamglass are the average density and the compressive strength. 70x70x70 mm in size. The samples were obtained after burning of raw mixtures in heat proof steal moulds. Two series were carried out: with use as gas-agent the limestone ($Ca CO_3$) and the coke. The quantity of ash used as a replacement was 10%, 20%, 30% by mass of mixture. It was established that when using the limestone as a gas-agent in amount of 1-3% from the mass of mixture and at the temperature interval of 770-880°C it's possible to produce foam glass with average density of 390-590 kg/m^3 and compressive strength of 3,4-3,8 N/mm^2. When using coke as a gas-agent in amount of 2-3% of mass at the temperature interval 820-880°C it's possible to produce foam glass with average density of 360-485 kg/m^3 and compressive strength of 2,8-3,2 N/mm^2.(see Table 1).

It can be concluded that it is better to use coke as gas-agent it is possible to produce foam glass with lower average density compared with limestone. Mixtures with quantity of ash 15%, 30%,45% from their mass and pure glass (without gas-agents) were investigated to define the temperature interval of emolliating the mixtures. The simples-cylinders with 20 mm in diameter and 20 mm in height were formed from raw mixtures on hydraulic press under pressure of 5,0 N/mm^2. The simples on the ceramic plate were placed into the furnace, which has been heated up to necessary temperature and were held out 1 hour.

Table 1 Influence of quantity of Al and sort and quantity of gas-agent on properties of foam glass

COMPONENTS OF RAW MIXTURES,% by mass		GAS-AGENTS, % by mass	PROPERTIES OF FOAM GLASS	
glass	Al		the average density, kg/m^3	the compressive strength, N/mm^2
90	10	CaCO$_3$, 1%	390	3,4
80	20		560	5,3
70	30	CaCO$_3$, 2%	590	3,8
90	10		360	2,8
80	20	coke, 2%	435	3,0
70	30		485	3,2

The tests were carried out at temperatures 750, 825, 850, 900°C. Volume modifications and density of samples was fixed after the end of burning.

It was established that presence of Al causes expansion. It is confirmed by increased of volume of samples and expanding coefficient. It was fixed after the end of burning by visual inspection and measuring sizes of samples. It confirms presence in ashes foam agents (particles of coke) and gives the possibility to use it in the technology of foam glass. At the same time the increase of quantity of ashes in raw mixtures results in the rise of temperature interval of emolliating process and viscosity of mixtures. It is confirmed by decrease of expanding coefficient (for instance, at temperature 850°C in mixture with contents of ash 15% the coefficient of expansion is 2.77; in mixture with contents of ash 30% - 1.17) (see Figure 1). If the content of ash is over 30% the effect of expansion is not observed. Therefore the mixture with ash content of 30% was selected for next research.

Using modifying admixtures

The addition of modifying admixtures into the raw mixtures is one of the solutions of the problem, which allows regulating (reducing) the viscosity of the silicate melts and improving the foam glass properties. As a rule they are the components, which contain alkalis in their chemical structure.

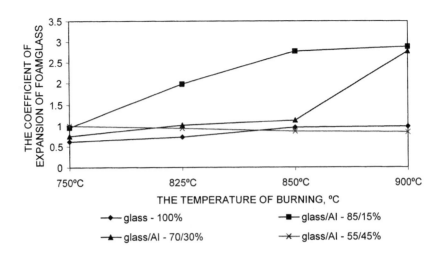

Figure 1   Influence of quantity of Al and temperature of burning on
expanding coefficient of foamglass

Such substances have lower fusion temperature than the basic ones and form low viscosity melt in which more refractory components of mixture are dissolved. In such way the reducing of viscosity of melts, producing of greater amount of liquid phase, decrease of temperature of sintering of mixtures is reached. The last condition causes more uniform porous formation and production of foam glass with improved properties. To study the possibility of decrease of melt viscosity the mixture with contents of ash of 30% was modified by Na_2CO_3 and tested according to the previous procedure (above considered) at temperatures of 750, 800, 850, 900 and 950°C.

The coefficient of expansion of a modified mixture has increased into 25-27% in comparison with mixture without Na_2CO_3. That proves the possibility to use Na_2CO_3 as admixture for decrease of melt viscosity. It is necessary to note that the increase of coefficient of expansion can also be caused by gas emission at decomposition of Na_2CO_3 and its interaction with components of silicate melt. The increased density of pressed samples, which causes their fast heating, and intensive and fuller interaction between particles of mixture, can also be very important. The density of mixture in mould is much less (0.9-1.1 g/cm^3 in compared with 1.45-1.5 g/cm^3 at pressed simples). That is why if producing foam glass in moulds the results can differ a somewhat.

Influence of modifying admixtures on properties of foamglass

To decrease the power consumption whilst producing foam glass by using Al and to improve its properties, the using as modifying admixtures a liquid glass and silicate aggregate (silicate lump) were held. Mixture contains glass and Al in proportion of 70/30 % by mass was used. Raw materials have been used: breakage of window glass, Al from Kiev waste-incineration factory "Energy"

The liquid glass properties were as follows:

- The silicate module – 2.85;
- Density –1.4 g/cm³.

The silicate aggregate by joint-stock company "Zaporozhflus" properties were:

- The silicate module –2.85;
- Na_2O –26.46%;
- SiO_2 –73.04%.

The coke was used as gas agent in amount of 2 % from the mass of mixtures.

Raw mixtures were prepared by grinding of components in a ceramic mill with subsequent mixing in ceramic mill up to value of specific surface of 4500-4700 cm²/g.

The basic characteristics of foam glass were defined on samples 70x70x70mm in size obtained by burning the raw mixtures in heatproof metal forms. Roasting was carried out at temperatures 820,850,880°C for 25 minutes.

Figure 2 it can be seen that the average density of foamglass varies depending on the use as a modifying agent liquid glass or silicate aggregate in compared to mixtures without modifying admixtures.

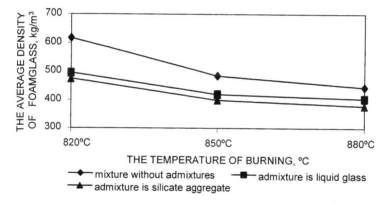

Figure 2 Influence of modifying admixtures on average density of foamglass obtained by using Al

Figure 3 shows the dependence of coefficient of expansion of foamglass on factors mentioned above. The characteristics of foam glass (density and coefficient of expansion) obtained by using a silicate aggregate are better on 7-9% than while using a liquid glass.

The characteristics of foam glass (density and coefficient of expansion) obtained by using as a modifying admixture the silicate aggregate are better on 15-18% than characteristics of foam glass obtained without modifying admixtures (for raw mixture with glass and ash resulting from incineration of household waste in proportion 70/30 % by mass).

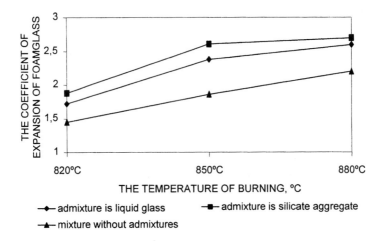

Figure 3. Influence of modifying admixtures on coefficient of expansion of foamglass obtained by using AI.

Further research is necessary to increase the efficiency of using of modifying admixtures in a direction of improvement of foam glass properties. The use of AI in foamglass technology is efficient technological solution for utilizing of dangerous wastes and environmental protection.

CONCLUSIONS

The results and comparison them with previous ones enable to come to conclusion about efficiency of using liquid glass and a silicate aggregate as modifying admixtures in foam glass production technology. It is more efficiently to use the silicate aggregate. It is confirmed by decrease of average density and increased of expanding coefficient of foamglass, obtained by using AI.

REFERENCES

1. SHILL F. Foamglass (obtained and using). Moscow, 1967, pp 307.

2. DEMIDOVICH B.K. Foamglass. Minsk, 1975, pp 275.

3. DEMIDOVICH B.K., SADCHENKO N.P. Foamglass – technology and using. Moscow, 1990, pp 25.

4. KITAYGORODSKIY I.I. The technology of glass. Moscow, 1967, pp 564.

5. KUKOLEV G.V. The physical chemistry of silicates. Moscow, 1966, pp 463.

6. GORYAINOV K.E., GORYAINOVA S.K. The technology of heat-insulating materials. Moscow, 1982, pp 276.

7. ARTAMONOVA M.V. The chemistry and chemical technology of silicate materials. Moscow, 1983, pp 144.

8. BEZBORODOV M.A. The viscosity of silicate systems. Minsk, 1975, pp 351.

9. APPEN A.A. The chemistry of glass. Moscow, 1970, pp 351.

10. PAVLUSHKIN N.M. The chemical technology of glass. Moscow, 1983, pp 432.

11. KRUPA A.A., GORODOV V.S. The chemical technology of ceramic materials. Kiev, 1990, pp 398.

12. PLEMYANNIKOV M.M., KRUPA A.A. The chemistry and termal-physics of glass. Kiev, 2000, pp 559.

INNOVATIVE MATERIALS FOR CONCRETE—AN OVERVIEW

C M S Kutty

Context Data Services

India

ABSTRACT. This paper profiles the development of innovative materials for applications in making concrete and concrete products world wide. India is a major producer of concrete. Approximately 102 million tons of cement and 10 million tons of steel goes into its production .The heavy dependence on these materials is going to be a major problem when we confront accelerated demand for concrete in future. The key issues raised are :Spiraling prices ,sustained availability of key inputs, eco-bio conservation, energy management and affordability, Concerns have been raised that the conventional materials used in the production of concrete do not respond to the emerging pattern of demand such as (i) the possibility of obtaining many variations in shape and surface patterns (ii) potential for industrialization, (iii) materials that can stand up to aggressive environment improved crack control (iv) enhanced alkalinity for corrosion resistance (v) high strength to weight ratio and improved tensile strength. These demands have awakened planners and researchers to identify non-conventional materials for the production of concrete. This paper is an overview of the efforts made in India and overseas for the development of materials for concrete endowed with the properties highlighted above.

Keywords: Tensile strength, Eco-bio conservation, Industrialisation, Corrosion resistance, structural debilitation.

Dr C M Sankaran Kutty is Director, Context Data Services, Mumbai, India. The organization offers consultancy services in the area of Industrial and Market Research, Executive training programme for civil engineers, publication of status reports on various inputs to Construction Industry etc. Dr. Kutty holds a Doctoral degree in the area of the Economics of Housing Construction. He has undergone short term courses in the area of Construction Management at the University of Michigan, U.S.A and University of Reading, U.K. He has published several papers in India and Abroad. He teaches various disciplines in the area of Construction Management in various institutions in Mumbai.

INTRODUCTION:

The rising level of aspirations and demographic pressure places heavy demand on infrastructure projects and housing. Concrete structures and products form a substantial portion of these built facilities. Countries world wide, especially the developing countries face the following problems in the mass usage of concrete. The important among them being:-

- Structural debilitation due to corrosion of reinforcing steel
- The spiraling prices of conventional materials that go into the production of concrete

Developing countries have been identified as endowed with innovative materials with promising prospects for partial or total substitution. These materials are abundantly available and research work at various institutes have proved their economic viability while ensuring the desired structural and quality parameters .This paper looks into these innovative developments with possible application in India.

VEGETABLE FIBRE REINFORCED MATERIALS

Status: Vegetable fibers are abundantly available in developing countries which make them convenient materials for brittle matrix reinforcement.

The Composites made from Vegetable Fiber has the following Matrix

Blast furnace slag based cement mortar (1:1.5);
Binder BFS+Lime +Gypsum (0.88: 0.02:o.10)
Reinforcement : 30mm long chopped coir fibers (2% by volume).

Relatively poor mechanical and durability performance acts as a major problem in its wider application. The solution to this problem lies in the protection of the fiber by coating them or sealing dry composite to avoid the effect of the water.

Product development

Hollowed load bearing wall panels were developed in Brazil with the above mentioned composite. This was the result of extensive research carried out by *Vahan Agopyan of Escola Politechnica & Vanderly M. John , Institute de Pesquisas Technologycas* , Brazil A prototype of a dwelling embryo was assembled with these panels in a low cost housing settlement located in the city of So Paulo. This prototype has been in use since 1989 by the local community as a Nursery, Short term house and community center.

Bagasse Ash as a pozzolanic material

This agriculture residue is thrown as a waste in many countries. In India the current availability is about 2 million tonnes per annum The Central Building Research Institute, Roorkee, India has carried out extensive evaluation work of Bagasse Ash from different places. Thirteen samples were analysed and observed very favourable compressive strength. Table 1 clearly supports these findings.

Table 1   Compressive strength of various pozzolanas

POZZOLANAS	COMPRESSIVE STRENGTH
29.30 kg/cm^2	Lime: Bagasse Ash (1:2)
5.60 kg/cm^2	Fly ash
6.3 kg/cm^2	Cinder
15.61 kg/cm^2	Surkhi

Practical Applications:

Concrete under floor, rendering and plastering solid concrete blocks and hollow concrete blocks. The features of the Hollow Concrete blocks are listed below:

(1) Dimension 45x22.5x11.5 cm; (2) Compressive strength 40.5 kg/ cm^2 ;   (3) Drying shrinkage 0.045%; (4) Water absorption 185 kg/m^3; (5) Moisture movement 0.03%

CERAMIC AGGREGATE FROM ALUMINIUM WASTE

Laboratorio Ricerca Tecniche Industriali , Itali has developed a process to manufacture Light weight aggregate and light weight concrete from Aluminium Waste.

The properties of the products evidenced by the Institute are indicated in Table 2.

Table 2   Properties of the Products

A: AGGREGATE	B: CONCRETE
Granules volume mass 0.6 kg/lt	Volume mass (dried) 1.1. kg/lt
Inhibition coefficient 24h>1%	Water absorption (7 days) $\geq$ 10%
Compression resistenace ,5.0 MPa	Compression Resistance $\leq$25 MPa

Application of Silica fume in Tunnel Construction

Tunnel Engineering demands concrete with favourable properties such as high strength, corrosion resistance and water prevention etc.,It has been established by studies that construction cost is directly proportional to its diameter squared (*Lu Ji Guang,Shanghai Institute of Building Sciences & Liu De Fang , Shanghai Tunnel Engineering Co, China)*

Study of Yanan Road Tunnel

The Institute carried out a study of Yanan Road Tunnel, China. One third of Yanan Road Tunnel (490 meters) is made of silica fume concrete with high strength. Through Differential Thermal Tests, X ray diffraction and Pore structure Analysis , it was established that the content of Ca (oh)2 in cement stone of Silica fume Concrete gets reduced owing to the Pozzolanic activity of Silica fume and micro aggregate effect, Its pore structure is improved, concrete formation is compacted and strength of concrete and other properties are raised.

Test Results conducted at the Institute

50 MPa normal concrete and 70 MPa silica fume concrete were subjected to test. The results obtained is provided in Table III.

Table 3 showing the relative strength of silica fume concrete vis-à-vis normal concrete.

Table 3   Average Strength

GRADE OF SEGMENT	NO OF SAMPLES	COMPRESSIVE STRENGTH (3d)	COMPRESSIVE STRENGTH (28d)
50 MPa normal concrete	280	30,0 MPa	60.8 MPa
70 MPa Silica fume Concrete	20	35.0 MPa	73.4MPa

Impermeability

450 pieces of segment were spot checked under 0.8 MPa water pressure. All were found to be up to the standard.

Structural Test

In comparison with 50 MPa normal concrete , 70 MPa silika fume concrete has demonstrated increased compressive strength and shear strength of end ribs and deferred the occurance of shear cracks on end ribs , This feature offers the possibility of reducing the wall thickness of segments ranging between 9 to 18 %, implying substantial cost reduction.

Aggregates for Concrete from Fly Ash

Turning waste into opportunities

India is facing a paradoxical situation. The country is facing increasing shortage of coarse aggregates. Unbridled quarrying of rock formations for construction activities are threatening its fragile eco-system. On the other hand, mega thermal plants that dot the Indian landscape are facing the serious problem of waste disposal in the form of Fly Ash.

The search for a balancing act demands a judicious exploration of the beneficial application of this waste. Studies at the *F-U vogdt Institute of structural engineering & Material Strength ,Technical University of Berlin, Germany* provided promising possibilities for producing aggregates for concrete fly ash.

Methods of Production

- Pelletizing
- Briquetting

World wide production of Ashes and Slags and their utilization rate are indicated in Table 4.

Table 4   World wide production of Ashes and Slags and their utilization rate

NO	COUNTRY	TOTAL ANNUAL PRODUCTION OF SLAG & ASH, 10^3T	UTILIZATION IN CEMENT CONCRETE	
			Annual Utilization, 10^3T	Share in total Production, %
1	CIS	10000	200	2.02
2	USA	6400	510	8.00
3	China	4227	100	2.30
4	Poland	2000	220	11.00
5	Czech	1630	110	6.70
6	U. K	1600	400	25.00
7	Germany	3000	255	17.00
8	India	1000	40	4.00
9	S.Africa	1000	in developing stage	--
10	Romania	700	30	4.30
11	France	520	127	24.40
12	Hungary	510	16	3.10
13	Bulgaria	500	10	2.00
14	Japan	400	50	12.50
15	Australia	350	65	18.50
16	Canada	300	50	16.20
17	Denmark	100	44.5	44.50
18	Belgium	65	25	38.50
19	Holland	45	15	33.50

Source: Journal of Indian Building Congress : Volume 5 , No 1 , 1998.
Cost Comparison: 1 ton of Cement: Indian Rs.3000/- per ton., Fly Ash: NIL

Uses of Polymers in Concrete

Reinforced Concrete is a durable Construction material in Building and civil engineering projects. During the past thirty years or so it has become increasingly evident that reinforcement corrosion has become a major factor in the deterioration of reinforced concrete members.. In fact this is more critical in the case of reinforced concrete structures such as (a) Bridges; (b) Marine Structures; (c) Chemical & Mineral Processing Plants; (d) Waste Water Treatment Plants. The implications are (a) Reduced service life for structure; (b) user inconvenience; and (c) higher maintenance cost.

Research findings and field applications have demonstrated the advantages of using synthetic fibers to overcome this problem. The fibers identified include the following viz. (a) Glass; (b) Kevlar; (c) Carbon & (d) Polyethylene.

Synthetic fibers are credited with the following favourable properties

- high strength to weight ratio;
- wide range of stiffness values;
- excellent corrosion resistance;
- inert to electro magnetic effects.

Immense Possibilities

Potential possibilities for the application of polymers in concrete have been documented. Chopped glass fibers and chopped carbon fibers have been used in concrete elements to replace conventional steel reinforcement and reduce component weight and also to impart improved tensile, flexural and impact properties.

Continuous unidirectional fibers

Continuous Glass fibers are available in fiber bundles or fiber bundles encapsuled in a resin matrix. Materials such as **epoxy, polyester vinyle ester** and **phenolic resins** are used for making resin matrix. Encapsuling the fiber bundle in a matrix protects the fibers against damage and the bars are easier to handle and install during construction.

Polymer Concrete Composites (PCC)

New composite materials with promising possibilities have been developed at the *Centre for building studies , concordia university, USA* and have found applications in the following areas : ✔ Bridge decking ✔ Tunnel support lining system desalting plants Beams, ✔ ordinary reinforced beams and post tensioned beams ✔ Underwater Habitats ✔ Dam Outlets , ✔ Offshore Structures , ✔ Ocean Thermal Energy Plants etc.

PCC has different variants such as Polymer impregnated concrete (PIC), Polymer Cement Concrete (PCC), Fiber Reinforced Polymer Concrete (FRPC) and Phase Change Material Concrete (PCMC) and can subserve the above referred segments. Typical properties of PCC are given in Table 5.

The following Polymer Latices have been identified for PCC Production :

a) Styrene –Butadiene copolymer

b) Ethylene –Vinyl Acetate copolymer

c) Acrylate Styrene copolymer; and

d) PVC

Table 5   Typical properties of PCC

	PCC	PIC	PIC(A)*	PC	PPCC
Composite strength,					
MPa	34-47	137-88	268-86	130-98	37-91
psi	5000	20000	39000	19000	5500
Tensile strength,					
MPa	2.41	10.34	14.47	9.65	5.51
psi	350	1500	2100	1400	800
Shear strength,					
MPa	861	....	...	>4481	>4 481
psi	125				
% water absorption	5.5	0.6	<0.6	0.6	....
Freeze thaw resistance	700/25	3500/2	''	1600/0	'''
Acid resistance**		10x	>10x	>20x	4x
Benefit/Cost	1	2	3	4	'''

 *   based on autoclaved concrete
 ** improvement factor

Use of Recycled Polymers in Concrete

A novel idea of generating wealth from waste is through recycling of polymers for application in concrete. A brief overview is given under:

Products used for recycling	**Polymer**	**Application in Concrete**
Milk Bottles	HDPE	Concrete Foam work
Motor Oil Containers		
Carbonated Beverage Bottles		
Piping Oil Containers	PTE	Composite bldg. Materials
Battery Cases, Diaper linings		
Foams, Cups, Trays, Food Containers	PVC, P P & P S	

Application of FRP Materials in Concrete:

Of late, FRP materials show promising signs of large scale applications in concrete. A detailed review is made here under:-

FRP Materials

They comprise a combination of fibrous reinforcement held in polymer matrix to plates, rods, tubes and structural profiles. The fibres used are carbon Aramid or glass. The resins used are Epoxy, Polyester, Vinyl Ester Phenolic. The main purpose of the matrix (resin) is to hold the fibres in the desired orientation, transmit load into the fibers in the desired orientation, transmit load into the fibers and protect the fibers from damage. For beams and bridges the technique generally involves bonding either unstressed or pre stressed carbon fibre reinforced polymer (CFRP) plates to underside or bottom flange of the beam. Alternatively reinforcing rods may be embedded. This has the effect of increasing the tensile strength of the lower part of the beam. For materials such as reinforced concrete and cast iron which are strong in compression, but weak in tension this is usually the most critical area. Pre-stressing of CFRP plates has the effect of further reducing the tensile stress on the bottom flange thereby increasing the load capacity. CFRP plates can be boned to the side of beams near the supports to increase the shear capacity. For such structures as columns, silos and cooling towers, the reinforcement is applied cross-wrapped or in bonds with the tensile capacity of the carbon, aramid or glass fibre acting as confinement.

Favorable Properties :

(a) Resistance to fatigue and creep; (b) durable; (c) good fire resistance as they are poor heat conductors.

Uses :

Strengthening of structural elements such as Beams, floors, columns, bridges, silos, cooling towers, chimneys etc.

CONCLUSIONS

Status Of The Use Of The Materials In India

The Technology for the applications of these materials into the making of concrete is available. Strangely enough, not much has been done to transform these benefits to commercial applications. Every effort should be made to transform these opportunities to the benefit of the masses so that there will be a sudden improvement in their affordability threshold. Attitudinal change is called for.

CONSTRUCTION MATERIALS MANAGEMENT ADOPTING GIS TECHNOLOGY

E Arunbabu

P Thirumalini

P Partheeban

Jaya Engineering College

India

ABSTRACT. This papers deals with the development of Geographical Information System (GIS) technology for the material management in the construction field. This study suggests the way in which the transportation of construction materials cost can be minimized. With the advancement in construction techniques, lot of materials has to be transferred from the retailers to the consumers (Construction sites). As GIS is a readily available spatial analysis tool, which gives unique and unparalleled insights into the natural and man made environments due to its strength to link the generic information with its location. Hence an attempt is made to graphically represent the construction materials shops combined with the graphical representation of construction sites so as to identify the optimal routes based on the distance between the shops and the cost of the various construction materials. It is proved that the GIS is best tool for materials management for the construction industry, which not only analysis the current scenario but also helps in projecting the future in other words, one can effectively use the GIS tool for past, present and future studies in the construction field.

Keywords: GIS, Retail Shops, Cost of materials, New Construction, Optimum.

Mr E Arunbabu, is a Lecturer within the Department of Civil Engineering, Jaya Engineering College, India.

Mrs P Thirumalini, is a Lecturer, Department of Civil Engineering, Jaya Engineering College, India.

Dr P Partheeban, is Professor and Head of Department of Civil Engineering, Jaya Engineering College, India.

INTRODUCTION

In India Construction industry is the second largest employer supporting 32 million people and 16 major core sector industries. The civil engineering is updated with computerized findings with regard to procurement of materials, manufacturing, strengthening and implementation. GIS is a tool for the management, query, visualization and analysis of spatially referred information. GIS is capable of handling spatial and non-spatial data. It also identifies the spatial relationship between Map features [2].

Proper planning in a construction project is one of the major contributing factors to the success of a project. The construction process can be subjected to the influence of highly variable and sometimes unpredictable factors. Therefore, any construction project has to be planned well, anticipating and solving the problems in that particular project to optimize the cost of construction and to commission the project in the shortest possible time [3]. With the rapid advances in science and technology resulting in new construction of complex nature, the need for the introduction of effective and improved planning techniques has been felt for quite sometimes.

GIS technology will lead to increased productivity in many industry operations and to greater accuracy and timeliness of information for both professionals and the general public [1]. If its full potential is realized, the market will move to higher levels of efficiency, which may ultimately drive down the cost of construction. There should also be a reduction in traditional information arbitrage of construction industry, where inside information or market knowledge historically has allowed for unusually large construction returns.

Recent innovations in GIS technology will not only increase the productivity of existing construction industry operations but will also lead to greater accuracy of information and access to information that previously was cost prohibitive or simply unavailable. These advances will provide a greater range of information with which a construction industry decision can be made, which will benefit consumers by leading to decisions that more closely address their needs and preferences. From the perspective of the industry, as GIS and related computer technology are integrated into market operations, fewer people may be needed to perform some of the traditional tasks such as residential brokerage and mortgage origination [6]. Somewhat as a counterbalance, however, the need for professionals with advanced technology skills, including GIS, will grow.

This article assesses the current and potential contribution of GIS to the residential and commercial construction activities efficiently and economically. GIS applications by major industry function are examined, including materials purchase, materials transportation, new building construction, cost of materials and market analysis. The article concludes the optimum use of cost specific to construction materials in GIS framework.

STUDY AREA CHARACTERISTICS

Thiruninravur is located at 30 km northwest of Chennai city has been used as a case study to demonstrate the capabilities of GIS in effective use of construction materials by civil engineers. This area comprises a total area of 5 km^2 (out of which 50 percent of them are residential). Population in this town Panchayat is 53,321 (27,973 male and 25,348 female). The birth rate and death rate is 233 and 165 per year, respectively.

Thiruninravur is an educational centre, which attracts hundreds of students each year. In and around the study area there are about 22 educational institution and 10 of them are colleges. In the last ten years, this place has experienced unusually strong population growth, increasing at 20 % per year. This growth is part of the pattern of long-term rural growth of the nation. It is the result of small industrial growth in rural towns, home construction, an increase in the number of urban people who choose rural areas for retirement and an increased ability to sustain a living through telecommunications linkages.

The challenge of population growth faced by Thiruninravur is common to the problem faced throughout the Chennai city. A total number of houses in this area are 9927 with a household ratio of 5.37, because this area is earlier rural in nature. The status of income level is varying from lower level to higher level. The rainfall in this area is moderate and the main source of the drinking water is from the borewell. The quality of the water is within the standard limits for drinking. The development of Chennai city is at saturated level in the west direction and hence at present development is taking place at Thiruninravur. So due to the above said reasons lot of construction activities are taking place in this study area.

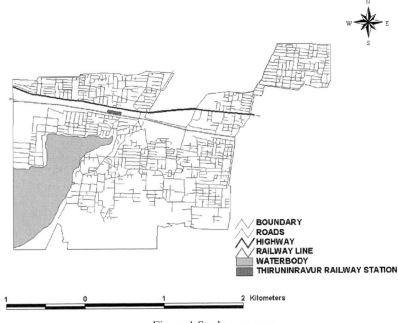

Figure 1 Study area map

DATA COLLECTION AND ANALYSIS

Secondary data such as a basemap of the Thiruninravur town Panchayat, population data, etc were collected from the town Panchayat office. Primary data such as location of material shops, location of the construction site were surveyed physically. As the construction materials cost varies in different shops and in different locations, the cost variation of each material was collected.

DEVELOPMENT OF GIS

The spatial and non-spatial data has been collected from the study area (Thiruninravur). The collected Thiruninravur Town Panchayat basemap was scanned and taken in to MapInfo environment to digitize the map, where it is digitized using the technique *Headsup Digitization* [4, 5]. The different features of the area are separated and stored in different layer so as to facilitate for varies analysis. The digitized map is exported as shape file format to ARC VIEW GIS software for analysis. Various themes included in this study are study area boundary, roads, existing building, buildings under construction, locations of shops, educational institutions and water bodies. Figure 1 shows the study area basemap along the Chennai – Thiruvallur Highway. Themes showing location of shops and new building constructions are point themes and road network and boundary are line themes. Water bodies are represented as polygon themes. The locations of construction material shops and new building construction sites are presented in Figure 2. The attributes of the various themes have been entered through data tables called *Attribute Tables*.

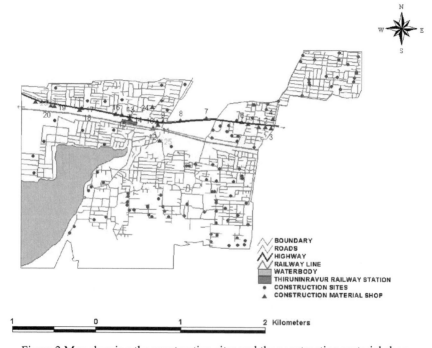

Figure 2 Map showing the construction sites and the construction material shop

RESULTS AND DISCUSSION

Buffers are created to find the radius of accessibility (with a radius of 0.5 km and 1.0 km for the new construction sites). These buffers were created in the Arc/View environment. Radius of accessibility can be fixed depending upon the housing construction with respect to retailer's locations. In this study all the shops are clustered in the Chennai – Thiruvallur Highway and the construction takes place on either side of roads.

Wherever the construction takes place they have to procure the materials from these retailers. On one side of the highway the construction sites were at a maximum distance of 1.0 km and other side it is 1.5 km. Hence to analyze the rate variation in the construction material shop the radius of accessibility of 0.5 km is fixed as a minimum radius. Buffers are created for the new building constructions sites with radius of accessibility of 0.5 km and 1.0 km are shown in Figure 3 and Figure 4.

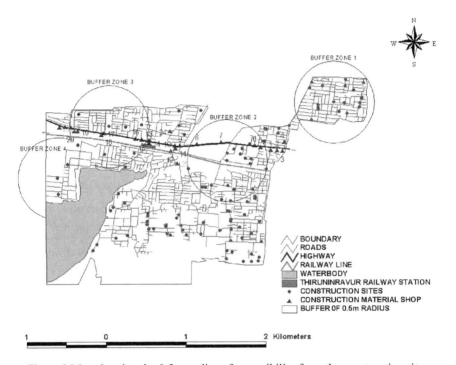

Figure 3 Map showing the 0.5 m radius of accessibility from the construction site

Based on the cost of materials with respect to shops and proximity of construction sites, optimal routes were identified for each 0.5 km buffer. Figure 5 shows the optimal route for the materials purchase for the site located within the radius of accessibility of 0.5 km. Similarly the optimal routes were identified from the material shop to the site based on the cost comparison between the shops falling within the radius of accessibility.

Critical areas are those areas, which cannot be covered within the radius of accessibility. That is, the site is more than 1 km from the highway, the sites are identified as critical sites. For these areas it is suggested for the retailers that a new shop can be established, so that the sites in the critical areas can procure the materials from the newly established shop.

Figure 6 shows the optimal route from the site to the shop 2. Even though the distance is lesser from the site to the other shops 3 and 4, when the cost comparison arises it is found that the cost of the materials is lesser in the shop 2. Hence the material is procured from the shop 2.

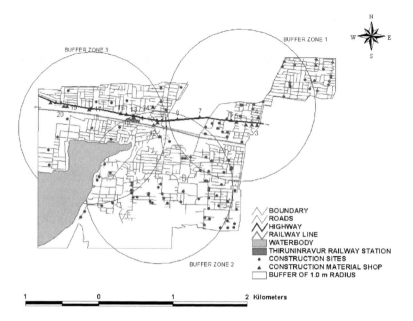

Figure 4 Map showing the 1.0 m radius of accessibility from the construction site

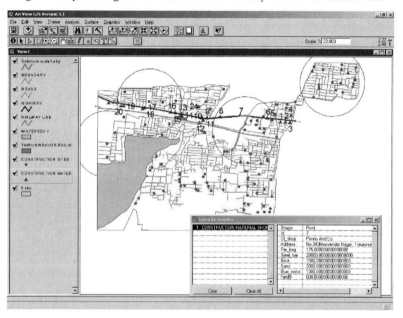

Figure 5 Arc view window shows the attribute table of shop 7 with the optimal route to reach the construction site

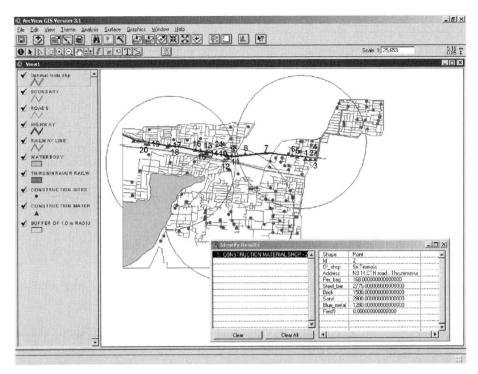

Figure 6 Arc view window shows the attribute table of shop 2 with the optimal route to reach the construction site

APPLICATIONS

The developed GIS provide excellent support for construction industries, which develops fast. The potential application of this GIS includes:

1. Local planning authorities or managers may use for their planning purposes or decision making

2. Engineers or executives officers can use this system to plan and execution

3. Service oriented organizations such as Red Cross, Lions Club and Rotary Club may use for general purpose

4. It will be useful for the maintenance personnel

5. It provides technical advice and assistance on the new house builders, contractors, real estate people, construction materials retailers and general public

CONCLUSIONS

The conclusion part of this paper gives information to the planners, Builders, Contractors Consultants Administrators and the Public on cost-effective method of utilizing the construction materials. Hence this study is more useful to those who are involved in construction industry since it gives information both spatially and non-spatially.

The advancement of software and hardware technologies has meant an improvement in the productivity and quality of construction engineering and planning in both public and private sectors. With the increase in population and registered contractors the numbers of constructions that are taking place in our country have increased tremendously. The following measures may be adopted to reduce the cost of construction:

- Maintaining the construction records

- Scientific analyses of materials and its transportation to the construction sites

- Data on construction materials cost

- Area under development or developed

To achieve all the above stated measure, greater use of GIS and Internet GIS can be used. GIS technology will continue to play a vital role in Construction field. Thus GIS becomes a primary repository of information that can be quickly accessed and viewed when required. There are virtually no limits to the scope of GIS analysis by combining any modes of analysis and query tools provided by a GIS we could derive answers to the most complex questions posed by any activity or field of study.

ACKNOWLEDGEMENTS

We sincerely thank Prof. A. Kanagaraj, Chaiman, Jaya Educational Trust for extending all support to complete this research work. We also express our gratitude to all the staff members and students who helped us to complete this work in successful manner.

REFERENCES

1. ABLE SOFTWARE CORPORATION, Users Manual for R2V for Windows 9X & NT, (1998).

2. ESRI, ArcView GIS, New York, 1999.

3. JAIN, K.C., AGGARWAL, L.N., Production Planning Control and Industrial Management, Khanna Publishers, 1995, pp 20-25.

4. JAMES A.K., Technical papers, XIX Annual ESRI Conference, USA, July 26-30, (1999)

5. MAPINFO CORPORATION, MapInfo Professional Users Guide, New York, 1996, www.esri.com

THE SIGNIFICANCE OF EMBEDDED ENERGY FOR BUILDINGS IN A TROPICAL COUNTRY

S P Pooliyadda
London Borough of Haringey
United Kingdom
W P S Dias
University of Moratuwa
Sri Lanka

ABSTRACT. A computerised relational database management system is used to represent and calculate the embedded energies and carbon coefficients of building materials. The embedded energy requirements are also calculated on the basis of the lowest quality energy (called a "bio-equivalent" basis), in addition to the more conventional basis of Tonnes of Oil Equivalent. Comparisons are then made between alternative materials for building components such as purlins, walls, roofs and windows. Among the more common construction materials considered, the lowest energy option is timber while the highest is steel, with concrete in between. Timber products have negative carbon coefficients as well, i.e. they store more carbon than is emitted in their use for house building. The walls and wall plastering contribute significant proportions of the gross energy of a building, indicating that the use of low energy materials for walls and the elimination of wall plastering will make significant contributions to reducing the embedded energy of buildings. The ratio between total embedded energy and annual operational energy for selected buildings ranges from 14 to 35 for the houses, while for an office building with air-conditioning loading it is around 5.

Keywords: Embedded energy, Building materials, Raw materials, Energy sources, Energy quality, Carbon emission

S P Pooliyadda, BScEng, MPhil, is currently a structural engineer with London Borough or Haringey and obtained both her degrees from the University of Moratuwa, Sri Lanka.

Prof W P S Dias, BScEng, PhD (Lond), DIC, CEng, MIStructE, FIE (SL), is a Professor in the Department of Civil Engineering at the University of Moratuwa, Sri Lanka. His research interests encompass Concrete Technology, Energy & Environmental Issues and Philosophy of Engineering. He serves as the Sri Lanka Representative for the Institution of Structural Engineers, U.K.

INTRODUCTION

Most of the construction industry energy considerations have been made with respect to the operational phase - e.g. energy saving buildings. In cold countries such as U.K. and U.S.A. the ratio between construction and annual operational energy ranges from around 3 to 6; in a temperate zone country that is a little warmer such as New Zealand [1], this ratio goes up to around 9. However in a tropical country such as Sri Lanka, the operational energy is likely to be very small as very little heating is required. This means that the energy embedded in the building assumes much greater significance.

The construction or embedded energy of building materials will vary from one country to another, depending on the sources of energy used for manufacturing. In Sri Lanka, for example, there is a wide range of energy sources used for building materials manufacture, from firewood for brick and tile production to fossil fuel and electricity for cement and steel production. An energy policy would clearly need to take into account such differences in quality of the energy sources [2] that are used for manufacture, and not merely the quantity.

OBJECTIVES

The objectives of the research described in this paper are as follows:
1.   To compare energy contents and carbon coefficients for alternative materials of construction.
2.   To find the gross embedded energy input to typical residential and office buildings.
3.   To find the influence of various building components on the embedded energies of residential buildings.
4.   To compare the significance of construction energy with that of operational energy in Sri Lanka.

ENERGY CONTENTS AND CARBON COEFFICIENTS

The energy embedded in a building material will in general comprise (i) production energy, (ii) energy to transport its component materials and (iii) the energy embedded in its component raw materials (with the energy to extract and transport energy sources being ignored). All these energies can also be classified according to input energy type as biomass, fossil fuel and electrical energy. Figure 1 shows this schematically for a brick wall element.

The authors describe elsewhere [3, 4] the assumptions used for the above calculations. The salient features are as follows: TOE (Tonne of Oil Equivalent) values for diesel and firewood are 1.05 and 0.38 respectively; 1 MWh of hydro electricity has a TOE value of 0.086; the Sri Lankan energy mix is 66% hydro and 34% thermal; carbon coefficients are 20.3 kgC/GJ for oil type fossil fuels and 54 kgC/GJ for thermal electricity. Wood fuels are assumed to be sustainable (i.e. zero carbon emissions) and timber used in construction has a negative carbon coefficient of 250 kg/m^3 of wood. The manufacture of cement and aluminium results in carbon releases of 142 kg and 130 kg/tonne respectively, apart from fuel related emissions.

Although it may seem that the least energy intensive material is the preferred option, there are other issues that have to be considered too. One of the main issues is regarding the differences in energy quality. For the purpose of this study the "quality" of energy is characterized by the concept that "a high quality energy or energy source can be used for a variety of end-uses with relative ease" [2]. Thus electricity (which has a wide variety of end-uses) has the highest quality while biomass the lowest (since burning is the only way it can yield its energy and because it cannot be easily transported and utilized like many fossil fuels).

In order to compare different energy qualities, the amount of the lower quality energy that is required to obtain a given quantity of the higher quality energy via established processes was assessed. All the energy requirements can then be compared on the basis of the total equivalent amount of lowest quality energy (i.e. biomass energy). The resulting energy values can be said to have a "bio-equivalent" basis. The "quality factor" between electricity and biomass energy was obtained as 5, and that between electricity and fossil fuel energy as 2.78; the quality factor between fossil fuel energy and biomass energy then worked out as 1.8, in order for the system to be consistent (i.e. $1.8 \times 2.78 = 5$) [3, 4].

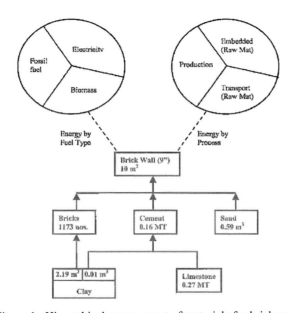

Figure 1   Hierarchical arrangement of materials for brickwork

COMPARISON OF BUILDING MATERIAL ALTERNATIVES

Purlins

The energy inputs and carbon emissions corresponding to purlins made out of 3 different materials, i.e. timber, steel and prestressed concrete, were analysed. The cross section of each purlin was selected so that it could carry a uniformly distributed load of 1 kN/m over a span of 3 m. The cross section required for timber purlins is 75×100 mm. The cross section required for steel purlins is a 75×75×6 mm angle section. The cross section required for prestressed concrete purlins is 60×120 mm (grade 40 concrete) with 3 nos. 6 mm diameter high tensile wires. The energy inputs corresponding to timber, prestressed concrete and steel purlins, which have similar spans and approximately similar bending capacities, are 29 MJ, 121 MJ and 654 MJ respectively in terms of TOE, indicating that steel is around 23 times more energy intensive than timber and that prestressed concrete is around 4 times more energy intensive than timber. However on a bio-equivalent basis, the ratios are 15 and 3 respectively. It was also found that 5.6 kg of carbon is stored in the timber purlin compared to 3.7 kg of carbon and 13.1 kg of carbon released to the atmosphere via prestressed concrete and steel purlins respectively.

Walls

The energy inputs and carbon emissions corresponding to 10 m^2 of 225 mm thick brickwork and 200 mm thick blockwork, both of which will have approximately similar strengths, are compared. The raw material requirement for 10 m^2 of brickwork is 1173 nos. of bricks, 0.16 MT of cement and 0.59 m^3 of sand. This is obtained from the Building Schedule of Rates (BSR) [5]. The raw material requirement for 10 m^2 of blockwork is 120 Nos. of 200×200×400 mm cement blocks, 0.04 MT of cement and 0.18 m^3 of sand. This is also obtained from the BSR [5]. The energy inputs corresponding to 10 m^2 of brickwork and blockwork are 10.89 GJ and 0.95 GJ respectively in terms of TOE, indicating that brickwork is around 11 times more energy intensive than blockwork. However on a bio-equivalent basis, the ratio is around 6. Net carbon emissions are similar, with 40.6 kg of carbon being released to the atmosphere via brickwork compared to 41.7 kg of carbon via blockwork.

Roofs

The energy inputs and carbon emissions corresponding to a 10 m^2 plan area of a calicut tile roof and an asbestos roof were compared. The raw material requirement for a calicut tile roof is 0.02 m^3 of air-dry treated timber purlins, 0.06 m^3 of air-dry treated timber planks (for the reepers) and 156 nos. of calicut tiles. These quantities are obtained from the BSR [5]. The raw material requirement for an asbestos roof is 0.11 m^3 of air-dry treated timber purlins and 12.29 m^2 of asbestos sheets. These requirements are also obtained from the BSR [5]. The energy inputs corresponding to a 10 m^2 plan area of a calicut roof and an asbestos roof are 4.1 GJ and 0.75 GJ respectively in terms of TOE, indicating that the calicut tile roof is around 5 times more energy intensive than an asbestos roof. However on a bio-equivalent basis, the ratio is only around 2. Furthermore 18 kg of carbon is stored for the calicut tile roof compared to 6.5 kg of carbon released to the atmosphere via the asbestos roof.

Windows

The energy inputs and carbon emissions corresponding to aluminium and timber windows of size 1250×1550 mm with two shutters were compared. The raw material required for an aluminium window is 15.7 kg of aluminium extrusions (7.96 kg for the frame and 7.74 kg for the shutter) and 1.9 m^2 of 3 mm thick glazing. To construct a timber window one needs 0.04 m^3 of rough purlins for the frame, 0.01 m^3 of rough planks for the shutter, 1.5 m^2 of wood painting and 1 m^2 of 3 mm thick glazing. The energy inputs corresponding to an aluminium window and a timber window of similar size are 2.44 GJ and 0.20 GJ respectively in terms of TOE, indicating that the aluminium window is around 12 times more energy intensive than the timber window. On a bio-equivalent basis, the ratio is around 10. Furthermore, 11.2 kg of carbon is stored for the timber window compared to 50.3 kg of carbon released to the atmosphere via the aluminium window.

DISCUSSION

The analyses carried out indicate that, among the more common construction materials considered, the lowest energy option is timber. For example, on the basis of TOE, timber purlins are around 23 times less energy intensive than steel purlins and 4 times less energy intensive than prestressed concrete purlins. Similarly timber windows are around 12 times less energy intensive than aluminium windows. This is consistent with other research findings [6]. However, because of the worldwide phenomenon of deforestation, the use of timber will be possible only with aggressive re-forestation and perhaps even the use of biotechnology.

Wood products have negative carbon coefficients as well, i.e. they store more carbon than is emitted in their use for construction. For example, for purlins of 3 m span, 5.6 kg of carbon are stored for the timber purlin compared to 3.7 kg and 13.1 kg of carbon released to the atmosphere via prestressed concrete and steel purlins respectively. For windows of size 1250×1550 mm, 11.2 kg of carbon is stored for the timber window compared to 50.3 kg of carbon released to the atmosphere via the aluminium window.

It is also seen that materials that use wood fuels (e.g. bricks and tiles) consume more energy. However, the use of wood fuels is more competitive when compared on a bio-equivalent unit basis. For example, the energy input corresponding to a calicut tile roof and an asbestos roof, which have a similar plan area of 10 m^2, is 4.1 GJ and 0.75 GJ respectively in terms of TOE, indicating that the calicut tile roof (where wood is the fuel used for the production of calicut tiles) is around 5 times more energy intensive than an asbestos roof; however on the basis of bio-equivalent units the ratio is only around 2. When the energy input corresponding to 10 m^2 of brickwork and blockwork is considered, brickwork (where again wood fuels are used for the production of bricks) is around 11 times more energy intensive than blockwork; however on the basis of bio-equivalent units the ratio is only around half that value. Furthermore, with respect to carbon emissions, wood fuels are considered to be self-sustaining, as the carbon released during burning is that which has been absorbed from the atmosphere and stored. If wood fuels are to be promoted however, fuel wood farming would have to be undertaken on a large scale. The use of wood fuel would also decrease dependency on fossil fuels.

The variety of energy sources and the consequences of raw material extraction would also result in differing environmental costs. For example, blockwork is 11 times less energy intensive than similar strength brickwork. Although the mining and transporting of river sand, which is a major raw material for blocks, may have a rather low energy cost, thus resulting in a low embedded energy for cement blocks, the environmental cost of sand mining may be high, giving rise as it does to river bed degradation and coastal erosion. Hence, future studies must account for a comprehensive environmental costing of building materials, using the techniques of environmental economics.

GROSS EMBEDDED ENERGY IN BUILDINGS

There was insufficient data to find the Sri Lankan energy requirement for finishing work such as tiling, sanitary fittings, electrical installations etc. Thus gross energies for the embedded energies of the buildings were calculated by neglecting these items. Therefore the actual gross construction (or embedded) energy requirement will be higher than the calculated value; however other researchers have shown that the contribution from the above items are not that large [1].

Variations on aspects of typical Sri Lankan single and two storey houses constructed according to National Housing Development Authority (NHDA) type plans [4] were studied, using different building materials to represent maximum and minimum energy intensive houses. These were compared in order to obtain a range for the embedded energy of a house. These high and low energy intensive construction material types are based on the comparisons made in the previous section. For the single storey house the gross embedded energy value was found to lie between 2.4 and 4.9 GJ/m^2, and for the 2 storey house between 3.5 and 5.7 GJ/m^2. For an air conditioned office 8 storeys building the actual gross energy was found to be 5.5 GJ/m^2.

A review of the literature revealed a very wide range in the gross embedded energy per unit floor area of a building calculated by different authors, even for the same building type. For residential construction, for example, the range was from about 1.2 to about 8 GJ per square metre [1]. It is seen that the embedded energies estimated in this study (i.e. between 2.4 and 5.7 GJ/m^2) lie in the above range.

Influence of Building Components on Embedded Energy

The contributions to the total embedded energy of the houses from various building elements made of minimum and maximum energy intensive materials was next determined. In the percentage ranges stated below, the first value corresponds to the contribution from the minimum energy material in the minimum energy house, while the second value corresponds to the contribution from the maximum energy material in the maximum energy house.

The influence from walls for the two storey house is 10% to 44%; for the single storey house it is 29% to 49%. This shows that the walls can contribute a significant proportion of the gross energy and that they have a wide range as well. This is because there is a large difference in energy contents between brick walls and block walls (as seen before). It also shows that choosing block walls instead of brick walls could make the greatest contribution to the reduction of gross embedded energy in residential buildings of 1-2 storeys. The influence from roofs for the two storey house is 4% to 7%, whereas it is 8% to 16% for the single storey house. It should be noted that this energy contribution from the roof is somewhat smaller than the corresponding cost contribution. The contribution from the floor slab for the two storey house was found to be around 7%, which is much lower than the cost contribution.

Although alternatives for plastering were not studied, the contribution from wall plastering is 29% to 13% for the single storey house and 32% to 17% for the two storey house; and that from the brick paved, cement rendered floor is 22% to 10% for the single storey house and 5% to 2% for the two storey house. This indicates that the elimination of plastering (if acceptable to house owners) can also contribute significantly towards reducing the embedded energy of houses. In this context un-plastered block walls may be more acceptable than un-plastered brick walls, because of the poor dimensional accuracy of Sri Lankan bricks.

Operational Energy

An indication of the annual operating energy requirement of the typical houses that have been used in this research was obtained by a 1998/1999 sample survey of household energy consumption. The houses that were surveyed were those constructed according to the NHDA type plans used for taking off BOQ quantities. The operational energy requirement was of two types namely, (i) electrical energy required for the lighting and other electrical equipment and (ii) fuel used for cooking. Most of the houses surveyed made use of Liquid Petroleum Gas for cooking, and the contribution from this is around half the contribution from electricity consumption.

For the air-conditioned office building the annual operational energy was found to be 3706 GJ or 964 MJ/m^2, based on an electricity consumption of 213,730 kWh over 4 months.

Comparison of Embedded and Operational Energy

The ratios between total embedded energy and annual operational energy for the above buildings ranges from 14 to 35 for the houses while for an office building with air-conditioning loading it is around 5.

This shows the large influence that air-condition loading has on the operational energy. The operational energy of the air conditioned building (964 MJ/m^2) is around 6 times higher than those for the houses.

Although air conditioning has a large contribution towards the annual operational energy of a building, the total number of air conditioned buildings is small for a developing country such as Sri Lanka. Nevertheless, the above results show that the focus of energy efficient design for buildings with air-conditioning has to be on the operational energy. On the other hand, for houses, which are largely not air conditioned, the way to promote energy efficiency is by reducing the embedded energy through the appropriate choice of building materials. This is borne out not only by the high ratios of construction to operational energy ratio obtained, but also by the fact that the ratios for the houses with low energy materials are around half those for the houses with high energy materials.

Figure 2 shows some comparative ratios between total embedded energy and annual operational energy. In the case of an air conditioned office building in Sri Lanka, the ratio of 5 obtained is similar to the ratios obtained for houses in the U.K. and New Zealand (where operational energy includes space and water heating) [2]. However, for houses without air conditioning in Sri Lanka the above ratio (varying between "Min" and "Max" in Figure 2, is very much higher than those obtained in temperate zone countries.

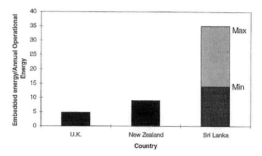

Figure 2   Comparison of Embedded and Operational Energy for Houses

CONCLUSIONS

1.   Among the more common construction materials considered, the lowest energy option is timber while the highest is steel, with concrete in between.

2.   Materials which use wood fuels (e.g. bricks and tiles) consume more energy. However, the use of wood fuels is more competitive when compared on a bio-equivalent unit basis.

3.   The energy contribution from walls for a typical two storey house is 10% to 44%; for a single storey house it is 29% to 49%. This indicates that choosing block walls instead of brick walls could make the greatest contribution to the reduction of gross embedded energy in residential buildings of 1-2 storeys. The contribution from wall plastering is 29% to 13% for the single storey house and 32% to 17% for the two storey house. Hence, the elimination of plastering can also contribute significantly towards reducing the embedded energy of houses.

4.   The ratio between total embedded energy and annual operational energy for the buildings selected ranges from 14 to 35 for the houses, while for an office building with air-conditioning loading it is 5. Hence, the focus of attention for energy efficient design should be on the operational phase for *air-conditioned* buildings. For *houses* in tropical countries however, which are generally not air-conditioned, energy efficiency is best achieved by the appropriate choice of construction materials.

REFERENCES

1. BAIRD, G., CHAN, S.A., Energy Cost of Houses and Light Construction Buildings, Report 76, New Zealand Energy Research and Development Committee, University of Auckland, Auckland, 1983

2. PERERA, K.K.Y.W., Energy Status of Sri Lanka, Institute of Policy Studies, Colombo, 1992

3. DIAS, W.P.S., POOLIYADDA, S.P., Quality based energy contents and carbon coefficients for building materials: system approach, Energy, Vol. 29, 2004, pp. 561-580.

4. POOLIYADDA, S.P., Energy Content and Carbon Emission Audit of Building Materials, M.Phil. Thesis, Dept. of Civil Engineering, University of Moratuwa, Sri Lanka, 2000

5. MINISTRY OF LOCAL GOVERNMENT, Building Schedule of Rates, Sri Lanka, 1988

6. HONEY, B.G., BUCHANAN, A.H., Environmental impacts of the New Zealand building industry, Research Report 92/2, Dept. of Civil Engineering, University of Canterbury, Christchurch, New Zealand, 1992

DESIGN AND EVALUATION OF SUSTAINABLE CONCRETE PRODUCTION

Z Li

Y Yamamoto
University of Tsukuba
O Takaaki
Hiroshima University
Japan

ABSTRACT. In this study, the environmental burden (EB) factors of concrete are discussed, land use change and forest's CO_2 absorption reduction, caused by gathering raw materials or disposing waste in landfill, are taken as EB factors for reflecting well the environmental benefits of recycling besides air emissions, waterborne releases, and energy resource consumption. Then, a method of evaluating environmental performance (EP) of concrete is proposed, which is based on life cycle impact assessment approach. The EP indicator of concrete is expressed by social cost caused by its life cycle. A social cost intensity database is also constructed for some of fuels, raw materials, and processes such as transportation, etc. Finally, an EP design method is suggested for maximizing the EP of concrete material on the condition that its mechanical properties and durability are guaranteed, that is, minimizing the environmental burden of designed concrete. Furthermore, an example of EP design is given to make the retaining wall concrete have the minimum EP indicator, using available environment-friendly raw materials.

Keywords: Concrete, Environmental performance indicator, Environmental performance design, Life cycle impact assessment, Waste utilization.

Zhuguo Li, Research scientist in Graduate School of System and Information Engineering at University of Tsukuba, Japan. His research interests include rheology and workability design of fresh concrete, development and evaluation of high performance concrete, environment-conscious design of construction materials, and environmental impact analysis of building and regional development.

Yasuhiko Yamamoto, Professor in Graduate School of System and Information Engineering at University of Tsukuba, Japan. His research interests include concrete materials, concrete construction, and risk management of concrete structure.

Taka-aki Ohkubo, Professor in Department of Social and Environmental Engineering at Hiroshima University. His research interests include building materials and components, advanced construction method and technology, and methods for maintaining and repairing buildings, etc.

INTRODUCTION

Concrete is one of main construction materials, and this situation will not change in the foreseen future. During the life cycle of concrete, it consumes various natural resources, including energy resources, water, land, and mineral resources, etc., and releases many kinds of polluting substances back to global/regional environment. Concrete has been identified as one of main factors of environmental burden (EB) in construction industry. There is no doubt that reducing concrete's EB is an urgent assignment for constructing an environment-harmonious society system.

Fortunately, past studies and practical experiences have shown that many kinds of wastes or by-products such as fly ash, blast furnace slag, sludge incineration ash, and recycled aggregate, etc. can be successfully recycled in concrete. This recycling not only can greatly decrease the EB of concrete itself, but also can alleviate the pollution of other industries. In recent years, the technologies of recycling wastes and co-products in concrete have gained remarkable development. It is expected to use various recycled materials to the full extent through environmental performance (EP) design for reducing the EB of concrete to a minimum.

In this study, an integrated assessment method of concrete's EP is firstly proposed, which is a life cycle impact assessment approach and gives EP indicator in form of social cost, and a database of social cost intensities of 14 EB factors is also constructed for various concrete's raw materials and processes such as transportation and final waste disposal. Then, an EP design method is suggested for minimizing concrete's EB by using environment-friendly materials. Finally, an application example of this EP design method is given for designing retaining wall concrete.

ENVIRONMENTAL PERFORMANCE OF CONCRETE

During the life cycle of a product or process or service, natural resources are consumed, and polluting substances are discharged to global/regional environment. These environmental inputs and outputs will inflict damages on human health, primary production, natural resources and biodiversity, accordingly they are usually called environmental burden (EB) of the product or process or service. As stated just above, environment-conscious design of minimizing concrete's EB should be performed on an equal footing with strength design and durability design. Hence, in this study, the integrated EB of concrete is considered as one of its performances, like as strength, and fluidity, etc., referred to environmental performance (EP). Proper evaluation of EP is the integral first step toward performing an EP design.

Environmental Burden Factors

Many natural resources including oil, limestone, clay, rock, sand, gypsum, etc., are consumed in the life cycle of concrete. Meanwhile, various pollutants, such as CO_2, SO_x, NO_x, CH_4, N_2O, suspended particulate matter (SPM), COD (Chemical Oxygen Demand), total P (T-P), total (T-N), and solid waste, are released back into the environment. Therefore, exhaustible energy resources including coal, oil, natural gas, and uranium ore, and air emissions as well as waterborne releases are taken as EB factors of concrete in this study, as shown in Table 1.

Moreover, gathering raw materials or disposing waste will cause the loss of habitat and natural environment resource in the working region. It is necessary to introduce an EB factor, called land use change, to reflect the EB caused by the disruption or reduction of habitat, especially in case of using recycled materials. For example, when part of land sand in concrete is replaced by JIS grade II fly ash, if not considering the effect of land use change, the integrated EB of fly ash concrete is almost the same to that of normal one because the energy consumption for collecting the fly ash is not less than that for gathering the land sand [1]. The areas of sites needed to acquire unit weight of rock or clay (ton), and to dispose unit volume of waste (m^3), are 0.02 m^2 and 0.20 m^2, respectively [2].

Table 1 EB factors considered and their embedded social cost intensities

EB CATEGORY	EB FACTOR	EMBEDDED SOCIAL COST INTENSITY (JAPANESE YEN ¥)
Atmospheric emissions	CO_2* (g)	$1.62 \cdot 10^{-3}$
	SO_x (g)	$5.57 \cdot 10^{-2}$
	NO_x (g)	$3.98 \cdot 10^{-2}$
	CH_4 (g)	$3.73 \cdot 10^{-2}$
	N_2O (g)	$4.80 \cdot 10^{-1}$
	SPM (g)	$3.08 \cdot 10^{-4}$
Waterborne releases	COD (g)	$6.40 \cdot 10^{-4}$
	T-N (g)	$8.25 \cdot 10^{-2}$
	T-P (g)	$9.74 \cdot 10^{-1}$
Energy resource consumption	Oil (g)	$1.65 \cdot 10^{-3}$
	Coal (g)	$4.54 \cdot 10^{-4}$
	Natural gas (g)	$1.29 \cdot 10^{-3}$
	Uranium (g)	1.16
Habitat loss	Land use (m^2)	80**

[Notes] * Forest's CO_2 absorption reduction is added up as an increase of CO_2 emission.
** The social cost for converting woodland to other uses, such as quarry and landfill.

At present, it is difficult to assess quantitatively and completely the loss of natural environment resource resulting form raw materials acquisition or final waste disposal. Since rock acquisition or landfill construction usually takes place in relatively low mountainous areas in Japan, the EB caused by the loss of natural environment resource is always simply evaluated by a reduction in the area of leafy deciduous forest [3]. The reduction of forest area is 0.067 m^2/m^3 when disposing solid waste, or is 1.9×10^{-5} m^2/ton for quarrying rock. The operations of quarry and landfill last respectively 50 and 10 years on average in Japan, and the recuperative growth period of the leafy deciduous forest is set up to be 30 years. Therefore, CO_2 absorption ability of forest, being 1.87 kg CO_2 / m^2 per year, will lose roughly for 80 years when quarrying rock, or for 40 years due to final waste disposal [3]. Forest's CO_2 absorption reduction is employed to represent roughly the EB related to the loss of natural environment resource.

In this study, solid waste discharge is not considered as an independent EB category, and the use of clay, aggregate, gypsum, and limestone, etc. is not regarded as EB factor of exhaustible resource consumption. However, the resulting energy resources consumption, air emissions, waterborne releases, forest's CO_2 absorption reduction and land use change are taken in account.

Environmental Performance Indicator

Environmental life cycle inventory (LCI) can quantify the inputs of energy and raw materials taken from the environment, and the outputs released back into the environment throughout the whole life cycle of evaluated object. However, it is difficult to understand the environmental impacts of these inputs and outputs and to make an EB comparison between different evaluated objects only by these LCI result data. A life cycle impact assessment is therefore necessary to translate the inventory data to impacts on the safeguarded subjects of human health, public assets, biodiversity, primary production capacity, and further integrate the damages of all the safeguarded subjects into a single indicator.

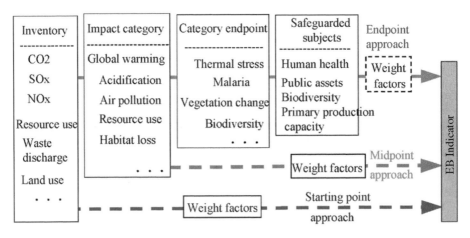

Figure 1 Three kinds of integrating LCA approaches

The integrating LCA approaches can be classified into three types: endpoint approach, midpoint approach and starting point approach, as shown in Figure 1 [4]. The endpoint approach can firstly give quantitative analyses of potential environmental impacts and damages to the safeguard subjects, resulting from resource uses and environmental releases, and then integrate all the damages into a single indicator. Therefore, the endpoint approach, also called as life cycle impact assessment (LCIA), is the most desirable LCA method.

In order to integrate the inventory data or the environmental impacts or the damages of the safeguarded subjects into an indicator, weight factors are of course needed. For making the integrated EB indicator (i.e. EP indicator) of concrete to easily understand and use in the environmental accounting of enterprise or local government, the weight factors in currency unit are adopted to make the obtained EP indicator in form of monetary loss, i.e. social cost.

In order to develop an LCIA approach applicable to Japan, a national research project, called Development of Assessment Technology of Life Cycle Environment Impacts of Products (also called briefly LCA Project,), had been conducted from 1998 to 2002 [5]. In the LCA project, a link was firstly established from EB factors to environmental impact categories, such as global warming, further to the safeguarded subjects including human health, public assets, biodiversity, and primary production capacity. The damage factors of various EB factors to the four kinds of Safeguarded subjects and the normalization values of the Safeguarded subjects in Japan are respectively obtained. Then external diseconomy, resulted from the damages of the safeguarded subjects, were clarified by doing a conjoint analysis, which estimates how much people would be willing to pay to preserve the safeguarded subjects. Finally, the weight factors of the safeguarded subjects and the social costs embedded in per unit of various EB factors were achieved. Social cost intensities of the EB factors considered in this study are shown in Table 1. Using these social cost intensities of EB factors, we can further estimate the EP of concrete in social cost after performing LCI.

If waste such as fly ash is recycled into concrete or enters into other product's life cycle, the potential environmental burdens caused by final waste disposal can be avoided. To reflect the environmental benefits of waste recycling, when assessing the EP of concrete, subtraction operation is performed for the potential social cost brought by the final disposal of the waste recycled in the concrete, as shown in Equation (1).

$$EPI = \sum_i [(\sum_j E_{ijv} - \sum_r E_{irw}) \times F_i] \qquad (1)$$

where EPI = EP indicator of concrete; i = EB factor; j = virgin material and process; r = recycled material; F_i = weight factor of EB factor i (embedded social cost intensity); E_{ijv} = magnitude of EB factor i of virgin material or process j; E_{irw} = magnitude of EB factor i caused by the final disposal of waste r.

ENVIRONMENTAL PERFORMANCE DESIGN APPROACH

Proposal of EP Design Method

EP design aims to minimize the EB of concrete caused during its life cycle. The boundary of the LCI, which evaluates the environmental inputs and outputs for assessing further the EP of concrete should encompass the acquisition of raw materials (limestone, clay, etc.), manufacture of intermediate materials (cement, aggregate, etc.), concrete production, construction, maintenance, demolition and final disposal. However, it is considered that there is no great difference in the environmental burdens associated with the demolition stage and final disposal stage between different concretes. The concrete works in the stage of maintenance (repairing, and strengthening, etc.) cannot be predicted quantitatively at present because they are different from achieved structure and locating environment. And if alternative concrete of lower EB are also ensured to be on the demands for the practical performances such as strength, workability and durability when EP design is performed, there is no great difference in the concrete works in construction and maintenance stages between alternative concrete and normal one. Since the purpose of EP design is to minimize the EB of concrete, in this study LCI boundary only covers the stages from the acquisition of raw materials to the production of concrete, as shown in Figure 2. The environmental burdens, caused by the production of transport vehicles and the constructions of factories for manufacturing the raw materials, are excluded.

The EP design method of concrete for minimizing the EB associated with the production stage by using recycled materials is proposed as shown in Figure 3. Firstly, the normal mixture not using recycled material is designed and its EP indicator is further assessed, of which practical performance indexes such as compressive strength chloride content meet the requirements of structural functions and locating environment. Then, mix proportions of concrete mixtures using different recycled materials available are calculated or selected, of which other performances except EPI are nearly on an equal footing with that of the normal mixture, and their EP indicators are estimated. The mixture (No.1) of the minimum EPI, or and other mixtures (No.2, No.3, …), if the EPI of No.2 or No.3 is greater than that of No.1 within 10%, is selected to conduct trial mix of concrete and modify the mix proportions for ensuring their practical performances to meet the demands. Since it is possible that the EPI of mixture No.1 is greater than that of the mixture No.2 or No.3 after the mix proportions are revised through trial mix test, the mixtures No.2 or and No.3 is also need to perform trial mix test. Finally, the mixture that has a minimum EP indicator is determined to be final result of concrete design, and its environmental advantage is estimated by calculating the difference in EP indicator from the normal mixture.

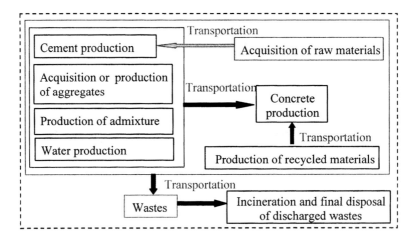

Figure 2 LCI system boundary for EP design of concrete

In Japan, the design guidelines have been established for the concrete using environment-friendly materials, including mineral admixtures of fly ash, silica fume and blast furnace slag powder, fine aggregates of blast furnace slag, electric furnace oxidation slag, copper slag, ferronickel slag and fly ash, blast furnace slag coarse aggregate, artificial fly ash coarse aggregate, and blended Portland cements of fly ash cement, silica fume cement and blast furnace slag cement, etc. The mix proportions design of concrete using recycled material can follow these guidelines to conduct. It has been proved experimentally that eco-cement made from trash incineration ash, recycled aggregate, waste glass, dissolution dust, limestone powder, sludge incineration ash, sludge incineration ash, swage-sludge slag, trash molten slag, pulp-sludge ash, cullet, and rice hush ash, etc. can be also recycled in concrete [6, 7]. The concrete design for using these recycled materials may refer to the related literatures. If using any kind of blended Portland cement as binder, the mineral admixture such as fly ash is not considered to replace some of Portland cement for reducing further concrete's EB.

Social Cost Intensity Database

Table 2(a) shows the EB factor intensities and social cost intensities for purchased electricity and some of fuels, raw materials and processes. The air emissions and waterborne releases, caused by electricity, light oil, and superplasticizer, are obtained respectively by Input-Output Table analyses [8, 9] and the measurement and statistics [10]. The air emissions and waterborne releases caused by with incineration are actual survey results [5, 11]. The air emissions, associated with final waste disposal, are estimated using "Bottom-Up Method" based on the energy requirements [5], but the waterborne releases are actual survey results [5]. The consumption amounts of the energy resources and the habitat loss areas, associated with electricity, light oil, incineration, final waste disposal and superplasticizer, are estimated by the Bottom-Up Method, based on the energy requirements [5] and the amount of final disposal waste [10, 11], respectively.

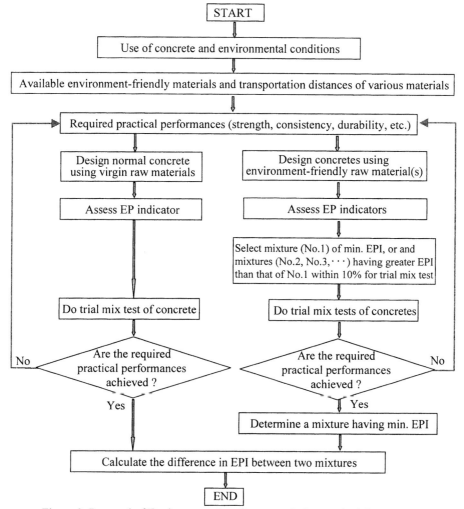

Figure 3 Proposal of Environmental performance design method for concrete

106 Li, Yamamoto, Takaaki

Table 2(a) EB factor intensities and social cost intensities of electricity, fuels and process

INPUTS AND OUTPUTS		ELECTRICITY (kWh)	LIGHT OIL (MJ)	INCINERATION (kg)	FINAL WASTE DIPOSAL (kg)	TRANSPORTATION (10t diesel truck, t×km)	MIXING STAGE OF CONCRETE (M³)
Atmospheric emissions (g)	CO_2	5.64×10^{-1}	7.80×10^{-2}	1.27	3.61×110^{-2}	78.3	7.65×10^{3}
	SO_x	4.61×10^{-1}	1.14×10^{-1}	0.76	5.74×10^{-4}	1.05×-10^{-1}	9.01
	NO_x	6.15×10^{-1}	5.17×10^{-1}	1.24	4.10×10^{-1}	1.19	35.5
	CH_4	5.43×10^{-1}	5.00×10^{-3}	5.27×10^{-3}	1.83×10^{-2}	2.41×10^{-2}	2.93
	N_2O	3.86×10^{-3}	3.16×10^{-3}	4.52×10^{-3}	1.34×10^{-5}	1.40×10^{-3}	2.08×10^{-1}
	SPM	7.40×10^{-3}	1.03×10^{-3}	7.53×10^{-2}	2.97×10^{-3}	9.65×10^{-2}	3.43×10^{-2}
Waterborne release (g)	COD	1.50×10^{-4}	3.46×10^{-4}	9.72×10^{-4}	9.30×10^{-2}	3.60×10^{-4}	3.87
	T-N	5.10×10^{-3}	3.35×10^{-4}	6.09×10^{-3}	1.13×10^{-2}	3.39×10^{-4}	5.77
	T-P	7.00×10^{-4}	2.60×10^{-5}	8.31×10^{-4}	2.09×10^{-4}	2.62×10^{-5}	3.04×10^{-1}
Energy resource consumption (g)	Oil	26.00	21.80	11.30	2.84×10^{-1}	21.90	1430
	Coal	57.20	2.49×10^{-3}	11.60	2.67×10^{-1}	2.54×10^{-3}	271
	Natural gas	6.93	3.36×10^{-1}	1.41	3.24×10^{-2}	3.38×10^{-1}	52.90
	Uranium ore	4.25	1.85×10^{-3}	8.68×10^{-1}	1.99×10^{-2}	1.89×10^{-4}	20.10
Land use change	Habitat loss (m²)	1.35×10^{-5}	1.93×10^{-7}	1.93×10^{-6}	0.005	2.56×10^{-8}	4.46×10^{-4}
Social cost intensity (¥)		—	—	—	—	2.19×10^{-1}	41.2

Table 2(b) EB factor intensities and social cost intensities of raw materials

INPUTS AND OUTPUTS		SUPER-PLASTI-CIZER (kg)	PORT-ABLE WATER (kg)	LAND SAND (kg)	CRUSHED STONE (kg)	PORT-LAND CEMENT (kg)	SLAG CEMENT B (kg)	FLY ASH CEMENT B (kg)	SLAG POWDER * (kg)	FLY ASH (JIS II, kg)	FLY ASH (JIS I, kg)	FLY ASH (JIS IV, kg)	SILICA FUME (kg)	RECY-CLED AGGRE-GATE (JIS I kg)	RECY-CLED AGGRE-GATE (JIS III, kg)	WASTE MELTING SLAG AGGRE-GATE ** (kg)
Atmospheric emissions (g)	CO_2	1.22	2.81×10^{-1}	1.87	5.68	856.00	466.00	690.00	-9.13	-9.13	8.62	-80.00	-29.55	-65.10	-6.96×10^{-2}	435.00
	SO_x	1.58	2.31×10^{-4}	2.50×10^{-3}	7.09×10^{-3}	9.30×10^{-2}	6.18×10^{-2}	7.76×10^{-2}	5.74×10^{-2}	5.74×10^{-2}	7.19×10^{-2}	-5.74×10^{-4}	4.55×10^{-2}	6.71×10^{-3}	3.04×10^{-3}	4.29×10^{-1}
	NO_x	2.73	3.15×10^{-4}	1.26×10^{-2}	2.94×10^{-2}	1.43	8.10×10^{-1}	1.16	-3.32×10^{-1}	3.32×10^{-1}	-3.13×10^{-1}	-4.10×10^{-1}	-3.48×10^{-1}	-3.97×10^{-1}	-4.02×10^{-1}	6.63×10^{-2}
	CH_4	9.41×10^{-1}	2.67×10^{-4}	2.30×10^{-4}	1.84×10^{-3}	1.34×10^{-2}	7.43×10^{-3}	1.04×10^{-2}	5.00×10^{-2}	5.00×10^{-2}	6.71×10^{-2}	-1.83×10^{-2}	3.60×10^{-2}	-1.07×10^{-2}	-1.50×10^{-2}	4.84×10^{-1}
	N_2O	3.15×10^{-1}	1.96×10^{-6}	6.82×10^{-5}	1.71×10^{-4}	3.95×10^{-3}	1.97×10^{-3}	2.95×10^{-3}	4.73×10^{-4}	4.73×10^{-4}	5.94×10^{-4}	-1.34×10^{-5}	3.73×10^{-4}	6.46×10^{-5}	3.38×10^{-5}	3.59×10^{-3}
	SPM	1.96×10^{-1}	3.83×10^{-5}	5.62×10^{-3}	4.87×10^{-3}	3.89×10^{-2}	2.32×10^{-2}	3.20×10^{-2}	-2.04×10^{-3}	-2.04×10^{-3}	-1.81×10^{-3}	-2.97×10^{-3}	-2.23×10^{-3}	-2.86×10^{-3}	-2.92×10^{-3}	3.22×10^{-3}
Water-borne release (g)	COD	2.93	1.12×10^{-4}	1.06×10^{-3}	1.06×10^{-3}	2.14×10^{-1}	2.04×10^{-1}	2.17×10^{-1}	-9.30×10^{-2}	-9.30×10^{-2}	-9.30×10^{-2}	-9.30×10^{-2}	-9.30×10^{-2}	-9.30×10^{-2}	-9.30×10^{-2}	-1.15×10^{-1}
	T-N	2.55	9.31×10^{-5}	7.09×10^{-4}	7.23×10^{-4}	7.95×10^{-2}	7.65×10^{-2}	8.05×10^{-2}	-1.07×10^{-2}	-1.07×10^{-2}	-1.05×10^{-2}	-1.13×10^{-2}	-1.08×10^{-2}	-1.13×10^{-2}	-1.13×10^{-2}	-9.28×10^{-3}
	T-P	1.01×10^{-2}	1.03×10^{-5}	8.08×10^{-5}	8.27×10^{-5}	9.90×10^{-3}	9.50×10^{-2}	1.01×10^{-1}	-1.21×10^{-4}	-1.21×10^{-4}	-9.93×10^{-5}	-2.09×10^{-4}	-1.39×10^{-4}	-2.00×10^{-4}	-2.05×10^{-4}	3.94×10^{-4}
Energy resource consumption (g)	Oil	53.20	1.31×10^{-2}	7.41×10^{-1}	1.18	18.57	11.45	14.57	2.99	2.99	3.81	-2.84×10^{-1}	2.32	2.45×10^{-1}	3.73×10^{-2}	23.91
	Coal	50.10	2.83×10^{-2}	7.40×10^{-3}	1.61×10^{-1}	95.63	56.21	76.25	6.93	6.93	8.73	-2.67×10^{-1}	5.45	5.29×10^{-1}	7.23×10^{-2}	53.03
	Natural gas	6.07	3.49×10^{-3}	8.10×10^{-3}	3.66×10^{-5}	4.50×10^{-1}	2.84×10^{-1}	3.38×10^{-1}	8.40×10^{-1}	8.40×10^{-1}	1.06	-3.24×10^{-2}	6.61×10^{-1}	6.67×10^{-3}	1.14×10^{-2}	6.43
	Uranium ore	3.73	2.10×10^{-3}	5.50×10^{-4}	1.20×10^{-2}	4.22×10^{-1}	2.55×10^{-1}	3.30×10^{-1}	5.15×10^{-1}	5.15×10^{-1}	6.48×10^{-1}	-1.99×10^{-2}	4.05×10^{-1}	3.92×10^{-2}	5.34×10^{-3}	3.94
Habitat loss (m^2)	Land use change	8.20×10^{-6}	3.87×10^{-9}	2.09×10^{-5}	2.09×10^{-5}	-1.84×10^{-5}	-9.34×10^{-5}	-4.91×10^{-5}	-1.18×10^{-4}	-1.18×10^{-4}	-1.18×10^{-4}	-1.18×10^{-4}	-1.33×10^{-4}	-1.00×10^{-4}	-1.00×10^{-4}	-1.24×10^{-4}
Social cost intensity (¥)		7.04	2.99×10^{-3}	6.94×10^{-3}	2.87×10^{-2}	2.12	1.22	1.71×10^{3}	1.11	5.73×10^{-1}	7.62×10^{-1}	-1.81×10^{-1}	4.08×10^{-1}	-8.42×10^{-2}	-1.32×10^{-1}	5.38

*: Blaine specific surface area is 400 cm²/g; **: Electric type municipal solid melting slag fine aggregate

Based on the obtained EB intensity data of electricity, light oil, incineration and final disposal, and the energy requirements shown in References 1 and 12, the environmental burdens, associated with various raw materials, diesel truck transport, and mixing process of concrete, are then estimated by using the Bottom-Up Method, as given in Tables 2. Also, the calculations are made for the land use area and forest's CO_2 absorption reduction caused by mineral acquisitions and final waste disposal, the portion of forest's CO_2 absorption reduction is included into CO_2 emission. The final disposal waste here includes not only the wastes discharged during the manufacturing process of raw material itself [10], but also those from the generating processes of electricity and fuel used. Obtained EB factors' intensities and social cost intensities of some of raw materials and processes for producing concrete are as shown in Table 2(b).

EXAMPLE OF ENVIRONMENTAL PERFORMANCE DESIGN

As an example of EP design of concrete, environment-conscious design of retaining wall concrete is carried out. The performance requirements of concrete and available raw materials are shown in Tables 3 and 4, respectively. According to the related guidelines of concrete design, it is possible to use recycled aggregate and blended Portland cement or replacing some of ordinary Portland cement by fly ash or slag powder in this kind of concrete. Six series of mixtures containing different environment-friendly materials are designed besides a normal mixture using virgin materials, as given in Table 5. These seven series of mixtures have nearly the same fluidity, strength and durability. Their EP indicators are shown in Figure 4.

Table 3 Quality requirements of designed concrete

USE OF CONCRETE	LOCATION	DESIGN REFERENCE STRENGTH	GRADE OF USE PERIOD PLANNING	EXPECTED TEMPERATURE DURING CONSTRUCTION	STANDARD DEVIATION OF STRENGTH	SLUMP	AIR CONTENT
Retaining wall	Tokyo	21 N/mm^2	Standard	>20°C	2.5 N/mm^2	80 mm	4.5%

Table 4 Properties and transportation distances of available raw materials

WATER	CEMENT			ADMIXTURE			AGGREGATE		
	Portland cement	Slag cement	Fly ash cement	Slag powder	Fly ash	Chemical admixture	Land sand	Crushed stone	Recycled coarse aggregate
Potable water	Ordinary type D: 100 ρ: 3.15	B type ρ: 3.0 4 D: 100	B type ρ: 2.97 D: 100	Blast furnace Slag Blaine value II : 4000 cm^2/g D: 150	JIS grade g II ρ: 2.2 0 D: 150	Water-reducing AE agent D: 200	ρ: 2.60 F_m: 3. 1 D: 30	D_{max}: 25 ρ: 2.65 ρ_v: 1570 V_a: 60.4 F_m: 6.65 D: 30	JIS grade 1 D_{max}: 25 ρ: 2.32 ρ_v: 1420 V_a: 60.4 F_m: 6.56 D: 50

[Notes] D: Transportation distance by diesel truck of 10 ton (km), D_{max}: Maximum size (mm), ρ: Absolute dry density (g/cm^3) ρ_v: Bulk specific gravity (kg/m^3), V_a: Ratio of absolute volume (%), F_m: Fineness modulus.

Table 5 Mix proportions of designed mixtures

SERIES	W/B (%)	s/a (%)	W (kg)	C (kg)	ADMIXTURE (kg)	CRUSHED STONE (kg)	RECYCLED COARSE AGGREGATE (kg)	LAND SAND (kg)	AE AGENT (kg)
N	54.1	44.2	146.0	OPC, 270	0.0	1059.8	0.0	840.8	0.68
W-BB	52.4	44.8	141.6	BB, 270	0.0	719.0	308.2	833.2	0.68
W-FB	51.9	44.7	140.1	FB, 270	0.0	719.0	308.2	831.6	0.68
W-BG30	50.1	44.0	140.8	OPC, 200	BG, 85.7	719.0	308.2	808.1	0.71
W-BG50	35.2	40.9	139.3	OPC, 200	BG, 200	719.0	308.2	711.0	1.00
W-FA20	41.6	42.6	140.4	OPC, 270	FA, 67.5	719.0	308.2	762.8	0.81
W-FA30	37.8	41.8	136.7	OPC, 289	FA, 72.4	719.0	308.2	737.8	0.90
N*	54.8	44.1	148.0	OPC, 270	0.0	1059.8	0.0	835.6	0.68
W-BB*	52.9	44.7	143.2	BB, 270	0.0	719.0	308.2	829.6	0.68

[Notes] W: mixing water, B: binder, s/a: ratio of sand to aggregates, C: cement, OPC: ordinary Portland cement, BB: Blast furnace slag cement (B type), FB: flay ash cement (B type), BG: Blast furnace slag powder, FA: flay ash, *: mixture revised through trial mix test.

According to Figure 4, in case of producing concrete that has lower design reference strength grade like this example, if using fly ash or a large amount of slag powder to replace part of ordinary Portland cement (series W-FA20, W-FA30 and W-BG50), on the contrary, EP indicator of concrete increases, because water-binder ratio has to be decreased and Portland cement content must be maintained at the minimum limit for satisfying the durability requirements such as neutralization resistance. Series W-BB, using B type blast furnace slag cement and JIS grade 1 recycled coarse aggregate, has the minimum EP indicator (390 Japanese yen/m^3), and the second magnitude of EP indicator is greater than that of W-BB beyond 10% (W-FB: 528 Japanese yen/m^3). Therefore, trial mix tests of series N and W-BB are conduced to revise their mix proportions for attaining surely the required fluidity, strength and durability. The revised mix proportions are shown in the lower section of Table 5 (series N* and W-BB*). The EP indicators of N* and W-BB* are respectively 673 and 396 Japanese yen/m^3, as shown in Figure 4. Hence, W-BB* is selected to be final result of concrete design, of which EP indicator is smaller than that of normal one by 41.2%.

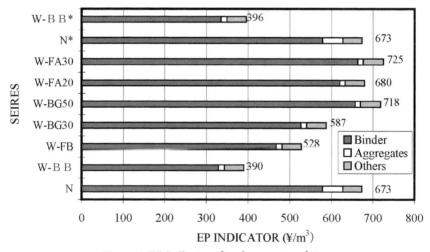

Figure 4 EP indicator of each concrete mixture

CONCLUSIONS

This paper firstly gave a discussion on the environmental burden factors of concrete material that should be considered, the introduction of land use change and forest's CO2 absorption reduction as EB factors can reflect well the environmental benefits of recycling. Then, a method of evaluating environmental performance of concrete was suggested, which is based on life cycle impact assessment. The obtained EP indicator of concrete is expressed by social cost caused in its life cycle. A database of social cost intensity was also constructed for some of raw materials of concrete and processes. Finally, a design approach of environmental performance design was proposed for concrete production to minimize its environmental burden. An example of EP design was further given to minimize the EP indicator of retaining wall concrete by optimizing sorts and contents of used raw materials. Moreover, it was concluded that the environmental performance of concrete with lower strength grade is not always improved by using waste in large quantities on the condition of guaranteeing its durability.

REFERENCES

1. LI, Z., YAMAMOTO, Y., SAGISAKA, M., Environmental life cycle assessment of concrete containing fly ash, Proc. of 6th CANMET / ACI International Conference on Fly Ash, Silica Fume, Slag and Natural Pozzolans in Concrete (ACI SP-221), U.S.A, 2004.5, pp 573-588.

2. ANBE, W., II, R., Ecosystem impact assessment of land use, Proc. 2nd Workshop on Environmental Impact Assessment Based on Estimation of Damage-Impact Assessment of Ecosystem (Tokyo), 2002.7, pp 13-30.

3. NAKANO, K., MIURA, H., WADA, Y., Evaluation of environmental impact reduction by waste asphalt concrete recycling, Journal of the Japan Society of Civil Engineers, No.559/VII-2, 1997.2, pp 81-89.

4. ITSUBO, N., INABA, A., Assessment of Environmental Impact of Manufacturing Steel Considering Physical Damage to Human Health, Material Transactions, The Japan Institute of Metals, Vol. 44, No. 1, 2003.1, pp 167-172.

5. JAPAN ENVIRONMENTAL MANAGEMENT ASSOCIATION FOR INDUSTRY, Report on Development of Assessment Technology of Life Cycle Environment Impacts of Products, 2003.3, pp 29-918.

6. JAPAN CONCRETE INSTITUTE, Proceedings of the JCI symposium on application of industrial and municipal wastes to concrete materials, JCI-C 56, 2002.9, pp 1-188.

7. UCHIKAWA, H., OBANA, H., Environment-compatible cement – eco-cement made from municipal solid waste. Concrete Journal, Vol. 34, No. 4, 1996.4, pp 57-63.

8. ARCHITECTURAL INSTITUTE OF JAPAN, Guide to LCA of building, Maruzen Publishing Division, Tokyo, 2003.2, pp 1-155.

9. NANSAI, K, MORIGUCHI, Y and TOHNO, S. Embodied energy and emission intensity data for Japan using Input-Output Tables-inventory data for LCA. National Institute for Environmental Studies, Japan, 2002, pp.3EID1-3EID27.

10. TSURUMAKI, M., NOIKE, T., Study on the LCA evaluation of wastewater treatment. Journal of the Japan Society of Civil Engineers, No.643/VII-14, 2000.2, pp 11-20.

11. ECO-MANAGEMENT RESEARCH INSTITUTE, Investigative report of environmental burden intensity data associated with waste treatment, Technical Report, 2000.2, pp 5-60.

12. SANO, S., TANIMURA, M., YOSHIDA, A., ICHIKAWA, M., Life-cycle assessment of reinforced concrete, Annual Report of Taiheiyo Cement, No. 142, 2002, pp 111-122.

GREEN BUILDINGS AND FLYASH CONCRETE –
THE COMMERCE CITY, COLORADO PROJECT

M Reiner

K L Rens

A Ramaswami

University of Colorado at Denver

United States of America

ABSTRACT. Current "green building" guidelines establish a goal for locally sourced or manufactured materials, particularly for using recycled materials. This is a qualitative environmental goal, relating to the energy associated with the transportation of the product and not the energy of production, termed embodied energy. A quantitative environmental goal would require a life cycle analysis (LCA) to measure the embodied energy that would account for production, the insulating properties, and the ability to be again recycled over the effective life-time of the building. A quantitative environmental life cycle analysis of the primary construction materials in building construction would provide a definitive green building assessment. The Sustainable Youth Zone center (SYZ) project located in Commerce City, Colorado is a planned Zero Net Energy building for after school educational programs. Building material selection is evaluated based on current quantitative and qualitative green criteria including; Building for Environmental and Economic Sustainability (BEES), Leadership in Energy and Environmental Design (LEED). However, in an attempt to determine if the building will actually be green, a life-cycle analysis and cost (LCA/LCC) analysis was completed on concrete as the primary construction material considered. Fly ash is a by-product of coal burning power plants and can be substituted for a large portion of the Portland cement. High volume fly ash (HVFA) concrete was selected for analysis as fly ash is a recycled material, therefore reducing the impacts resulting from extraction and processing of new virgin materials. A "mini-mix" (MM) testing program was also conducted to evaluate HVFA mixes for additional full-scale testing.

Keywords: LCA, LCC, HVFA, BEES, Mini-mix, Fly-ash, LEED

Mark Reiner, P.E., P.G., is a PhD Candidate, University of Colorado at Denver, Department of Civil Engineering, Denver, Colorado.

Kevin Rens, P.E., is Associate Professor, University of Colorado at Denver, Department of Civil Engineering, Denver, Colorado.

Anu Ramaswami, is Associate Professor, University of Colorado at Denver, Department of Civil Engineering, Denver, Colorado.

INTRODUCTION

In May of 2003, the Urban Sustainable Infrastructure Engineering Project (USIEP) was initiated at the University of Colorado, Denver (CU Denver) to work in the multi-disciplinary area of sustainable urban infrastructure development. A citizens' group from Commerce City, Colorado approached CU Denver USIEP for assistance in engineering and architecture for a planned Youth Center. This project has become known as CU Denver's Sustainable Youth Zone Project (SYZ – pronounced "SIZE"). The overall goal for the design is an integration of building materials, energy, water, and waste sub-systems to achieve a zero net energy consuming building and a *Platinum* certification from the U.S. Green Building Council's Leadership in Energy and Environmental Design (LEED).

As concrete is typically the most massive individual element of a building, a comparison of ordinary Portland cement (OPC) concrete, high volume fly ash (HVFA) concrete, and concrete with slag as a partial cement substitute was selected for evaluation. In order to quantify the environmental and economic cost impacts of foundation material selection, a life-cycle analysis (LCA) and life-cycle cost (LCC) assessment was conducted. Also, considered are the current building codes and structural engineering requirements for using HVFA concrete in the SYZ building for slab-on-grade, basement walls, and other structural elements. A "mini-mix" (MM) testing program was conducted to evaluate promising HVFA mixes for additional full-scale testing.

The primary LCA/LCC software used for the analysis of HVFA is the U.S. National Institute of Standards and Technology (NIST) Building for Environmental and Economic Sustainability (BEES) model version 3.0, 2003. This paper presents the regulations, LCA/LCC results, and results of the MM testing.

The Environmental Impacts of Concrete

Selecting materials based not only on performance, but also the energy required for extraction, production, and transportation, can have a significant influence on the economic and environmental impact on the environment. The total world-wide cement demand is expected to rise to about two billion tonnes by the year 2010 and the current annual world-wide production of coal ash is estimated to be over 700 million tonnes [1]. Although cement typically constitutes 10% to 15% of concrete by weight, cement production is responsible for most of concrete's environmental impacts. The manufacture of Portland cement accounts for 6% to 7% of the worldwide CO_2 emissions, adding the greenhouse gas equivalent of 330 million cars driving 12,500 miles per year [2]. Even for residential houses, concrete accounts for 11% to 25% of the total embodied energy of the structure and is the largest contributor to global warming among the individual building elements [3].

HVFA Concrete as a "Green" Construction Material

According to a generally accepted definition, HVFA is constituted by a minimum of 50% fly ash, a low water content (130kg/m^3), less than 200 kg/m^3 cement content, and a low water-cementicious ratio (less than 0.4). The lesser volume requirements for cement and water and the high performance of HVFA contribute to meeting increasing demands for concrete in the future at reduced or no additional environmental impacts or cost. In addition, as of Sept. 2,

1993 in the U.S., fly ash is a Subtitle D solid waste (under RCRA) and is required to be disposed of in monofills. The use of high percentages of fly ash in concrete would further lessen the regional ecological impacts from the disposal of large quantities of fly ash.

Potential LEED credits for HVFA

As the aggregate is mined locally to the proposed SYZ building (much within twenty kilometers) and the cement kilned regionally (approximately 100 kilometers), it is anticipated that the use of HVFA concrete may contribute to receiving four credits: Recycled Content (5% post-consumer + ½ post-industrial) Credit 4.1, Regional Materials: Credit 5.1 (20% regionally), Credit 5.2 (50% regionally), and an innovation in design credit.

Beyond LEED

The criterion for assigning LEED credits is based on assigning relative environmental impacts to a single life-cycle stage. This can lead to the selection of materials that may have a benefit of being locally produced, for instance, but may have a higher total embodied energy. In order to make material selections for the elements of a building based not only on performance, but also the amount of embodied energy, a LCA-assisted building design should be performed to quantify environmental impacts and lifetime economic costs of the individual design elements of the building.

Regulatory and Other Barriers for HVFA

Although there is no limit of fly ash content dictated in the Universal Building Code (unless exposed directly to de-icing salts where a maximum of 25% is dictated), there is a perception among engineers and architects that concrete construction codes are "prescriptive" in the sense that there is a maximum permissible limit of fly ash. The confusion arises from the governing standards as, for instance, ASTM C 595 limits the proportion of the pozzolan in the cement to 40 percent by mass while a new, performance-based cement standard ASTM C 1157 does not limit the type and the content of components in the blended cement. The UBC generally refers to ASTM C 618 in regards to fly ash content. It states that the optimum amount of fly ash or natural pozzolan for any specific project is determined by the required properties of the concrete and other constituents of the concrete and is to be established by testing. However, the time and cost of obtaining the performance data and putting it in the hands of the decision-makers is a significant barrier to commercializing this innovative technology [4].

HVFA CONCRETE LIFE-CYCLE ANALYSIS

The BEES Version 3.0 software is a LCA/LCC tool that contains environmental and economic performance data for nearly 200 products across 23 buildings. The LCA assessment is a "cradle-to-grave" approach that considers raw material acquisition, product manufacture, transportation, installation, operation and maintenance, and ultimately recycling and waste management. For this paper, BEES has been applied to evaluate the effects of the inclusion of fly ash and slag into concrete manufacture. However, only fly ash contents of

15%, 20%, and 35% (for slabs) has been included in the BEES database. The manufacturing data for concrete products in BEES are from the Portland Cement Association LCA database. Concrete mixes modeled in the BEES software are limited to compressive strengths of 21 MPa, 28 MPa, and 34 MPa (3 ksi, 4 ksi, and 5 ksi).

The BEES methodology measures environmental performance using an LCA approach in accordance with ISO 14040 of four steps: 1) goal and scope definition spell out the purpose of the study and its breadth and depth; 2) inventory analysis step identifies and quantifies the environmental inputs and outputs associated with a product over its entire life cycle; 3) impact assessment characterizes these inventory flows (inputs and outputs) in relation to a set of environmental impacts; and 4) interpretation step combines the environmental impacts with the goals of the LCA study. Economic performance is separately measured using the ASTM International standard LCC approach.

Goal and Scope

When used in concrete, slag and fly ash are cementitious materials that can act as a partial substitution for portland cement without significantly compromising compressive strength. Fly ash from modern thermal plants generally does not need any processing and is considered a waste product and is therefore considered to be an environmentally "free" input material [5]. Slag was considered in this assessment as it is also a waste product of the steel industry and considered to be environmentally free. But unlike fly ash, slag has a higher embodied energy as it must be processed by quenching and granulating prior to inclusion in concrete. The authors selected the analysis of the individual element slab-on-grade, under the BEES substructure major group element category, as it is the only element that evaluates 35% fly ash content. The functional unit for concrete quantities is 0.09 m^2 (1 ft^2) and a product service lifetime of 50 years.

Fly ash definition

The pozzolanic properties of a good-quality fly ash are not governed so much by the chemistry but by the mineralogy, low carbon content, high glass content, and 75% or more particles finer than 45 μm [1]. Fly ash reacts with any free lime left after the hydration to form calcium silicate hydrate, which is similar to the tricalcium and dicalcium silicates formed in cement curing. The two main types of fly ash used for concrete additives in the U.S. are low-calcium, bituminous-coal fly ash (ASTM Class F) and a high-calcium, subbituminous-coal fly ash (ASTM Class C). Class F is more common and is used in the BEES model.

Inventory Analysis

The energy requirements for the materials composing OPC, HVFA, and slag concrete in BEES are the following:

- Coarse and fine aggregate: BEES conservatively assumes that all aggregate is crushed and the energy assumed for production and a 85 km round-trip is 155 kJ/kg (66.8 Btu/lb).

- Portland cement: BEES data for cement manufacture is based on the average ASTM C150 Type I/II cement. The energy is based on a weighted average of manufacturing processes, a 100 km round-trip, and assumes 5,320 kJ/kg (2,280 Btu/lb).
- The transport energy for all materials in concrete production is by truck, consuming 1.18 kJ/kg*km (0.818 Btu/lb*mi).
- Slag requires processing and is assigned an energy 465 kJ/kg (200 Btu/lb).
- There is no energy assigned to fly ash other than 100 km round-trip transport.

Impact Assessment

Of the twelve environmental impacts evaluated in BEES, eight impacts had significant values for OPC, HVFA, and slag substitution. The results are shown in Figure 1.

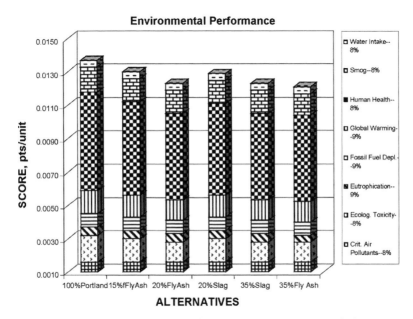

Figure 1   Environmental Performance of OPC, Fly Ash Concrete, and Slag Concrete

Interpretation

The BEES output for overall environmental performance indicates substantial indicator reductions for increasing volumes of fly ash and slag as a substitution for portland cement, particularly in the human health and global warming criteria. However, fly ash provided the greatest environmental benefits. The 11.7% reduction in overall environmental performance for 35% fly ash, when compared to OPC, was linearly extrapolated to estimate a 26.6 % reduction in environmental performance for 70% fly ash inclusion. If these scores are each first multiplied by the quantity of functional units to be used in a particular building the environmental impacts may then be compared and can provide insights into selecting building elements.

Economic Performance

The real interest rate (discount factor) of 3.5 percent was obtained from OMB Circular A-94, 2004 (http://www.whitehouse.gov/omb/circulars/a094/a94_appx-c.html). All costs for the alternatives during the period of analysis are in dollars with 2002 as the common base year. Again, fly ash provided the greatest reduction in lifetime costs, as shown on Figure 2. The 10.3% reduction in economic performance 35% fly ash, when compared to OPC, was linearly extrapolated to estimate a 21.3 % reduction for 70% fly ash inclusion.

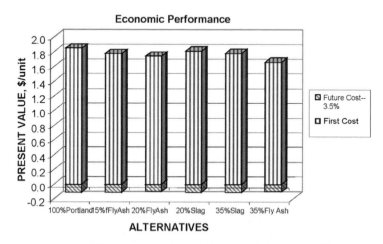

Figure 2   Economic Performance of OPC, Fly Ash Concrete, and Slag Concrete

HVFA MINI-MIX DESIGN AND RESULTS

The "mini-mix" (MM) approach substitutes the ASTM C-128 sand moisture cone for the ASTM C-143 concrete slump cone as a measure of workability. MM tests are mixed in the same proportions as a full-scale mix, but without the coarse aggregates. The process was selected for determining an appropriate HVFA mix as the effect of variables on concrete systems can be rapidly assessed. For this report, 12 MM batches were mixed over a range of fly ash substitution for OPC (Ordinary Portland Cement) and water reducing admixtures that would produce the required early and long-term strength for potential use as pre-cast and structural items in the SYZ building.

The mixes were divided into three ranges of total cementitious content (low-L, medium-M, and high-H). The total cementitious material for the three sets was 327-kg/m³ (550 pounds per cubic yard) for the low (L), 369-kg/m³ (620 pounds per cubic yard) for the medium (M), and 410-kg/m³ (690 pounds per cubic yard) for the high (H). Each range of cementitious content was further divided into four total fly ash replacement percentages of 40, 50, 60, and 70. The fly ash used in the mix was Class F obtained from the Coal Creek power plant in Wyoming. The fly ash particle diameters were 29.09 percent finer than 10 μm and 67 percent finer than 45 μm. In addition, Glenium 3020 HES (now known as Euclid), a water reducing agent was added to the mixes. The proportions used in the 12 MM tests are shown in Table 1.

Table 1   Mix Proportions for the 12 Mini-mixes

MIX CODE	MATERIAL, kg/m³						AIR CONTENT, %	W/C RATIO
	Cement	Fly Ash	WRA	Fine Agg.	Coarse Agg.	Water		
H70	122.8	286.5	12.3	777	1071.1	122.8	0.6	0.3
H60	163.7	245.6	16.4	769.8	1093.9	115.8	0.7	0.28
H50	204.7	204.7	20.8	765.3	1083.4	128.6	0.6	0.31
H40	245.6	163.7	24.6	767.4	1082.3	132.5	0.6	0.32
M70	110.3	258.1	11.1	825.7	1069.7	118.9	0.8	0.32
M60	148.3	221.9	14.8	821.4	1073.8	117.8	1.2	0.32
M50	183.9	183.9	18.3	831.4	1063.1	121.5	1.3	0.33
M40	220.7	147.1	22.1	782.8	1102.4	124.7	1.2	0.33
L70	97.9	228.4	9.8	854.5	1072.1	124.6	0.6	0.38
L60	130.5	195.8	13.1	862.9	1072.1	124.6	0.6	0.38
L50	163.1	163.1	16.2	869.5	1060.8	122.6	1.4	3.8
L40	195.8	130.5	19.4	868.8	1059.2	123.6	1.7	0.38

[1] The value represents an equivalent quantity for a full-scale mix. Coarse aggregate not used in mini-mix tests.
[2] Computed air content based on gravimetric analysis.

Measuring compressive strength using the MM procedure is conventionally obtained by 50.8-mm by 101.6-mm (2-in by 4-in) cylinders. But due to the number of samples, and that the MM procedure is a precursor to the full-mix design, 50.8-mm (2-inch) squares were tested for compressive strength. We anticipated that the MM strength results would be approximately 15 percent to 20 percent higher than a full-scale test using coarse aggregate and breaking 101.6-mm by 203.2-mm (4-inch by 8-inch) cylinders. The main criteria for selecting the MM mixes to be carried on for full-scale testing were those with the highest percentage of Portland replacement and achieved 1-day compressive strengths of 10.35 MPa (1500 psi) for pre-cast work, and 28-day strengths of 27.6 MPa (4000 psi) for structural concrete. The compressive strength test results for the 12 mixes at 1, 3, 7, 28, and 56-day breaks are shown on Figure 3.

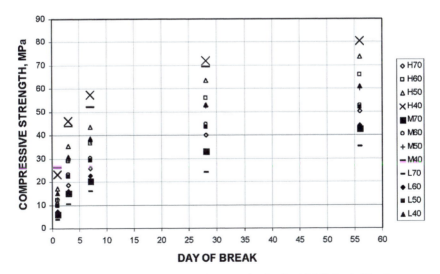

Figure 3   Compressive Strength Test Results for the 12 Mini-mix Batches

The performance of these mixes at early-age strength showed promising results. All of the tests met the 1-day break strength of 10.35 MPa (1500 psi), except for L70, M70, and L60. However, all tests except L70 met the 28-day strength of 27.6 MPa (4000 psi). The early-age compressive strength and the high volume of fly ash content led to the selection of four mixes for full scale testing using 101.6-mm by 203.2-mm (4-inch by 8-inch).

Much of the unexpected very high strengths are likely the result of low w:c ratios, the use of smaller cylinders and the lack of non-homogeneity usually found in larger inert aggregate. At early ages, the fly ash does not provide a chemical strength matrix, but it does act as a solid filler and a cement paste hydration site for further strength development. At later ages the fly ash does form additional calcium silicate which results in increasing strength. Although the strengths increase with decreasing fly ash content, we anticipate that the trend would not continue as a higher w:c would be required to compensate for decreased workability due to the floc associated with higher portland contents. Full-scale testing of H70, H60, M60, and M50 are planned to be performed at the University of Colorado – Denver structural lab starting Fall, 2004.

ACKNOWLEDGEMENTS:

- Bud Werner, Principal CTL Thompson, Materials Testing Division, Denver, Colorado.
- Tom Fox, ISG Resources, Tukwila, Washington.
- Andy Burke, Civil Engineering Department, University of Colorado – Denver.

REFERENCES

1. MALHOTRA, V.M., AND MEHTA, P.K., "High-Performance, High-Volume Fly Ash Concrete: Materials, Mixture Proportioning, Properties, Construction Practice, and Case Histories." Supplementary Cementing Materials for Sustainable Development, Inc, Ottawa, Canada, 2002

2. MEHTA, P.K., "Role of Flyash in Sustainable Development," Concrete, Flyash and the Environment Proceedings, December 8, 1998, pp. 13-25.

3. COLE, R.J., 1999. Energy and greenhouse gas emissions associated with the construction of alternative structural systems. Building and Environment 34: 335-348.

4. CIVIL ENGINEERING RESEARCH FOUNDATION. New Materials and Technologies Available for Use in Industrial Infrastructure: An Overview, Prepared for the DOE, Office of Energy Efficiency and Renewable Energy, March 25, 2003

5. U.S. NATIONAL INSTITUTE OF STANDARDS AND TECHNOLOGY (NIST) Building for Environmental and Economic Sustainability (BEES) model version 3.0, 2003.

6. U.S. Green Building Council's Leadership in Energy and Environmental Design (LEED)

CONCRETE MIXES FOR DESERT ENVIRONMENTS

B M El-Ariss

United Arab Emirates University
United Arab Emirates

ABSTRACT. This paper reports the results of experimental investigations carried out on the effects of concrete mixes with small or even null quantities of coarse aggregates on the mechanical properties of concrete in desert environments. It also presents the results of a statistical analysis aimed at optimizing concrete mix design for desert regions. A laboratory experiment using 72 samples of 12 different concrete mixes with three different curing methods was carried out. The influences of the water/cement ratio, coarse aggregate/total aggregate ratio, total aggregate/cement ratio, and curing methods (air curing, oven curing, and water curing) on the compressive strength of concrete were characterized and analyzed. Mathematical parabolas were developed for concrete strength as a function of the reduced coarse aggregate quantity. Based on the statistical analysis, recommendations are provided on the optimum concrete mix using desert sand.

Keywords: Concrete mix, Aggregates, Compressive strength, Curing.

Dr B El-Ariss, Assistant Professor, United Arab Emirates University, Department of Civil and Environmental Engineering, U.A.E.

INTRODUCTION

Sand/aggregate ratio and water/cement ratio are two basic parameters for the design of concrete mixes. These two ratios are mainly determined by using tables available in the literature, which are fairly arbitrary. In desert regions, such as those in the United Arab Emirates (UAE), there are special problems associated with concreting. These problems concern the availability of coarse aggregates. On the other hand, concreting in such regions has an advantage due to the presence the fine aggregates everywhere. Several experiments have been conducted to analyze the effects on concrete compressive strength of various factors such as water/cement ratio, cement content, supplementary cementitious materials, chemical admixtures, coarse/fine aggregate ratio, and temperature [1-3]. However, little research work has focused on the analysis of the measured experimental data and on modelling the effects of the reducing the quantity of coarse aggregates and increasing the quantities of fine aggregates. Tables available in the literature are no longer suitable for the design of concrete mixes with a small coarse aggregate content. Orr [4] used a factorial experiment to investigate the effects of cement temperature, mix temperature, cement composition, and water/cement ratio on the consistency and strength of concrete. He found that interactions between the factors were significant and hence concluded that it is possible for conventional single factor experiments to mask true effects on the response of concrete. Jerath and Aabbani [5] used a multiple nonlinear regression model to derive the relationships among various variables such as slump, maximum aggregate size, fineness modulus of sand, and water/cement ratio. However, only two independent variables were involved in the multiple nonlinear regressions. Abbasi et al. [6] used a reduced factorial experimental technique to investigate the simultaneous variations of the controlling factors.

In this research, the effects of reducing the coarse aggregate quantity and increasing the fine aggregate and cement quantities on concrete compressive strength were studied and analyzed. Empirical formulas for predicting concrete compressive strength as a function of reduced coarse aggregate quantity are derived and presented.

EXPERIMENTAL BACKGROUND

Test Program

Intensive laboratory testing of 900 samples of 60 different concrete mixes was carried out. The test variables include the water/cement ratio (W/C), total aggregate/cement ratio (TA/C), coarse aggregate/total aggregate ratio (CA/TA), and curing. Different values of W/C ratio, 0.4 and 0.5, were selected so that most common degrees of workability were considered. Three methods of curing were used to cure the concrete samples (air, oven, and water). No additives or superplasticizers were added to the concrete mixes.

Materials and Procedure

A crushed coarse aggregate and desert sand as fine aggregate were obtained from UAE. The aggregates were tested for specific gravity and absorption (ASTM C 127), and fineness modulus (ASTM C 33). The specific gravity was 2.45, absorption value was 4.97%, and maximum size was 19 mm. The fineness modulus of the sand was 2.74, specific gravity was 2.68, and absorption value was 1.6%. Type I ordinary Portland cement was used in all the mixes. The aggregates, water, and cement were mixed for 2 min and were cast in standard 150 x 150 x 150 mm cubes in the laboratory. The cubes were vibrated for 2 min using a vibrating table. The specimens were de-moulded after 24 h and kept immersed in water tanks, in ovens at 50°C, and in the air at room temperature. Three cubes for each mix from the three different curing methods were tested for compressive strength at 3, 7, 14, 21, and 28 days.

Test Results

A summary of the 3, 7, 14, 21, and 28-day compressive strengths for the 60 mixes with water cement ratios (W/C) of 0.4 and 0.5, from the three different curing methods is given in Table 1. The relationships between the compressive strength versus reduced coarse aggregate content, total aggregate to cement ratio (TA/C), and coarse aggregate to total aggregate ratio (CA/TA) for mixes 1-6 (with $W/C = 0.4$) only are shown in this paper graphically in Figures 1-3.

DISCUSSION OF RESULTS

Effect of Curing

The effect of curing on the compressive strength of concrete is significant as shown in Figures 1-3. These figures show comparisons of the measured compressive strengths at various material quantities for W/C = 0.4 and for different curing methods. It is clear that the strength is the least when oven curing is applied and is the largest when water curing is used. When air curing is applied, the compressive strengths are somewhere in between those of the oven and water curing strengths. These results are comparable to previous observations [3].

Effect of TA/C and CA/TA

The influences of TA/C and CA/TA on the compressive strength are also highly significant according to Figures 2 and 3. This is confirmed by the variation of the compressive strength with TA/C and CA/TA. Figures 2-3 show the measured test results for all different aggregate and cement quantities. Both sets of curves have a similar pattern for six different values of CA/TA or mixes. This means that the strength does not change much when the quantity of the coarse aggregates is reduced to zero. Similarly, it can be seen in Figure 1-3 that the curves, regardless of the curing method, exhibit a similar pattern for the six different mixes, which means that increasing either the fine aggregates, cement, or both will yield the same behavior. This agrees with Soudki and El-Salakawy [7].

Effect of W/C

The results of the test show that the W/C ratio did not affect the compressive strength much as long as it was kept constant throughout the different concrete mixes, as shown in Figures 1-3 for W/C ratios of 0.4 ($W/C = 0.5$ not shown in this paper). This is expected for well-controlled laboratory-made specimens.

Table 1 Summary of measured compressive strengths (MPa) x 10

| | | STRENGTH (x10 MPa) AT: | | | | |
		3 Days	7 Days	14 Days	21 Days	28 Days
Mix #1	Air Curing	388	438	470	480	490
	Oven Curing	367	415	418	436	440
	Water Curing	417	595	625	655	680
Mix #2	Air Curing	360	418	444	477	485
	Oven Curing	365	403	427	453	460
	Water Curing	425	575	640	650	665
Mix #3	Air Curing	377	437	450	467	480
	Oven Curing	392	411	427	430	435
	Water Curing	440	540	565	600	645
Mix #4	Air Curing	329	423	443	450	462
	Oven Curing	330	357	375	390	420
	Water Curing	385	487	590	605	635
Mix #5	Air Curing	295	420	421	426	450
	Oven Curing	320	343	355	363	380
	Water Curing	350	510	535	590	595
Mix #6	Air Curing	285	392	393	452	467
	Oven Curing	311	347	363	365	397
	Water Curing	330	530	560	575	615
Mix #7	Air Curing	390	510	520	525	530
	Oven Curing	420	465	470	485	490
	Water Curing	463	625	665	690	765
Mix #8	Air Curing	383	480	495	540	537
	Oven Curing	395	441	443	469	482
	Water Curing	454	600	616	670	705
Mix #9	Air Curing	427	457	493	515	518
	Oven Curing	426	443	453	465	475
	Water Curing	460	540	620	650	666
Mix #10	Air Curing	385	457	490	522	525
	Oven Curing	390	441	447	450	455
	Water Curing	470	530	575	625	630
Mix #11	Air Curing	398	475	527	535	565
	Oven Curing	450	476	475	482	495
	Water Curing	471	560	627	640	690
Mix #12	Air Curing	400	453	505	515	545
	Oven Curing	395	422	430	435	440
	Water Curing	436	530	570	595	645

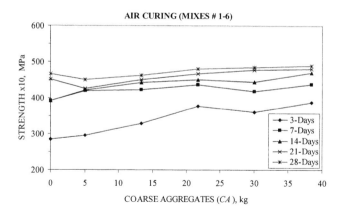

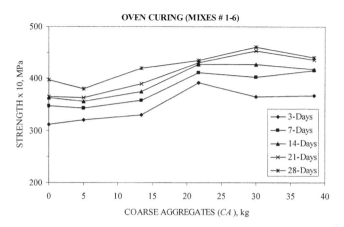

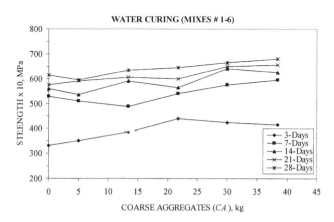

Figure 1 Relationships between measured compressive strength versus coarse aggregate quantities *CA* (*W/C*=0.4)

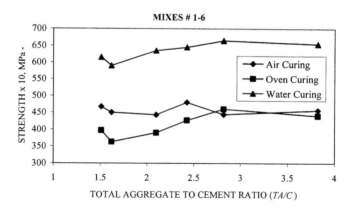

Figure 2 Relationships between measured compressive strength versus *TA/C* (*W/C*=0.4)

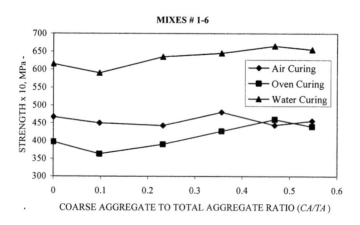

Figure 3 Relationships between measured compressive strength versus *CA/TA* (*W/C*=0.4)

CONCLUSIONS

In this paper, concrete mixes in various proportions with reducing quantities of coarse aggregates were used. An empirical relationship was derived for the compressive strength as a function of reduced coarse aggregate quantities. The following can be concluded:

- Empirical formula for predicting the 28 day compressive strength when reducing the coarse aggregates was achieved and formulated as follows:

$$f_{c(MPa)}' = 0.00005x(CA)_{(kg)}^2 - 0.018x(CA)_{(kg)} + 12.345$$

  where CA is the coarse aggregate content in kg.

- Reducing the amount of the coarse aggregates will reduce the cost of the concrete and will reduce transport trips; therefore reducing the pollution.

- For economical mix designs it is advisable to keep the cement content constant, increase the sand content, and decrease the quantity of coarse aggregates to achieve a comparable compressive strength.

ACKNOWLEDGEMENTS

This work was financially supported by the Research Affairs at the UAE University under a contract no. 06-01-7-11/03. The author would like to thank the UAEU for funding this research. The author would also like to thank Engineers Mohammed Iqbal Martini and Ismail Said Ahmed Ismail and Mr. Faisal Adb El-Wahab for their help in the laboratory work and interpreting the results.

REFERENCES

1. AMERICAN CONCRETE INSTITUTE (ACI), Hot weather concreting. ACI 305R-91, Detroit, 305R1–305R20, 1996

2. BURG, R.G., The influence of casting and curing temperature on the properties of fresh and hardened concrete. Res. and Devel. Bull. RD113T, Portland Cement Association, Portland, Oregon, 1996

3. NEVILLE, A.M., Properties of concrete, 4th Ed., Wiley, New York, 1997

4. ORR, D.M.F., Factorial experiments in concrete research. American Concrete Institute, Materials Journal, Vol. 69, No. 10, 1972, pp. 619–624

5. JERATH, S. and AABBANI I., Computer-aided concrete mix-proportioning. American Concrete Institute, Materials Journal, June–July 1983, pp. 312–317.

6. ABBASI, A.F., MUNIR, A. and WASIM, M., Optimization of concrete mix proportioning using reduced factorial experimental technique. American Concrete Institute, Materials Journal, Vol. 84, Jan.–Feb., 1987, pp. 55–63

7. SOUDKI, K.A, EL-SALAKAWY, E.F. and Elkum, N.B., Full factorial of optimization of concrete mix design for hot climates. Journal of Materials in Civil Engineering, November-December 2001

CONTRIBUTION OF MIXED CEMENTS IN PRODUCTION OF MORE DURABLE AND SUSTAINABLE CONCRETE

R Rosković   **J Beslać**

Civil Engineering Institute of Croatia in Zagreb

D Bjegovic

University of Zagreb

Croatia

ABSTRACT. This paper presents the results of an investigation into cement and concrete production, using varieties of mineral additions. By combined application of mineral additions in cement, the effects of mixed cement production are shown as a way of participation in sustainable development through improvement of mechanical characteristics of cement, conservation of natural resources and reduction of atmospheric pollution. The possibility of enhancing durability properties of concrete by using mixed cements is expected as well.

Keywords: Concrete, Mixed cements, Sustainable development

Ružica Rosković, is engineer of chemistry and technology with the master of science degree and in the course of getting the doctor of science degree. She has 18 years of experience, predominately in the cement production where she was technical manager (in Nasicecement Factory) for 8 years. Now she is head of chemical laboratory in Civil Engineering Institute of Croatia in Zagreb. She has been involved in chemical engineering design, production management and quality assurance, especially in the field of sustainable cement production and consumption. In this field she published several papers.

Prof. Dubravka Bjegovic, Ph.D.C.E., is Dean at Faculty of Civil Engineering, University of Zagreb, and professor at Department for Materials, Zagreb, Croatia. She has written more than 100 papers dealing with theoretical and practical aspects of concrete technology and durability of concrete and reinforced concrete exposed to aggressive environment. She is the associated member of Croatian Academy of Technology, Zagreb, Croatia, New York Academy of Sciences, New York, USA and NACE International, Houston, Texas, USA.

Jovo Beslać, is a civil engineer with a PhD degree and nearly 40 years of experience in concrete construction and technology. In this field he was engaged in many domestic and international projects, published several books and more than 70 papers. For several years he worked as professor for civil engineering materials in Architectural and in Civil Engineering Faculty in Zagreb. Now he is scientific advisor in Civil Engineering Institute of Croatia in Zagreb.

INTRODUCTION

In the process of cement-production large amounts of raw material, which by heat treatment in rotary-furnaces give clinker, are consumed. For its production, large amounts of fossil-fuels are consumed and a significant amount of CO_2 is released to the atmosphere. Current efforts of the cement-industry are directed towards the use of alternative energy sources and also the substitution of as significant a proportion of clinker in cement as possible, resulting natural resources being saved and contamination of the atmosphere being decreasing [1-4].

Approaches to the production of clinker for the purpose of the reduction of energy - consumption and CO_2 emission are different and dependent on the regional position of a factory- the availability of natural and secondary raw materials and regulations of particular countries regarding environmental protection, etc. [5-9].

Although the production of cement with additions is known and applied for a long time, today there is an increased trend of further development, with the purpose of the reduction of CO_2 emission and the saving of raw material. There are several ways of producing such cement depending on the type and quantity of addition. These include cement-production by joint grinding of all components, cement-production by extra addition to ground cement, or cement-production by mixing of individually ground components.

Each way of production consumes thermal and electric energy, and in the end this is reflected in CO_2 emission. Simultaneously, every type of cement contributes to certain characteristics of concrete and determines its purpose [8]. Mixed cement is commonly a combination of Portland cement and mineral additions in different quantities. Mineral additions can be natural materials which have hydraulic activity, they can be added as fillers (tuffs, limestone), or they can be materials which emerge as by products of some industrial processes (slag, fly ash, silica fume).

Research into the production of mixed cements to boost cement production and reduce expenses are a matter for history [8]. Later research has been directed towards finding means of improving features in modification of hydration systems [8-10]. Chemical and mineralogical composition together with microstructure of cement are very important parameters for the durability of concrete. However, the physical characteristics of cement such as fineness and water demand are also important for a specific consistency of fresh concrete, and porosity all effect on durability of concrete.

Certain improvements are realized by some mineral additions, such as slag, fly ash, silica fume etc. Regarding the increase of durability of concrete, almost all effect an improvement, mostly by pozzolanic reaction with the consumption of Ca $(OH)_2$ and by the decrease of pore volume [5, 8, 9, 11, 13].

Due to specific properties of particular materials in cement, the possibility of combined use of these materials and theirs mutual interaction is being researched in recent years and is the subject of our research.

With combined use of more than one addition it is possible to exploit the benefits of one to diminish the shortcomings of another and obtain the required at performance cement [14]. The benefits of mixed cement are shown on Figure 1.

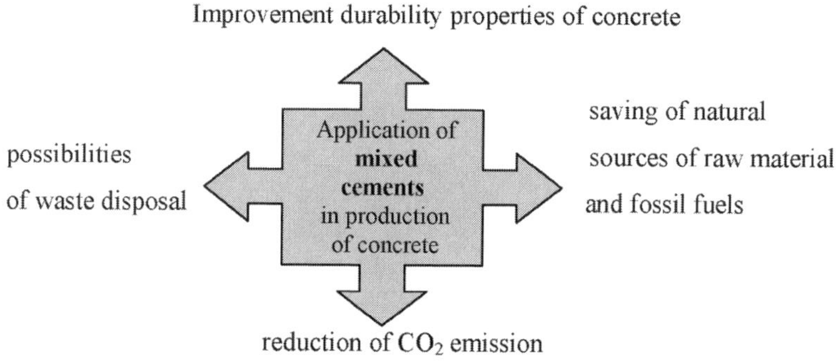

Improvement durability properties of concrete

possibilities
of waste disposal

Application of
mixed
cements
in production
of concrete

saving of natural
sources of raw material
and fossil fuels

reduction of CO_2 emission

Figure 1 Contribution of mixed cement in sustainable development

EXPERIMENTAL PROCEDURES AND RESULTS

The reference cement (RC) with chemical composition as given in Table 1 was used to which were added slag, fly ash and limestone in quantities of 7, 14, 21, 28 and 35% by mass. Mixed cements were produced by combining separately ground components (slag and limestone) or components in their original form (fly ash) with the reference cement.

The chemical compositions and particle size distributions of the materials are shown in Tables 1 and 2 [15]. The influence of the quantity and type of a particular addition on cement quality was tested through physical – mechanical investigation. The standard consistency and compressive strength were established according to standard test methods [16, 17] and presented in Figures 2 and 3.

Table 1   Chemical structure of components

PARAMETER	REFERENCE CEMENT %	SLAG %	FLY ASH %	LIMESTONE %
L.O.I	1.36	-	0.90	42.10
CaO	63.00	37.10	8.10	51.70
SiO_2	18.46	36.90	52.50	1.70
Al_2O_3	5.40	12.20	19.12	0.82
Fe_2O_3	3.49	0.60	9.83	0.59
MgO	2.78	9.20	2.40	2.72
$CaCO_3$	0.68	-	-	92.30
SO_3	3.10	0.21	1.80	-
$Na_2O + K_2O$	1.07	2.14	0.53	0.22

Table 2   Particle size distribution the materials

SIZE OF PARTICLE μm	REFERENCE CEMENT %	SLAG %	FLY ASH %	LIMESTONE %
> 89.9	4.4	2.7	15.9	3.2
66.9 – 89.39	7.6	7.6	10.4	7.2
42.9 – 66.9	14.2	15.1	12.9	4.9
32.0 – 42.9	10.2	9.4	7.4	3.1
20.5 – 32.0	15.4	14.5	11.9	7.9
9.8 – 20.5	22.6	22.1	17.6	15.6
1.9 – 9.8	25.1	27.0	22.7	43.0
< 1.9	0.5	1.6	1.2	15.1
Total	100.0	100.0	100.0	100.0

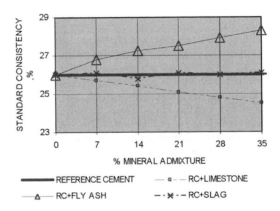

Figure 1   The effect of type and quantity of particular addition on standard consistency

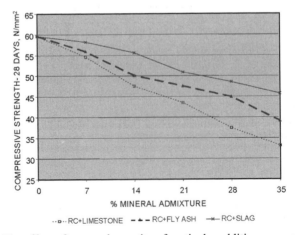

Figure 2   The effect of type and quantity of particular addition on compressive strength of cement after 28 days

Taking into account the influence of the additions, several standard industrial combinations of mixed cement were made.

The results of several industrial combination of mixed cement were analyzed and further examinations were made on two mixed cements:

a) Portland mixed cement with 20% of slag + 10% of fly ash and 5 % of limestone of mass, referred to as PCM (35) and,
b) Mixed cement with 20% of slag + 20% of fly ash and 10 % of limestone by mass, referred to as CM (50).

Table 3 Properties of mixed cements

CEMENT TYPE	FLEXURAL STRENGTHS N/mm^2	COMPRESSIVE STRENGTHS N/mm^2	STANDARD CONSISTENCY %
PCM (35)	7.60	47.12	26.20
CM (50)	6.35	33.80	27.10

With industrially prepared cements (PCM (35) and CM (50)), two mixtures of concrete with a target cement quantity of 400 kg/m^3 are made and marked CMC1 and CMC2. The next two types of concrete, as the reference concrete, were made in the same way as the concrete with the reference cement was made with a sufficient reference cement content to reach equal 28 day strength. These concrete were identified as CRC1 and CRC2. The concrete mix compositions and 28 days compressive strength results are shown in Table 4. For each type of cement, the clinker proportion, relative cement content, and relative clinker content are calculated and given in Table 4.

Table 4 Properties of concretes

MIX	CMC1	CRC1	CMC2	CRC2
Cement Type	PCM (35)	RC	CM (50)	RC
Cement, kg/m^3	399.0	340.0	405.0	320.0
Clinker, kg/m^3	244.0	319.6	190.4	300.8
Index of cement consumption	1.17	1.00	1.26	1.00
Index clinker proportion	0.76	1.00	0.63	1.00
Compressive strength - -After 28 days, N/mm^2	34.8	35.0	32.1	32.0

From Table 4 there are clear differences between the relative cement content and one relative proportion of clinker. Each reduction of clinker in cement is reflected by proportionate reductions in CO_2 emissions [4].

Comparing concretes with a cement content of 35% by mass (CMC1) and concrete with a cement content of 50% by mass (CMC2), it can be seen that concrete with a cement content of 50 % displays ones a small reduction in strength and gives advantages with respect to the production of sustainable concrete, as shown in Table 5.

Table 5 Contribution of mixed cement in concrete

	CMC1	CMC2
Index cement consumption	1.00	1.015
Index clinker proportion	1.00	0.78
Index strength change	1.00	0.92
Index change in - -chloride penetration	1.00	0.51

After the mechanical testing both concrete mixtures (CMC1 and CMC2) were tested for chloride penetration, as one criterion for durability judgments.

Tests were made according to ASTM C 1202-97: Standard Test Method for Electrical Indication of Concrete's Ability to Resist Chloride Ion Penetration, and the results of the tests are shown in Table 6.

Table 6 Chloride Ion Penetrability Based on Charge Passed

	CHARGE PASSED Coulombs	CHARGE PASSED CRITERIA OF ASTM C 1202 Coulombs	CHLORIDE ION PENETRABILITY
CMC1	3.562	2.000 – 4.000	Moderate
CMC2	1.832	1.000 – 2.000	Low

Promising results were obtained from Chloride Ion Penetrability testing. Concrete made of cement with 50% mixed additions (CMC2) in relation to concrete made of cement with 35% mixed additions (CMC1) gives almost twice the resistance to chloride penetration.

CONCLUSIONS

By producing cement, containing combinations of additional materials, it is possible to produce concrete with improved durability characteristics and similar mechanical characteristics, leading to significant savings of natural raw materials and fuels, and also reducing CO_2 emissions.

Results of experimental investigations show that concrete made of cement with 50% mixed additions (CMC2) in relation to concrete made of cement with 35% mixed additions (CMC1) gives:

- Savings in clinker (fossil fuels, natural raw materials) of 22%;

- Reduction in CO_2 emissions of 22%;

- Almost two times greater resistance to chloride penetration.

REFERENCES

1. POPOVIĆ, K, ROSKOVIĆ, R AND BJEGOVIĆ, D. Cement production and sustainable development, Građevinar, 55 (2003) 4 pp 201-206 (in Croatian)

2. HORTON, R. Factor Ten Emission Reductions: The key to Sustainable Development and Economic Prosperity for the Cement and Concrete Industry, CANMET/ACI Inter. Conf. Sustainable Development of Cement and Concrete, San Francisco, 2001, pp. 1-13

3. Reference Document on Best Available Techniques in the Cement and Lime Manufacturing Industries, European Commission, Institute for Prospective technological Studies, Seville, 2000

4. ROSKOVIĆ, R AND BJEGOVIĆ, D. Role of mineral additions in reducing CO_2 emission, Cement and Concrete Research (accepted, in press 2004)

5. ONUMA, E, ICHIKAWA, M AND SANO,S. Umweltbelastung von Zement und Beton-Bewertungsprobleme und Verbesserungsmöglichkeiten, ZKG (2000)10, pp.594-601

6. MARCHAL, G. Managing emissions, International Cement Review 1(2002), pp. 73-75

7. POPOVIĆ, K. Reducing CO_2 Emission into the Atmosphere-Achievements and Experience of Croatian Cement Industry, CANMET/ACI Inter. Conf. Sustainable Development of Cement and Concrete, San Francisco, 2001, pp. 73-84

8. GHOSH, S.N. Cement and Concrete Science Technology, Vol-1, Part-I, Akademia Book International, New Delhi, 1999, 494pp

9. HEWLETT, C.P. Lea's Chemistry of Cement and Concrete, 4d edn, Reed Educational and Professional Publishing Ltd, London, 1998, 1057pp

10. SARKAR, S L.& GHOSH, S N. Mineral Additions in Cement and Concrete, CRC Press LLC Vol-4, Akademia Book International, 1993, 565pp

11. MALHOTRA, V M. Supplementary Cementing Materials for Concrete, Ottawa, CANMET (Editor: Malhotra, V.M) ,1987

12. MOIR, G K. Gaining acceptance, International Cement Review, 2003, 3, pp.67-70

13. IRASSAR, E F. BONAVETTI, V L, MENENDEZ, G, DONZA, H AND CABRERA, O. Mechanical Properties and Durability of Concrete Made with Portland Limestone Cement, CANMET/ACI 3th Inter. Conf. Sustainable Development of Cement and Concrete, San Francisco, 2001, pp. 431-450

14. ISAIA, G.C. Synergic Action of Fly Ash in Ternary Mixtures with Micro silica and Rice Husk Ash, Proceedings of the 10th International Congress on the Chemistry of Cement, Vol.3.: Additives Additions Characterization Techniques, Gothenburg, 1997, 3ii110, 8pp

15. EN 196-2, Methods of testing cement-Part 2: Chemical analysis of cement

16. EN 196-1, Methods of testing cement-Part 1: Determination of strength

17. EN 196-3, Methods of testing cement-Part 3: Determination of setting time and soundness

CONCRETE MIXING AND CURING:
THE INFLUENCE OF WASTE WATER ON THE
CHARACTERISTIC STRENGTH OF CONCRETE

B Mohammed

A A Waziri

B A Umdagas

University of Maiduguri

Nigeria

ABSTRACT. This study was an attempt to study the possibility of re-cycling waste-water for concrete production/curing as well as determining the influence of waste-water on the compressive strength of concrete. Waste-water samples from four different sources/effluents; abattoir (A), vegetable-processing (B), industrial (C) and domestic (D) were collected and tested/analysed in terms of their physiochemical properties. A 'pure' water (E) was also collected from a water treatment plant to serve as a control in the experiment. The five water samples (A – E) were used in mixing/production and curing of concrete cubes. Using 150×150 mm moulds, fifteen concrete cubes were cast (three from each water sample) and cured for 28 days each. The cubes were subjected to crushing and the loads obtained were used in calculating their compressive strength. Results show values of 13.6 N/mm^2 for concrete A; 19.33 N/mm^2 for B; 24.53 N/mm^2 for C; 15.73 N/mm^2 for D and 26.67 N/mm^2 for E. Comparison of the results shows closeness for samples C and E. This is probably explained by the fact that the industrial effluent was treated to a large extent before disposal. Based on these findings, a number of conclusions were made.

Keywords: Concrete mixing, Curing, Wastewater-influence, Characteristic strength.

Babagana Mohammed is BSc, MSc (Agricultural Engineering) and M.I.L.R Registered Engineer, and holds Corporate Memberships of a number of professional bodies. Lecturer, Head of Department, and Consultant. Member, national EXCO of the Nigerian Society of Engineers since 1999.

Abubakar A Waziri BEng (Agric Engineering). Currently a researcher/postgraduate student at the University of Maiduguri, Nigeria.

Buba A Umdagas: HND, B.Eng. (Civil Engineering) and MBA. Currently Principal Technical Officer, Department of Civil Engineering, University of Maiduguri, Nigeria. Consultant. Has more than 10 years of work experience in research, projects and teaching.

INTRODUCTION

Concrete is the most widely used construction material in the world. It consists of Portland cement, water, aggregate (sand) and various chemicals or mineral admixtures [1, 2]. The properties of concrete can be divided into that of fresh and hardened concrete [3, 4, 5, 6]. The properties of fresh concrete are important only in the few hours of its history, whereas the properties of hardened concrete assume an importance, which is retained for the rest of the life of the concrete. The important properties of hardened concrete are strength, deformation under load, durability, permeability and shrinkage. Concrete exists in many types (plain, reinforced, pre-stressed and lightweight).

Curing (either by ponding, immersion, spraying/fogging or membrane), is defined as a procedure for ensuring the hydration of the Portland cement in newly placed concrete. It generally implies control of moisture loss and sometimes of temperature. In other words, it simply means keeping the water in the concrete where it can do its job of chemically combining with the cement to change the cement into a tough "glue" that will help develop strong, durable concrete [7]. Good curing means keeping the concrete damp and above 10°C until the concrete is strong enough to do its job. Curing would increase concrete strength, increase concrete abrasion resistance, lessen the chance of concrete scaling (shrinkage), reduce the possibility of surface dusting and concrete cracking [8, 9, 10, 11].

The ideal water for concrete mixing and curing is that water fit for drinking. Water for concrete should be free from such impurities as suspended soils, organic matter and dissolved salts, which may adversely affect the properties of the concrete. Due to the abundance, cheapness and easy accessibility of turbid water or wastewater [12], many people used it on concrete without prior testing of the water to investigate its properties, and the effects of these properties on hardened concrete. All these problems are as a result of insufficient pipe-borne water for domestic use. Waste-water is characterized in terms of its physical, chemical and biological properties. Every community produces both liquid and solid wastes. From the standpoint of sources of generation, waste-water may be defined as a combination of the liquid or water-carried wastes discharged from residences, institutions, commercial and industrial establishments, and surface and storm water. There are two main types of waste-water: domestic waste-water (containing sanitary wastes) and industrial waste-water (including sewage from treatment plants) [13, 14, 15, 16].

The overall objective of the study was to determine the influence of waste-waters on the characteristic strength of concrete.

MATERIALS AND METHODS

Materials

The different materials used during the experimental programme were ordinary Portland cement, complying with BS 12 (British Standard) for general concrete work, aggregates (coarse and fine; *Bama* gravel ranging from medium gravel to medium sand in size and mostly spherical in shape) and *Ngadda* sand (containing some percentage of silt) [17], waste-water samples, 26 litres each, (abattoir (A), vegetable-processing (B), industrial (C) and domestic (D)) and a treated/'pure' water (E) was also collected from a water treatment plant to serve as a control in the experiment.

The equipment used include moulds, slump cone, tamping rods, measuring cylinders, smoothing steel, weighing balance and pans, shovel, curing containers (baths), set of sieves, bricklayer hand trowel, trays, specific density bottle, universal material testing machine, spectrophotometer, turbidimeter, engine oil and wire brush.

Methods/Procedure

Four samples of waste-waters from four different sources were used in this research. Their physiochemical properties were determined in the laboratory (Table 1). Using a mix-ratio of 1:2:4, concrete was prepared and 15 concrete cubes (3 using each waste-water sample) cast in moulds of 150 mm × 150 mm. They were cured for 28 days. A sample of treated/'pure' water was used for a similar preparation and casting. This served as a control to the four others. The water-cement ratio used was 0.47. The experimental procedures included sieve analysis, determination of specific gravity, mixing of concrete, slump test, casting of cubes, curing of cubes and crushing of cubes.

Table 1 Physiochemical analysis of the five water samples (A – E), and the World Health Organisation (WHO) Standards

S/No.	TEST PARAMETER	A	B	C	D	E	WHO STANDARDS
1.	Temperature (°C)	30.1	28.8	28.6	28.8	27.0	25-30°C
2.	Total Dissolved Solids (TDS)	2474	592	466	1664	0.0	250 mg/l
3.	Electrical conductivity	4920	1166	912	3330	280	1000 ms/cm
4.	Turbidity	1240	78.0	16	24.0	0.0	0-5 NTU
5.	Colour (pt/co)	7925	2600	58	155	58	0-5 units
6.	pH	6.50	4.50	90	8.00	7.0	7.0
7.	Iron	1.25	0.50	0.00	0.20	0.00	0.3-1.0 mg/l
8.	Manganese	0.00	0.00	0.00	0.10	0.00	0.05 mg/l
9.	Sulphate	1.50	30.0	0.00	40.00	19.0	200-250 mg/l
10.	Magnesium	6.0	11.0	12.0	4.0	5.0	37-150 mg/l
11.	Calcium	12.0	28.0	42.0	38.0	28.0	75.270 mg/l
12.	Nitrate	363	83.6	0.44	0.00	0.00	50-100 mg/l
13.	Sodium chloride	528	64.35	42.9	67.65	3.30	-
14.	Chloride	32.0	39.0	26.0	41.0	2.00	50-250 mg/l
15.	Dissolved oxygen (DO_2)	8.00	6.20	5.6	9.10	0.00	0-15 mg/l

Strength Testing Procedure

For the determination of crushing strength, a universal testing machine was used to crush the cube specimens at 28 days. Three specimens were prepared for each mix. Each sample was wiped dry and weighed before being placed centrally in the load frame under a 50 mm diameter top plate. A metal plate measuring 250 × 120 × 2 mm was placed on top and bottom of the samples to allow for even distribution of the load on the surface of the cube. The load was then applied at a constant rate of displacement. The bottom plate advanced at 1.02 mm/min. Following the failure of a specimen, the maximum load was recorded and crushing strength was re-valuated on the basis of: P = F/A (where P = crushing strength, N/mm^2; F = failure load, N; A = cross-sectional area of specimen, mm^2) [18].

Calculations

i.   Cross-sectional area of cube = 150 mm × 150 mm = 22500 mm^2

ii.   Volume of cube = 0.15m × 0.15m × 0.15m = $3.375 \cdot 10^{-3}$ m^3

iii. Compressive strength = Load at Failure (N) / Area of cube (mm^2)

iv.   Density of cube = Weight of cube (kg) / Volume of cube (m^3)

RESULTS AND DISCUSSIONS

The result of compressive strength of concrete cubes mixed and cured with abattoir waste-water (sample A) reveals that the water has an adverse effect on the strength of concrete. It is clear from the result that sample A has the least strength in the curing age under consideration. This was attributed to the composition of the waste-water (i.e. high sulphate content, acids, silts, clays, turbidity etc). The presence of high sulphate (up to 150 mg/l) in the abattoir water may lower the strength of the concrete. This is so because most sulphate solutions react with calcium hydroxide, Ca(OH)$_2$, and calcium aluminate, C_3A, to form compounds that have limited solubility in water [19] and their volume is greater than the volume of the compounds of cement paste from which they originate. This increase in volume within the hardened concrete contributes to the build-up of internal stress and the break down of its structure [7]. The higher acidity of the abattoir waste-water may also lower the compressive strength of the concrete. The compressive strength of the curing age of sample A is 13.6 N/mm^2 and that of control sample (Sample E) is 26.67 N/mm^2. The compressive strength of sample A constitutes 51 % only of sample E. Therefore, from the observations made, it can be deduced that abattoir waste-water has adverse effect on the properties of concrete with special consideration to compressive strength.

The vegetable-processing waste-water (sample B) gained strength of 2.0 N/mm^2 between 7 and 14 days and only a marginal (0.7 N/mm^2) for the rest of the curing period. This shows that the rate of gain in strength is slow, decreases with age, and is not linear. The turbidity of the vegetable-processing waste-water is high due to clay and silt particles. The presence of large numbers of micro-organisms may have lowered the strength of the concrete and the pH level of 4.5 which indicated acidic nature of the water may also have affected the strength. The compressive strength of the curing age of sample B is 19.33 N/mm^2 as against 26.67 N/mm^2 for the control (sample E). The compressive strength of sample B constitutes 72 % of sample E in the curing age under consideration. However, non-uniformity in compaction (manual/hand compaction) may result in high void spaces, low density and the rate of permeability may also be higher. This allows more waste-water in the void spaces, which affect hydration and may lead to break down in its structure. It can be concluded that sample B has a low effect on the compressive strength of concrete.

The industrial waste-water (sample C) showed adequate strength at the curing age, which is 92 % of the control sample E. A graph of compressive strength against age is a straight line, which indicates little or no effect of the water sample on the concrete strength. The presence of sulphate, turbidity and alkalinity are possibly the source of lower strength of concrete.

Sample D (domestic waste-water) has lower strength than samples B, C and E but stronger than sample A. The lower value of the compressive strength may be as a result of presence of dissolved solids, urine and other impurities, which do not favour the hydration of cement

[20]. Chloride is responsible for brackish taste in water and is an indicator of sewage pollution because of the chloride content of urine, thus causing low strength and surface dampness of the concrete structure, as observed. The compressive strength of control D is 15.73 N/mm^2, which is only 59 % of the compressive strength of sample E. Sample E being the treated/'pure' water has the highest compressive strength of 26.67 N/mm^2 as compared with the other samples of waste-waters. A graph of compressive strength against ages is a straight line, which indicates the fitness of the water for concrete works.

The chemical composition of cement is also considered to be an important factor in strength gain of concrete [20]. Grading is also of vital importance in the proportion of concrete mixes, as it is often more economical to use the locally available material, even though it requires a richer mix, provided it would produce concrete free from segregation. The longer the period during which concrete is kept in water, the greater its final strength. It is normally accepted that concrete made with ordinary Portland cement and kept in normal curing conditions will develop about 75 % of its strength in the first 28 days (7). In general, water fit for drinking, such as treated water, is acceptable for mixing concrete. The impurities that are likely to have an adverse effect when present in appreciable quantities include silt, clay, acids, alkalis and other salts, organic matter and sewage. These are normally absent in treated water.

CONCLUSIONS

Curing is a process of preventing the loss of moisture from the concrete while maintaining a satisfactory temperature regime. This can be excellently achieved by using clean water. When waste-water is used for curing of concrete, the properties of concrete will be affected with special consideration to compressive strength. In this study five water samples were used for production and curing of concrete. Four of which were obtained from different sources of waste-waters (i.e. abattoir, vegetable-processing, industrial and domestic). A treated/'pure' water from a water treatment plant served as the control. Fifteen cubes were cast, three from each water sample, and were tested for compressive strength at the age of 28 days. The concrete cubes cured with abattoir waste-water were found to display compressive strength with a value of 13.6 N/mm^2 suggesting that the water (sample A) had an adverse effect on concrete properties, strength in particular. Domestic waste-water (sample D) also indicated a low value of compressive strength of 15.73 N/mm^2, and vegetable-processing waste-water (sample B) had a value of 19.33 N/mm^2. Industrial waste-water (sample C) showed little effect on the compressive strength giving a result of 24.53 N/mm^2, which is 92 % of that of the treated/'pure' water, which is 26.67 N/mm^2. From the experiments, the results obtained, and subsequent observations, it can be concluded that waste-water has an effect on the properties of concrete. Consequently, it can be deduced that surface contact between these waste-water samples and concrete structures either at retaining walls (hydraulic channels) or at the foot of structures, could reduce the strength of such structures, which in the long run reduces their life span. The erodibility of the surfaces of such structures in contact with these samples particularly waste-water from industrial, abattoir and domestic sources may be high and could have serious effects. Therefore, good water should be used in mixing and curing of concrete.

REFERENCES

1. BLACKLEDGE, G.F., Concrete Practice, Cement and Concrete Association, London, 1984.

2. JACKSON, N., DHIR, R.K., Civil engineering Materials. Macmillan Press, London, 1996.

3. MANNING, G.P., Design and Construction of Foundations, Cement and Concrete Association, London, 1972.

4. McKAY, W.B., Building Construction, Longman Limited, London, 1970.

5. NEVILLE, A.M., Properties of Concrete, Pitman, London, 1981.

6. REYNOLDS, C.E., STEEDMAN, J.C., Examples of the Design of Buildings to CP110 and Allied Codes", Cement and Concrete and Aggregates Association, London, 1978.

7. WAZIRI, A.A., An Assessment of Some Selected Properties of Wastewater Cured Concrete. A Final Year Project. Department of Agricultural Engineering, University of Maiduguri, Nigeria, 2003. 76 pp.

8. ASSELANIS, J.G., AITCIN, P.C., MEHTA, P.K., Effect of curing conditions on the compressive strength and elastic modulus of very high-strength concrete, Cement, Concrete and Aggregates Association, London. Vol. II, No. 1, 1989. pp 80–83.

9. AYODEJI, Y K. Effect of High Mixing Temperature 38°C on the Compressive Strength of Concrete made with *Bama* Gravel of Mix Proportion 1:½:4 and Water-Cement Ratio w/c (0.5, 0.55, 0.65), Final Year Project, Department of Civil and Water Resources Engineering, University of Maiduguri, Nigeria, 1999.

10. BALAMI, Y.G., IZAM, Y.D., Soil Bitumen as a walling material. Journal of Environmental Science, Vol. I, No. 2, Faculty of Environmental Sciences, University of Jos, Nigeria. 1998. pp 33.

11. BARRY, R., The Construction of Buildings, Blackwell Science Limited, United Kingdom, 1998.

12. JONES, A.N., A Survey of Waste-water Characteristics in Maiduguri, Final Year Project, Department of Civil and Water Resources Engineering, University of Maiduguri, Nigeria, 1999.

13. CLESCERI, L.S., GREENBERG, A.E., Standard Methods for the Examination of Water and Waste Water, American Public Health Association, USA, 1989.

14. HARCH, Water Analysis Handbook, Harch Company, London, 1987.

15. HAMMER, M.Y., Water and Waste-water Technology, John Wiley and Sons, Inc., New York, 1977.

16. METCALF, EDDY, Waste-Water Engineering, Treatment, Disposal and Reuse, Tata McGraw-Hill Publishing Company Limited, New Delhi, 1995.

17. SULE, M., Compressive Strength of *Bama* Gravel Concrete with Varying Crushing Stones, Final Year Project, Department of Civil and Water Resources Engineering, University of Maiduguri, Nigeria, 1991.

18. CONTROLS Testing Equipment for the Construction Industry, Controls Testing Equipment Limited, England, 1992.

19. TEBBUTT, T.H.Y., Principles of Water Quality Control, Pergamon Press Limited, England, 1997.

20. LAYA, S.S., Comparative Analysis of the Crushing Strength of Concrete Mixed with Turbid, Saline and Non-turbid Water, Final Year Project, Department of Civil and Water Resources Engineering, University of Maiduguri, Nigeria, 2002.

POLLUTION, WASTE AND RECYCLING

SUSTAINABILITY OF CEMENT AND CONCRETE INDUSTRIES

T R Naik

University Of Wisconsin Milwaukee

United States of America

ABSTRACT. Sustainability is important to the well-being of our planet, continued growth, and human development. Concrete is one of the most widely used construction materials in the world. However, the production of Portland cement, an essential constituent of concrete, leads to the release of significant amount of CO_2, a greenhouse gas (GHG). The production of one tonne of Portland cement produces about one tonne of CO_2 and other GHGs. The environmental issues associated with GHGs, in addition to natural resources issues, will play a leading role in the sustainable development of the cement and concrete industry during this century. For example, as the supply of limestone decreases it will become more difficult to produce adequate amounts of Portland cement for construction. Once there is no more limestone, and thus no Portland cement, all of the employment associated with the concrete industry as well as new construction projects will be terminated. Therefore, it is necessary to look for sustainable solutions for future concrete construction.[1]. Sustainable concrete should have a very low inherent energy requirement, be produced with little waste, be made from some of the most plentiful resources on earth, produce durable structures, have a very high thermal mass, and be made with recycled materials [2]. Sustainable constructions have a small impact on the environment. They use "green" materials, which have low energy costs, high durability, low maintenance requirements, and contain a large proportion of recycled or recyclable materials. Green materials also use less energy, resources, and can lead to high-performance cements and concrete. Concrete must keep evolving to satisfy the increasing demands of all of its users. Designing for sustainability means accounting for the short-term and long-term environmental consequences in the design.

Keywords: By-products, Environment, Portland cement, Pozzolan, Sustainable concrete.

T R Naik, FACI, FASCE, is a Professor of Structural Engineering and Academic Programme Director of the UWM Center for By-Products Utilisation (UWM-CBU) at the University of Wisconsin-Milwaukee. He is a member of several ACI committees and was also Chairman of the ASCE Technical Committee on Emerging Materials (1995-2000)

INTRODUCTION

According to the World Commission on Environment and Development sustainability means "Meeting the needs of the present without compromising the ability of the future generations to meet their own needs" [3]. The sustainability of the cement and concrete industries is imperative to the well-being of our planet and to human development. However, the production of Portland cement, an essential constituent of concrete, leads to the release of a significant amount of CO_2 and other greenhouse gas (GHGs). The production of one ton of Portland cement produces about one ton of GHGs [4]. The environmental issues associated with CO_2 will play a leading role in the sustainable development of the cement and concrete industry during this century. One of the biggest threats to the sustainability of the cement industry is the dwindling amount of limestone in some geographical regions. Limestone is essential to the production of Portland cement. As limestone becomes a limited resource, employment and construction associated with the concrete industry will decline. Therefore, those involved with these industries must develop new techniques for creating concrete with minimal use of limestone. Concrete production is not only a valuable source of societal development, but also a significant source of employment. Concrete is the world's most consumed man-made material. It is no wonder that in the U.S.A. alone, concrete construction accounted for 2,000,000 jobs in 2002 [5]. Moreover, concrete is second only to water consumption in all materials consumed worldwide. About 3.5B yd^3 of concrete was produced in 2002 worldwide. This equals to more than ½ yd^3 of concrete produced per person worldwide. Therefore, to create not only sustainable societal development, but also to sustain employment, such as batch plant operators, truck drivers, ironworkers, laborers, carpenters, finishers, equipment operators, and testing technicians, as well as professional engineers, architects, surveyors, and inspectors, the concrete industry must continue to evolve with the changing needs and expectations of the world.

WHAT IS SUSTAINABILITY

Limestone is used to manufacture Portland cement. Currently, Portland cement is the most commonly used material (Figure 1). Entire geographical regions are running out of limestone resource to produce cement. And major metropolitan areas are running out of materials to use as aggregates for making concrete. Sustainability requires those in the construction industry to take the entire life- cycle, including construction, maintenance, demolition, and recycling of buildings into consideration [2, 6].

A sustainable concrete structure is one that is constructed such that the total societal impact during its entire life cycle is minimal. Designing with sustainability in mind includes accounting for the short-term and long-term consequences of the structure. In order to decrease the long-term impact of structures, the creation of durable structures is paramount.

Building in a sustainable manner and scheduling appropriate building maintenance are significant in the "new construction ideology" of this millennium. In particular, to build in a sustainable manner means to focus attention on the effects on human health, energy conservation, and physical, environmental & technological resources for new and existing buildings. It is also important to take into account the impact of construction technologies and methods when creating sustainable structures [2]. An integrated sustainable design process can reduce the project costs and operating costs of the development.

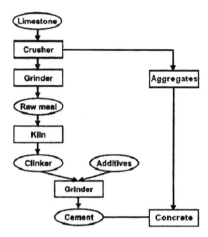

Figure 1 Limestone for Cement and Concrete (modified from [6]).

The production of one ton of Portland cement releases approximately one ton of CO_2 & other greenhouse gases (GHGs) into the atmosphere [4]. There are many challenges associated with Portland cement production. Of these, energy and resource conservation, the cost of producing Portland cement, and GHG emissions are the most significant.   Therefore, supplementary cementing materials such as fly ash and slag should replace larger amounts of Portland cement in concrete. However, before any construction occurs, all aspects of the building materials to be used should be evaluated.

In order to build structures and infrastructure that is cost efficient, environmentally friendly, and durable, the impact of the building materials on local and worldwide air conditions must be examined [2]. CO_2 emissions are expected to rise by about 50 % by 2020 from the current levels due to Portland cement production (Table 1).

Table 1 CO_2 Emissions in 2002; Total CO_2 emissions worldwide – 21 billion tons [4].

COUNTRY	PERCENT, CO_2
U.S.A.	25
E.U.	20
Russia	17
Japan	8
China	>15
India	>10

For each ton of Portland cement clinker, 1.5 to 10 kg of NOx is released into the atmosphere. In 2000, worldwide cement clinker production was approximately 1.5 billion tons. That means that in 2000 between 23 and 136 B kg of NOx was produced to make Portland cement clinker [4]. If the challenges associated with reducing CO_2, NOx, and other GHGs are to be met, then the concrete industry must develop other materials to replace Portland cement. Use of blended cements and chemical admixtures must be significantly increased for sustainability of the cement and concrete industries.

CONCRETE

For over 200 years, concrete has been accepted for its long-lasting and dependable nature. In addition to durability and dependability, concrete also has superior energy performance, is flexible in design, affordable, and is relatively environmentally friendly [7]. It can be expected that concrete will be needed to both increase industrialization and urbanization while protecting the environment. To do this, the concrete industry should consider recycling industrial by-products such as fly ash safely and economically. When industrial by-products replace cement, even up to 70 %, in concrete, the environmental impact improves along with the energy efficiency, and durability of concrete [8].

Concrete is a building material that is not only strong and durable, but can also be produced in ways that are environmentally friendly, and architecturally moldable in esthetically pleasing forms [9]. With sustainable concrete structures and infrastructure, the concrete industry can develop a sustainable future for generations yet to come. Furthermore, buildings that are constructed to be both durable and environmentally safe often lead to higher productivity because the buildings generally lead to better air quality and, therefore, higher productivity [10]. One example of the advantages of sustainable concrete is buildings constructed with concrete that have reduced maintenance and energy costs. Another is concrete highways, which reduce the fuel needed for heavily loaded trucks. A third example of the benefits of sustainable concrete construction is illustrated in insulating concrete homes, that have energy reductions of up to 40 % [7].

The cement and concrete industries can make substantial contributions to sustainable developments by creating and adopting technologies that can reduce the emissions of greenhouse gases. The cement and concrete industries could contribute to meeting the goals and objectives of the 1997 Kyoto Protocol [3]. Those involved with the manufacture of Portland cement would have a huge impact on the sustainable development of the concrete industry as a whole.

Several characteristics apply to innovative concrete products. First, they are produced with precast or cast-in-place reinforced concrete elements that are made with Portland cement and pozzolanic materials that includes renewable and/or recycled components. Second, innovative concrete products are constructed to enhance the performance of concrete elements, which may also contain recycled concrete as aggregates. High performance materials are intended to reduce cross-sections and the volume of concrete produced. They are also intended to increase the durability of concrete structures to minimize the maintenance needs of the concrete construction, and limit the amount of non-renewable special repair materials that need to be used in the maintenance of the concrete [10].

Concrete producers are creating sustainable solutions for many market sectors including agriculture and construction. In agriculture, integrated waste management solutions have been developed that convert manure into biogas, nutrient rich fertilizer, and reusable water. Industrial, commercial, and institutional buildings are being constructed so that they are more energy efficient, have better air quality, and necessitate less maintenance [2].

U.S. foundries generate over 7 million tonnes (8 million tons) of by-products. Wisconsin alone produces nearly 1.1 million tonnes (1.25 million tons) of foundry by-products, including foundry sand and slag. Most of these by-products are landfilled. Landfilling is not a desirable option because it not only causes a huge financial burden to foundries, but also makes them liable for future environmental costs, liability, and restrictions associated with landfilling [11]. In addition, in 1996 the USA produced 136 million tons of construction and demolition (C & D) debris; about 1.5 kg per person per day. About 25 to 40% of landfill space is C&D debris [12]. If this trend continues, the cost of landfilling will continuously increase, as will the potential health and environmental risks of landfill materials. Furthermore, the cost of landfilling is escalating due to shrinking landfill space and stricter environmental regulations. One of the innovative solutions appears to be high-volume uses of foundry by-products in construction materials [11].

A study was reported in 1999 whose aim was to evaluate the environmental impact of Controlled Low Strength Materials (CLSM) incorporating industrial by-products such as coal fly ash and used foundry sand [11]. The results demonstrated that excavatable flowable slurry incorporating fly ash and foundry sand as a replacement of fly ash up to 85% could be produced. In general, inclusion of both clean and used foundry sand caused a reduction in the concentration of certain contaminants. The use of foundry sand in flowable CLSM slurry, therefore, provided a favorable environmental performance.

PORTLAND CEMENT

Portland cement is not an environmentally friendly material; its manufacture creates greenhouse gas emissions; and, it also reduces the supply of limestone. As good engineers, we must reduce the use of Portland cement in concrete. We must use more blended cements.

The most energy intensive stage of the Portland cement production is during clinker production. It accounts for all but about 10 % of the energy use and nearly all of the GHGs produced by cement production. Kiln systems evaporate inherent water from the raw meal and calcine the carbonate constituents during clinker preprocessing [6].

Sources of CO_2 and GHG emissions in the manufacturing of Portland cement [4] are:
- from calcinations of limestone = ± 50 –55%;
- from fuel combustion = ± 40 –50%; and,
- from use of electric power = ± 0 –10%.

INNOVATIVE CEMENT PRODUCTS

While the embodied energy linked to concrete is low, supplementary cementing materials (SCM), especially fly ash, have been used by the concrete industry for over 70 years. Their use can contribute to a further reduction of concrete's embodied energy. When used wisely and judiciously, SCM can improve the long-term properties of concrete. Fly ash can, and does, regularly replace Portland cement in concrete [4, 8, 13].

One process that is even more environmentally friendly and productive is blended cements. Blended cements have been used for many decades and are made when various amounts of clinker are blended and/or interground with one or more additives including fly ash, natural pozzolans, slag, silica fume and other SCM. Blended cements allow for a reduction in the energy used and also reduces GHGs emissions [4, 13].

Most innovative concrete mixtures make use of SCMs to partially replace cement. The advantages of using of blended cements include increased production capacity, reduced GHG emissions, reduced fuel consumption in the final cement productivity, and recycling of SCMs [6, 7]. The manufacture of Portland cement is the third most energy intensive process, after aluminum and steel. In fact, for each ton of Portland cement, about six million BTU of energy is needed [4, 11].

Although cement production is energy inefficient, there have been major initiatives that have reduced energy consumption [6]. Of these, the most significant has been the replacement of wet production facilities with dry processing plants. In addition the cement industry has also moved away from petroleum-based fuel use.

Despite these advances, there are still some shortcomings when energy use is evaluated for the concrete industry. Dry process cement plants use pre-heaters, which increase the alkali content of cement [6]. Thus, researchers need to continue to develop ways to control the alkali content without increasing the energy consumption levels of the cement [10]. Furthermore, current innovations and energy savings are linked to the actual amount of energy consumption by wet-to-dry-kiln conversions and the number of pre-heaters needed to complete the process [6].

For each million ton of capacity, a new Portland cement plant costs over 200 million dollars. The cost associated with the production of Portland cement, along with the CO_2 emissions and energy issues make it unlikely that developing countries will be able to employ such technology. Also, government regulations of GHGs will likely force the cement industry to create blended cements and use supplementary materials for blended cements in order to meet the societal development needs [4, 6].

To produce one ton of Portland cement, 1.6 tons of raw materials are needed. These materials include good quality limestone and clay. Therefore, to manufacture 1.6 billion tons of cement annually, at least 2.5 billion tons of raw materials are needed [14]. As good engineers, we must use environmentally friendly materials to replace a major part of portland cement for use in the concrete. In the USA these materials primarily are fly ash, slag, silica fume, natural pozzolans, rice-husk ash, wood ash, and agricultural products ash. These all can be used to supplement the use of cement in concrete mixtures while improving product durability.

One of the important benefits of the increased use of other cementitious materials is the reduction of GHG emissions. With a replacement of cement with other recyclable resources, up to 15% of worldwide CO_2 emissions would be reduced. A replacement of 50% of cement worldwide by other cementitious materials would reduce CO_2 emissions by 800 million tons. This is equivalent to removing approximately 1/4 of all automobiles in the world [4].

Fly ash availability in 2002 is estimated at 100 million tons; and, in 2010 it is 160 million tons. While Portland cement availability in 2002 is estimated at 80 million tons; and, in 2010 it is 100 million tons. The fly ash disposal challenge, and the limited availability of Portland cement have the same solution: replace large amounts of Portland cement with fly ash to create durable and sustainable concrete.

THE HANNOVER PRINCIPLES FOR DESIGN FOR SUSTAINABILITY

In 1991, as the planning of the World's Fair was underway, the City of Hannover, Germany asked William McDonough and Michael Braungart to create sustainability principles to guide the large-scale development of EXPO 2000 in Hannover. "The Hannover Principles - Design for Sustainability" also include directives concerning the use of water. Although these guidelines were created for the World Fair, they are still a good tool to guide current and future development around the world [2].

Designers, planners, government officials, and all those who participate in the construction of new buildings and infrastructure should use the Hannover Principles. A new design philosophy has developed from these principles and should be included in the proposed sustainable systems and construction in the future. There are a number of examples of societies that have created sustainable and environmentally friendly communities. There is hope that the Hannover Principles will inspire development and improvements that are committed to sustainable growth with practical limits to create a sustainable and supportive future for communities and the world.

Hannover Principles by William McDonough [2]: insist on rights of humanity and nature to co-exist; recognize interdependence; respect relationships between spirit and matter; accept responsibility for consequences of design; create safe objects of long-term value; eliminate the concept of waste; rely on natural energy flows; understand the limitations of design; and, seek constant improvement by the sharing of knowledge. The Hannover Principles are not "cast-in-concrete". They were devised to provide a tangible document that could evolve and be adapted as our understanding of our interdependence with nature becomes more important over time.

For sustainability consider your actions on [2]: materials (use indigenous materials); land use (protect and create rich soil); urban context (preserve open spaces); water (use rainwater and gray-water); wastes (recycle), air (create clean air); energy (use solar & wind energy; recycle waste energy); and, your responsibility to nature (create silence), and, for the future generations minimize or eliminate maintenance.

Materials are key to creating sustainable and responsible concrete designs. In order to ensure that the most effective and environmentally friendly materials are being used, the entire life-cycle of the structure should be taken into consideration. Material choice should include anticipation of the extraction, processing, transport, construction, operation, disposal, re-use, recycling, off-gassing and associated Volatile Organic Compounds (VOC) of materials [2].

According to McDonough, constructions should be flexible to allow and serve different needs (e.g. today's storage building can be tomorrow's school). Adapt materials that are sustainable in their process of extraction, manufacture, transformation, and degradation, as well as recyclability. Consider toxicity, off-gassing, finish, and maintenance. Recycling is essential. Make allowance for disassembly and reuse. Plan for reuse of the entire structure in the future. Minimize use of hazardous chemicals. Eliminate waste that cannot be part of naturally sustainable cycle. Any solid wastes remaining must be dealt with in a non-toxic manner. Life-cycle costs must be analyzed. Life-cycle cost analysis is a process to evaluate energy use and environmental impact during the life of the product, process, and/or activity.

This process must include extraction and processing of raw materials, manufacturing, transportation, maintenance, recycling, and returning to the environment. Evaluate and understand costs and benefits in both the short-term and long-term. Use recycled concrete for aggregates. For the sustainability of the cement and concrete industries use less water and portland cement in concrete production; and, use more blended cements and tailor-made chemical admixtures.

The devastation of air is a global problem, regardless of the locality in which the pollution is created [2]. The overall design of concrete structures must not contribute to atmospheric degradation. Those involved in the cement and concrete industries must evaluate ozone depletion and global warming throughout the construction and planning process. A major contribution to this effort will be the use of more blended Portland cement to minimize global warming.

Water resources are being depleted by various uses [15]. Therefore, potable water should be conserved to serve life-sustaining needs rather than infrastructural needs. Rainwater and surface run-off water can be used as a water conservation method by recycling these water resources in construction instead of using potable water. Gray water should be recycled and used for grass, shrubs, plants, trees, and gardens; as well as for concrete production [2]. Furthermore, mixtures with less water should be developed with new technologies to create mortar and concrete containing a minimal amount of water.

Benjamin Franklin said, over 200 years ago in Poor Richard's Almanac "When the well's dry, we know the worth of water," [16]. Many facilities may have requirements that can be completed with non-potable water. By using non-potable water, a significant amount of money can be saved by avoiding or reducing potable water purchases and sewerage costs. To be as effective as possible, non-potable water construction and building uses should be identified early to be the most cost-effective. Four ways to utilize and recycle water are to reuse water on site for repeated cycles of the same task, treat and reuse water on site for multiple purposes, use gray water (shower, sink, bath and laundry excess) water after solids have been eliminated, and collect non-potable water from sources such as rainwater, lakes, rivers and ponds for use in construction [15].

Energy efficiency, providing the same (or more) services for less energy, helps to protect the environment. When less energy is used, less energy is generated by power plants, thus reducing energy consumption and production. This in turn reduces GHGs and improves the quality of the air. Energy efficiency also helps the economy by saving costs for consumer and businesses. According to Mc Donough [2]: Use buildings' thermal inertia (e. g., concrete building's mass allows it to retain heat). Use day lighting and natural ventilation. Use wind power and solar power. Recycle waste energy. Judiciously use color materials on surfaces. Reduce heat-islands in buildings. Manage and moderate micro-climates of buildings.

WASTE MATERIALS

Contractors should reuse industrial by-products and post-consumer wastes in concrete. Post-consumer wastes that should be considered for use in concrete include glass, plastics, tires, and aluminum, steel & tin cans. To do this successfully, contractors must watch for harmful hydration reactions and changes in volume. The recycling of industrial by-products has been well established in the cement and concrete industries over the past decades [16]. The use of coal fly ash in concrete began in the 1930s, but volcanic ash has been reused for mortar and concrete for several millenniums in Egypt, Italy, Mexico, and India. The use of by-products such as rice-husk ash, wood ash, silica fume and other pozzolanic materials, in addition to coal fly ash, can help to reduce the need for Portland cement in addition to creating more durable concrete and reducing greenhouse gas emissions [4, 13, 17]. This will also contribute to the improvement of air quality, reduction of solid wastes, and sustainability of the cement and concrete industry [13].

In summary, for sustainability of the cement and concrete industries: use less Portland cement; use less water; use applications-specific high-quality, durable aggregates; and, use chemical admixtures. Fundamental laws of nature state that we cannot create or destroy matter; we can only affect how it is organized, transformed, and used. Obey the rules of nature: use only what you need and never use a resource faster than nature can replenish it. Resources are extracted from the earth by 20% more than the earth produces. Therefore, what is consumed in 12 months will take 14.4 months to be replenished. The use of sustainable development procedures will reduce that rate [18]. "The issue is not environment vs. development or ecology vs. economy; the two can be (and must be) integrated [19].

CONCLUSIONS

As Kofi Annan, U.N. Secretary General said in 2002, "We have the human and material resources needed to achieve sustainable developments, not as an abstract concept but as concrete reality" [18]. Professionals involved in the cement and concrete industries have the responsibility to generate lasting innovations to protect both the industries' future viability and the health of our environment. Large volumes of by-product materials are generally disposed in landfills. Due to stricter environmental regulations, the disposal costs for by-products are rapidly escalating. Recycling and creating sustainable construction designs not only contributes to reduced disposal costs, but also aids in the conservation of natural resources. This conservation provides technical and economic benefits. It is necessary for those involved in the cement and concrete industries to eliminate waste and take responsibility for the life cycle of their creations. In order to be responsible engineers we must all think about the ecology, equity, and economy of our design [2].

We must apply forethought into direct and meaningful action throughout our development practices. Sustainable designs must be used as an alternative and better approach to traditional designs. The impacts of every design choice on the natural and cultural resources of the local, regional, and global environments must be recognized in the new design approaches developed and utilized by the cement and concrete industries.

REFERENCES

1. Concrete for the Environment, Nordic Network Concrete for Environment by SP Swedish National Testing and Research Institute, Boras, Sweden, June 2003, 8 pages.

2. McDONOUGH, W Partners. The Hannover Principles: Design for Sustainability, EXPO 2000, The World's Fair, Hannover, Germany, 1992.

3. UNFCCC COP9 Report, Delivering the Kyoto Baby, REFOCUS, International Renewable Energy Magazine, Kidlington, Oxford, UK, Jan/Feb. 2004, pp. 52-53.

4. MALHOTRA, V M. Role of Supplementary Cementing Materials and Superplasticizers, Proceedings of ICFRC International Conference on Fiber Composites, High-Performance Concrete, and Smart Materials, Chennai, India, January 2004, pp. 489-499.

5. United States House Resolution 394, March 2004.

6. WORRELL, E, AND GALTISKY, C. Energy Efficiency Improvement and Cost Saving for Cement Making, Lawrence Berkeley Nat.Lab., Pub. No. LBNL-54036, 2004, 62 pp.

7. "Concrete Thinking for a Sustainable Future." Cement Association of Canada, http://www.cement.ca/cement.nsf/internetE/28BAAE6AB42AB69C852567B60056B657?opendocument, May 2004.

8. NAIK, T R, KRAUS, R N, RAMME, B W, & SIDDQUE, R. Long-Term Performance of High-Volume Fly Ash Concrete Pavements, ACI Materials Journal, 100, 2, 03, pp.150-5.

9. Liquid Stone: New Architecture in Concrete, National Building Museum, Washington, D.C., 2004.

10. COPPOLA, L, CERULLI, T, AND SALVIONI, D. Sustainable Development and Durability of Self-Compacting Concretes, Eighth CANMET/ACI Int.Conf. on Fly Ash, Silica Fume, Slag and Natural Pozzolans in Concrete, Las Vegas, NV, 2004, pp. 29–50.

11. NAIK, T R, AND KRAUS, R N. The Role of Flowable Slurry in Sustainable Developments in Civil Engineering, Proceedings of the ASCE Conference on Materials and Construction - Exploring the Connection, Cincinnati, Ohio, 1999, 9 pages.

12. McKAY, D T. Sustainability in the Corps of Engineers, a paper presented at the Technical Session Sponsored by the ACI Board Advisory Committee on Sustainable Developments, Washington, D. C., March 2004.

13. MEHTA, P K. Greening of the Concrete Industry for Sustainable Development, ACI Concrete International, Vol. 24, No. 7, 2002, pp. 23-28.

14. WU, Z. Development of High-Performance Blended Cement, Ph. D. thesis, T. R. Naik advisor, Dept. of Civil Eng. and Mech., College of Engineering, 2000, 177 pages.

15. BOURG, J. Water Conservation, http://www.wbdg.org/design/resource.php?cn=0&cx=0&rp=50, April 27, 2004.

16. http://www.cbu.uwm.edu, June 30, 2004.

17. White Paper,ACI Board Advisory Committee on Sustainable Development, Feb. 2004.

18. TIME Magazine, August 26, 2002, pp. A8.

19. Sustainable Developments: Planning our Future, Ricoh Co., TIME Magazine, Mar'04.

DEVELOPMENT OF SELF COMPACTING CONCRETE FOR PREFABRICATED STREET FURNITURE

Ravindra Gettu

Indian Institute of Technology Madras

India

B Barragán

Universitat Politècnica de Catalunya

Spain

R L Zerbino

National University of La Plata

Argentina

C Bernad   **M Bravo**   **C Cruz**

Universitat Politècnica de Catalunya

Spain

ABSTRACT. Self-compacting concrete is a relatively new construction material with a promising future in the prefabrication industry. In the present work, a previously-proposed experimental mix optimisation methodology is applied to develop a suitable concrete for manufacturing a street/park bench with a complex shape. Good results have been obtained, in terms of self-compactability, strength (20 MPa at the age of 24 hours) and aesthetic qualities.

Keywords: Self compacting concrete, Architectural concrete, Street furniture, Precast concrete.

Dr Ravindra Gettu was, until 2004, the Director of the Structural Technology Laboratory of the Universitat Politècnica de Catalunya (UPC). He is currently a Professor in the Department of Civil Engineering at the Indian Institute of Technology Madras in Chennai, India. His research interests include the application of self compacting concrete, and admixtures for structural concrete.

Dr Bryan Barragán is a Senior Researcher in the Department of Construction Engineering of the UPC, where he obtained his doctoral degree in 2002. His research interests include fibre reinforced concrete, self compacting concrete and experimental techniques.

Dr Raúl L Zerbino is an Associate Professor at the National University of La Plata and Researcher of the National Council for Scientific and Technical Research, Argentina. His research interests include fibre concretes, concrete technology and fracture mechanics.

Camilo Bernad is a technician at the Structural Technology Laboratory of the UPC.

Marina Bravo is a recent civil engineering graduate. She is currently a Researcher at the Structural Technology Laboratory of the UPC.

Claudia de la Cruz is working toward a doctoral degree in civil engineering at the UPC in the area of the development and utilization of self compacting concrete.

INTRODUCTION

Recent changes in the construction environment are demanding improved technology for the production of high performance concrete with far greater workability, high strength and durability. Self-compacting concrete (SCC) is designed to flow under its own weight, so it can easily fill heavily reinforced formwork or complicated forms, without mechanical compaction. The critical aspect of this technology involves attaining a high fluidity while preventing segregation of the concrete components so as to obtain good homogeneity.

Okamura proposed the concept of SCC in 1986 and the prototype for structural applications was first completed in 1988 [1]. Since then, various investigations have been carried out, though only a few large-scale construction companies have used the concrete in practical structures. The required level of self-compactability for casting depends on several factors, such as the type of construction, the selected placement and consolidation methods, the shape of formwork, and the congestion of the reinforcement. With the increasing use of congested reinforcements, slender elements and complex forms, there is a growing interest in specifying highly flowable concrete [2].

On the other hand, SCC has important benefits in the prefabrication of architectural elements, urban furniture and other non-structural elements. Accordingly, the main objectives of the present study have been the identification of a prefabricated element where the use of SCC would be more beneficial than conventional concrete, the application of a mix design methodology developed [3] at the Universitat Politècnica de Catalunya (Barcelona, Spain) in the fabrication of this element, and the casting of the element with the SCC.

The main conclusions that a high strength SCC can be developed that satisfies both the mechanical and aesthetical requirements of the element chosen, which was a street/park bench with a complicated form. The optimised concrete mix had a 43 % paste volume, with high filler and superplasticizer dosages.

DESCRIPTION OF THE CONCRETE ELEMENT

SCC can be used in the fabrication of different types of elements and structural components. The element chosen in the present work due to its complexity (see Figure 1) was a prefabricated modular concrete street/park bench produced by the Spanish company Escofet [4] with the name Silla-U. This model was designed by A. Viaplana and H. Piñón.

Figure 1   *Silla-U* produced by Escofet (from [4])

The degree of complexity in fabricating this element can be appreciated in the above photograph, especially in the shapes of the seat and the back, and the 55° angle between the leg and seat of the bench. Furthermore, the end of the curved back has a thickness of only 53 mm. The bench is normally cast upside down in a double-walled metal form so that all visible surfaces are moulded. The element is reinforced with a light steel mesh. With conventional concrete, significant vibration is needed to ensure the filling of the form, which obviously limits the life of the form, necessitates a heavy mould and makes the process labour intensive, in addition to being noisy and tiring for the workers involved. These aspects influenced the choice of this element as the target for the SCC application.

In a previous work, the applicability of SCC in the fabrication of the present element has been studied through the use of a high-strength concrete with CEM I 52.5 R cement, limestone aggregates, fly ash and a polycarboxylate-based superplasticizer. The results have been presented elsewhere [5, 6], where it was shown that the element can be fabricated without any vibration, yielding sufficient strength for being demoulded in 24 hours and satisfactory surface finish. Moreover, the use of fly ash in the SCC provided an added value to the product in terms of reduced environmental impact. Nevertheless, when the aggregates were exposed by water and acid treatment, there was not enough contrast between the colour of the aggregates and the matrix to make the surface aesthetically appealing. Therefore, as an extension of the previous work, the present study has been performed, where aesthetics were taken into account. In this context, the objective was to obtain the same surface texture and coloration as the element manufactured with conventional concrete.

MATERIALS AND MIX DESIGN USED

The material components and the mix design were chosen to be as similar as possible to, if not the same as, those used in the conventional concrete with which the elements are traditionally manufactured.

A cement of type CEM I 52.5 R was chosen since a high initial strength was needed in order to demould and manipulate the element (i.e., lift the element out of the mould and turn it over) as early as 24 hours.

Crushed granite aggregates were used mainly for aesthetic reasons. When the surface of the element is washed with water and acid, as soon as it is demoulded, the aggregates are exposed to reveal a strong contrast between the dark (grey) coarse aggregates and the lighter coloured hardened cement paste. Three fractions of aggregates were employed: 0-2.5 mm, 2.5-6 mm and 6-15 mm. The maximum aggregate size was limited to 15 mm in order to avoid blockage of the concrete during its flow within the mould. Three aggregate fractions were chosen instead of two (i.e., using sand with a grain size range of 0-6 mm) in order to facilitate the incorporation of a higher quantity of the finer aggregates, which generally leads to better flowability of the concrete.

SCC needs a paste or mortar with significant viscosity in order to provide the cohesion required to prevent the segregation of the coarse aggregates. This can be achieved by employing fine fillers in the concrete and/or viscosity-enhancing admixtures. Here, crushed white marble was incorporated in the concrete to provide the high fines content as well as to make the paste lighter in colour. The material used had a fairly uniform grain distribution with a maximum size of about 40 microns. The Blaine specific surface area of the crushed marble was 3580 cm^2/g, which is almost the same as that of the cement used (i.e. 3400 cm^2/g).

The superplasticizer is an essential component of an effective SCC. Here, a polycarboxylate-based commercial product, which also has a de-airing agent to improve the quality of the moulded surfaces, was used. The admixture had a density of 1.09 kg/litre and a solids content of 38%. The water content of the liquid product was taken into account in the water/cement ratio of the concrete.

The mix was designed following the procedure developed previously at the UPC for high strength SCC [3]. In this procedure, the cement paste system is first optimized for high fluidity and moderate cohesion using the Marsh cone and mini-slump tests. The water/cement ratio was set to be 0.45, which is the same as that used in the conventional concrete. The objective of the first step is to determine the maximum superplasticizer dosage for different dosages of filler (i.e., crushed marble). For each filler dosage (defined as a ratio of the cement weight), the Marsh cone test is performed with increasing superplasticizer dosage, sp/c (defined as the ratio between the solids content of the superplasticizer and cement, by weight). As seen in Figure 2(a), 1000 ml of paste is poured into the Marsh cone and the time taken for 500 ml to flow out of the 8 mm aperture is measured for each sp/c. The dosage beyond which the flow time does not decrease significantly is taken as the saturation dosage [3, 7]. For each filler dosage and the corresponding saturation dosage of superplasticizer, a mini-slump test is performed (see Figure 2b), where a mould is filled with paste and lifted to let the paste spread over a base plate [8]. Following previous works [3, 5, 6], the optimum filler dosage is taken as that which gives a final spread of about 180 mm with the time needed for the spread diameter to reach 115 mm approximately equal to 2 seconds. In the present case, the optimum paste composition is defined by a filler/cement dosage of 0.6 and a superplasticizer/cement dosage (i.e., sp/c) of 0.8%.

(a)   (b)

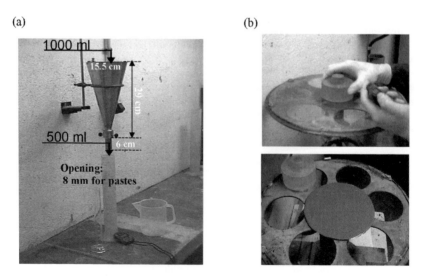

Figure 2   (a) Marsh cone test and (b) Mini-slump test

The aggregate proportions of the concrete are determined using dry uncompacted mixes of the three fractions and choosing the combination that yields the lowest void content [3, 6]. For the aggregates used here, the best packing density was obtained for the combination of the fractions 0-2.5 mm: 2.5-6 mm: 6-15 mm of 45:25:30. These proportions were maintained in the concrete mix.

After fixing the paste composition and the aggregate proportions, the only parameter that needs to be determined in order to complete the mix design is the paste volume. The optimum paste volume is take to be the minimum paste content that leads to self-compactibility. The self-compactability was evualated using the slump flow and V-funnel tests. In the slump flow test [9], the traditional conical mould of the slump test is filled with the SCC (without any compaction) and then lifted to let the concrete spread over a base plate. The final spread (D_F) and the time taken for the spread to reach a diameter of 50 cm (T_{50}) are determined. In the V-funnel test [10], the funnel with rectangular section and a bottom aperture of 65×75 mm is filled with concrete and the time taken for it be emptied is measured. The target values for these parameters were a final spread of 650-700 mm and T_{50} of 2-6 seconds in the slump flow tests, and a V-funnel flow time of 5-15 seconds. The choice of these ranges was made based on the characteristics of the element and the fabrication procedure; that is, the concrete had to flow steadily through the mould but as slow as possible in order to let the air escape.

After several trials, it was seen that the concrete with a paste content of 42.5% by volume, was that which provided adequate self-compactability. In addition to the tests of the fresh concrete, panels were cast with the SCC and the surface quality was assessed by the technicians of the manufacturing plant.

FABRICATION OF PROTOTYPE IN THE PLANT

Once the mix was considered to be appropriate, a batch of 250 litres was made in the prefabrication plant using a forced-action vertical-axis mixer. The mix composition and the parameters obtained in the self-compactability tests are given in Table 1. Note the relative proportions of the 0-2.5 mm, 2.5-6 mm and 6-15 mm aggregate fractions were 45%, 25% and 30%, respectively.

Table 1 Composition and properties of the concrete.

SCC FABRICATED IN THE PLANT

Composition (kg/m^3)					Properties			
Cement	Water	Super-plasticizer	Marble filler	Aggregates	Slump flow		V-funnel flow time (s)	24-hour compressive strength (MPa)
					D_F (cm)	T_{50} (s)		
406	177	8.4	243	1500	70	2	6	20

The mould was filled with the SCC by letting flow slowly out of a bucket. It was observed that the concrete flow was uniform and steady, without any indices of segregation (see Figure 3). No external vibration was needed to fill the mould (see Figure 4). The top of the concrete surface was levelled off manually with very little effort.

Figure 3   Filling of the mould with SCC

Figure 4   Mould after being filled

When conventional concrete is used in the fabrication of the same element, the mould is filled several times and heavily vibrated. Additionally, a needle vibrator is inserted into the part (as seen in Figure 5) where the leg of the bench is moulded to avoid air pockets.

Figure 5   Needle vibration used in the compaction of the conventional concrete

During the filling of the mould with SCC, only two workers were needed compared with four workers for the conventional concrete. The filling operation took about 4 minutes with SCC and about 10 minutes with vibrated concrete. It is clear that not using vibration for the compaction of the concrete results in significant savings in terms of equipment, labour and mould life.

CONCLUSIONS

Conventional vibrated concrete was successfully replaced by self-compacting concrete in the fabrication of a prefabricated street bench. In addition to satisfying the mechanical and aesthetical requirements, the use of SCC led to the elimination of both external and needle vibration, resulting in lower noise levels and harmful effects on the workers, and the potential for savings due to lower energy consumption, lower labour demand, reduced depreciation of the moulds and phasing out of the vibration equipment.

ACKNOWLEDGEMENTS

Partial support for this work from the UPC research project MAT2003-05530 funded by the Spanish Ministry of Science and Technology and the technology transfer project funded by SIKA is appreciated. Cementos Molins, Omya and SIKA donated the materials used here. The prototype elements were fabricated at the Escofet plant in Martorell, Spain. The generous help and participation of the Escofet personnel in this project is gratefully acknowledged.

REFERENCES

1. OKAMURA, H, OZAWA, K AND OUCHI, M. Self-compacting concrete. Structural Concrete, Vol. 1, No. 1, 2000. pp. 3-17.

2. CIMBETÓN. Self-compacting concretes: Monograph on SCC strucures, B.52A, Centre d'Information sur le Ciment et ses Aplications, France, 2004, 151 p.

3. GOMES, P C C, GETTU, R, AGULLÓ, L AND BERNAD, C. Experimental optimization of high-strength self-compacting concrete. Proc. Second Intnl. Symp. on Self-Compacting Concrete, Eds. K. Ozawa y M. Ouchi, COMS Engineering Corp., Kochi, Japan, 2001, pp. 377-386.

4. ESCOFET. Product Catalogue, http://www.escofet.com, 2003.

5. GETTU, R, COLLIE, H, BERNAD, C, GARCIA, T AND ROBIN, C. Use of high-strength self-compacting concrete in prefabricated architectural elements. Proc. Intnl. Conf. on Recent Trends in Concrete Technology and Structures, Vol. II, Eds. D.L. Venkatesh Babu, R. Gettu and R. Krishnamoorthy, Kumaraguru College of Technology, Coimbatore, India, 2003, pp. 355-363.

6. GETTU, R, GOMES, P C C, AGULLÓ, L AND JOSA, A. High-strength self-compacting concrete with fly ash: Development and utilization. Proc. Eighth CANMET/ACI Intnl. Conf. on Fly Ash, Silica Fume, Slag, and Natural Pozzolans in Concrete, ACI SP-221, Ed. V.M. Malhotra, American Concrete Institute, Farmington Hills, USA, 2004, pp. 507-522.

7. AGULLÓ, L, TORALLES-CARBONARI, B, GETTU, R AND AGUADO, A. Fluidity of cement pastes with mineral admixtures and superplasticizers – A study based on the Marsh cone test. Materials and Structures, Vol. 32, 1999. pp. 479-485.

8. KANTRO, D L. Influence of water reducing admixtures on properties of cement pastes - A miniature slump test. Cem. Concr. Aggregates, Vol. 2, 1980. pp. 95-102.

9. JSCE. Method of test for the slump flow of concrete. Standards of the Japan Soc. of Civil Engineers, F03, 1990.

10. EFNARC. Specification and Guidelines for Self-Compacting Concrete, EFNARC, Farnham, UK, 2002, 32 p.

DRYING SHRINKAGE AND MODULUS OF ELASTICITY OF SAND TOTAL LIGHTWEIGHT CONCRETES

M N Haque   **H Al-Khaiat**

University of Kuwait

Kuwait

O Kayali

University of New South Wales

Australia

ABSTRACT. The drying shrinkage (ε_s) of two total lightweight concretes (LWC) and two sand lightweight concretes (SLWC) of 35 and 50 MPa cube compressive strength and a normal weight concrete (NWC) of 50 MPa were monitored for a period of up to 3 years. In addition, their strength, stiffness (E), and corrosion potential have also been determined and reported. Whilst NWC 50 gave a drying shrinkage of approximately 700 microstrain, LWC 35 gave drying shrinkage of nearly 1000 microstrain. The results suggest that total and sand LWC shrank 20 and 5% more than the corresponding equal strength NWC, respectively. In spite of the differing values of shrinkage strains, compressive strength, modulus of rupture and modulus of elasticity of the concretes tested, the failure tensile strain capacity (extensibility) of all the concretes was somewhat similar. The results also suggest the ACI code equation to predict E value knowing the compressive strength gives a reasonable estimate of E for the lightweight concretes. Results of corrosion potentials indicated no likely corrosion activity within the test period. Also there was similarity between the values of corrosion potentials for lightweight and normal weight concrete.

Keywords: Drying shrinkage, Total lightweight concrete, Sand lightweight concrete, Normal weight concrete, Restrained shrinkage, Tensile strain capacity.

M N Haque, Professor, Department of Civil Engineering, College of Engineering and Petroleum, University of Kuwait, Kuwait.

O Kayali, Senior Lecturer, School of Civil Engineering, University of New South Wales, Australian Defence Force Academy, Australia.

H Al-Khaiat, Professor, Department of Civil Engineering, College of Engineering and Petroleum, University of Kuwait, Kuwait.

INTRODUCTION

Drying shrinkage of concrete and the cracking initiated by restrained shrinkage is a major weakness of concrete. The cracking induced by restrained shrinkage can degrade both the serviceability and durability of concrete structures. Whilst the drying shrinkage of most lightweight concretes (LWC) made using good quality manufactured LWA is often satisfactory, its behaviour under restrained shrinkage needs better understanding. This paper presents the drying shrinkage and some other relevant characteristics of both LWC and SLWC.

A testing program reports that 1 year drying shrinkage of LWCs varied approximately between 550 to 900 microstrain [1]. The study also reports the shrinkage behaviour of the two comparable LWC and NWC. The shrinkage of the LWC lagged behind early values of the NWC, equalled them at 90-110 days, and reached an ultimate value at one year approximately 14% higher. Berra and Ferrara [2] have also reported that total lightweight concretes resulted in a lower shrinkage rate, due to the presence of water in the aggregate particles: the value at infinite time was, however, higher. Nilsen and Aitcin [3] have also reported "the lightweight concrete mixture containing a high-quality expanded shale aggregate showed that it is possible to make a very-high-strength lightweight concrete with almost negligible drying shrinkage". In 1971, Pfeifer [4] concluded that in LWCs, when the percentage of natural sand fines was increased, drying shrinkage was reduced. Concretes containing 100% sand fines had 10 to 30% less drying shrinkage than those containing 33% sand fines. The comparative results of the drying shrinkage of total lightweight, sand lightweight and normal weight concretes of this study are discussed in the light of the modulus of elasticity, flexural modulus and hence the tensile strain capacities and the net potential of possible cracking induced due to restrained shrinkage in structural concrete.

Experimental Program

Total lightweight concretes (LWC) and sand lightweight concretes (SLWC) using a commercially available lightweight aggregate (LWA) and a normal weight concrete (NWC) were made to evaluate their strength, durability and shrinkage characteristics. The LWA used were oven dried at 105°C for 24 hours and sealed in containers. The LWC and SLWC were designed for a 28 day cube compressive strength of 50 and 35 MPa, hereafter referred to as LWC50, LWC35 and SLWC35 and SLW50, respectively. The NWC designed for 50 MPa cube compressive strength is designated as NWC50. For the NWC50, standard mix proportions, using local aggregates available in Kuwait, were adopted.

The coarse and fine LWA were soaked and mixed with about 30% of the mixing water for approximately 20 minutes in a pan mixer. Thereafter cementitious materials and the mixing water were added and mixed. The superplasticizer mixed with about 1 kg of water was added towards the final mixing stage. The mix quantities used for all the five concretes are included in Table 1.

The slump, bulk density, air content and temperature of each batch of concrete were determined. The average values of some of the characteristics of the fresh concrete are included in Table 2.

Table 1 Mix proportions (kg/m^3) of concretes tested

INGREDIENT	CONCRETE, kg/mm^3				
	LWC35	SLWC35	LWC50	SLWC50	NWC50
Cement	353	280	536	480	450
CSF	35.3	28	53.6	48	45
Water	280	195	294	200	219
Lytag Coarse[1]	625	700	567	726	1084[3]
Lytag Fine[2]	499	-	357	-	-
Sand[4] (normal weight)	-	570	-	345	571
Superplasticizer	3.5	3.0	6.5	5.5	5.0

[1,2]Fly ash based LWA
[3]Normal weight crushed gravel
[4]Washed and dried

Table 2 Some characteristics of the fresh and hardened concretes

CONCRETE	SLUMP (mm)	BULK DENSITY (kg/m^3)	CUBE COMPRESSIVE STRENGTH (MPa) (σ_c)			σ_r (MPa) 3 years	E (MPa) 1 year	$\varepsilon_{su} \times 10^{-6}$ 1 year
			28 day	1 year	3 years			
SLWC 35	100	1775	38.0	48.0	-	4.55[2]	26120	760
SLWC50	80	1800	49.5	64.5	-	5.50[2]	29040	680
LWC35	90	1795	36.0	48.5[1]	49.0	4.40	23210[3]	950 (1000)
LWC50	95	1815	51.0	62.0[1]	64.0	5.20	28250[3]	800 (835)
NWC50	85	2355	48.5	66.0[1]	67.5	7.90	36340[3]	650 (680)

[1] 2year results
[2] 1year results
[3] 9 month results
( ) 3 year results

The strength and durability characteristics of these five concretes have already been reported [5-7]. A summary of relevant results is included in Table 2. In this paper drying shrinkage of these five concretes is presented. The drying shrinkage of concrete was evaluated on 75 × 75 × 285 mm prisms from each mixture in accordance with ASTM C 157 [8]. After demoulding, shrinkage specimens were cured in water for 7 days. Then the shrinkage prisms were removed from the water tank, wiped off, transferred to control room, maintained at 50 ± 5% R.H. and 20 ± 2°C, and zero readings were taken. The drying shrinkage strains as given in Table 3 are the average of 3 specimens. As shown in the table, the readings were taken at 1, 3, 7, 14, 28, 91, 180, 270 and 365 days. The LWC35, LWC50 and NWC50 concretes were further monitored for a total period of 3 years. The SLWC35 and SLWC50 were monitored for a period of one year only.

Half-cell potential measurement was also done on reinforced concrete specimens of the five concretes tested. The tests were conducted according to ASTM 876-91 [9] by placing a half-cell on the concrete surface and connecting it, via a high-input resistance voltmeter, to the rebar. Two specimens per mix were cast for this purpose. The specimens were prismatic with dimensions of 75 × 75 × 285 mm. Two plain 10 mm diameter reinforcing bars were placed in each specimen with a concrete cover on two sides of the bar of only 10 mm. The specimens were cured for 1, 3 and 7 days before exposure on the seaside (referred to as 1SS, 3SS and 7SS curing and exposure regimes as detailed in Table 4). The half cell readings recorded in Table 4 are the average of two specimens.

Table 3 Shrinkage strain of total, sand LWCs and NWC ($\times 10^6$)

DAYS OF DRYING CONCRETE	LWC35	LWC50	NWC50	SLWC35	SLWC50
1	18	15	12	15	13
3	113	94	75	92	82
7	202	175	140	162	153
14	311	259	207	246	226
28	423	341	273	331	298
91	747	636	509	597	555
180	882	747	600	698	634
270	933	788	630	743	667
365	951	799	648	760	680
765	969	807	667		
1080	998	835	679		

Table 4 An estimate of failure tensile strain capacity (ε_{tc}) of the concretes

CONCRETE	σ_r (MPa) 28 days	E(MPa) 28 days	$\frac{\sigma_r}{E}=\varepsilon_{tc}$ $\times10^{-6}$ 28 days	$\frac{\sigma_r}{E}=\varepsilon_{tc}$ $\times10^{-6}$ ultimate	ε_s $\times10^{-6}$ 28 days	ε_s $\times10^{-6}$ at 1 year (& 3 years)	APPROX. FAILURE STRAIN CAPACITY $\times10^{-6}$
SLWC35	4.01	23782	169	174	331	760	338
SLWC50	4.65	26648	175	189	298	680	350
LWC35	3.80	21430	177	190	423	951(998)	354
LWC50	4.40	26130	168	184	341	799(835)	336
NWC50	5.70	31780	179	217	273	648(679)	358

RESULTS AND DISCUSSION

Drying Shrinkage

The shrinkage of the two LWC's and NWC50 were monitored for a period of 3 years whereas for the SLWC's for one year only as shown in Table 3. Obviously, drying shrinkage of the NWC50 is the least but that of the SLWC50 is not too different from the NWC50. The highest drying shrinkage recorded for the LWC35, at the age of 3 years is 1000 microstrain. It can be argued that the shrinkage characteristics of the SLWC's are satisfactory and would meet the specifications of a quality concrete [2-4].

The shrinkage response of all the five concretes can be explained and attributed to three main variables operative in this investigation. They are: total water content, paste content and the stiffness of the aggregate particles. For example, lower drying shrinkage of SLWC50 compared to LWC50 can be attributed to the lower paste content, water content and the use of normal sand of the higher modulus value. Of course, the same argument holds when shrinkage characteristics of SLWC35 and LWC35 are compared. In summary, drying shrinkage of LWCs and SLWCs, can be explained and attributed to the same factors as in NWCs. Further, the drying shrinkage values of all the five concretes tested are within values considered normal for concrete.

The drying shrinkage of NWC50 and LWC50, NWC50 and SLWC50 and, SLWC50 and LWC50 are plotted and compared in Figures 1-3, respectively. These graphs give a visual display of the deviation from equality. As shown in Figures 1, 2, and 3, approximately, LWC50 shrank 22% more than the NWC50, SLWC50 shrank 5% more than the NWC50 and LWC50 shrank 16% more than the SLWC50, respectively. These relationships can be used, as an approximation, to obtain an indicative value of drying shrinkage of total or sand LWC knowing the corresponding values for a similar strength NWC for the LWAs and the NWAs used in this study.

In addition to other disadvantages of drying shrinkage [10-11] in concrete structures, the one most deteriorating consequence is the cracking initiated due to restrained shrinkage. Factors like the modulus of elasticity (E), creep and tensile strength also influence the occurrence of cracking. A low modulus of elasticity value and a larger tensile creep reduce tensile stresses arising from restrained shrinkage. Of course a higher value of tensile strength would help minimise the occurrence of cracking [10].

Mehta has described the combination of these factors that are desirable to reduce the advent of cracking by a single term called extensibility [10]. Concrete is said to have a high degree of extensibility when it can be subjected to large deformations without cracking. Obviously, to minimize cracking, the concrete should shrink less, have low E, high tensile creep and high tensile strength.

Table 4 presents an approximate tensile strain capacity (extensibility) of the five concretes tested. Tensile strain capacity has been approximated by Houghton by dividing modulus of rupture, by the corresponding E value [12] and adopted by Mehta [10]. Accordingly, columns 4 and 5 of the table give of the concretes tested at 28 days and at 1 or 3 years (ultimate), respectively. Column 6 shows the measured drying shrinkage strains at 28 days while column 7 shows the longer term measured drying shrinkage strains at 1 year and 3 years. Column 8, which shows approximate failure strain capacity, is just double the value of tensile strain capacity at 28 days as included in column 4.

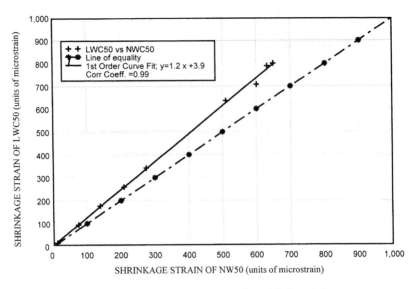

Figure 1 Shrinkage of normal weight and total lightweight concretes.

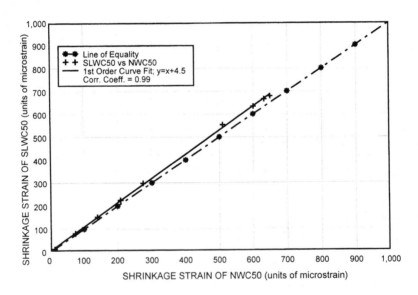

Figure 2 Shrinkage of normal weight and sand lightweight concretes.

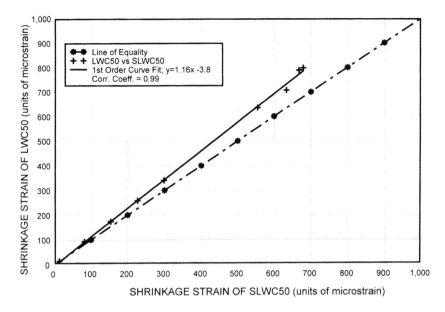

Figure 3 Shrinkage strain of sand and total lightweight concretes

This is based on experimental evidence presented by Altoubat and Lange [13] who suggested that "tensile creep relaxes shrinkage stresses by 50% and doubles the failure strain capacity". The main aim of this analysis is to see the extent of drying shrinkage and the potential of crack initiation in LWCs.   On the overall, it is concluded that the LWCs tested are comparable with known good quality structural concretes with all the other attendant advantages. Sand LWCs, however, seem to provide better protection against the occurrence of cracking caused by restrained shrinkage.   Nevertheless, long- term drying shrinkage measurement has shown that all the tested concretes may be liable to cracking if fully restrained.   Total lightweight concrete showed the highest tendency in this respect.   The authors acknowledge that cracking due to drying shrinkage can not be isolated from other influencing factors such as time-dependent stiffening and time-dependent gain in strength. Furthermore, the extent of constraint plays an important role if cracking is to occur.   In addition, allowing for tensile creep relaxation may increase the limit of constraint-free drying shrinkage strain by 50% [13]. The authors believe that more research is needed to confirm this figure.

Estimation of E value

The ACI [14] and AS 3600 [15] equation,

$$E = \rho^{1.5} \times (0.043) \sqrt{f_{cm}} \tag{1}$$

can be used to determine modulus of elasticity of concrete knowing its strength where:

ρ   =   density of concrete in kg/m^3
f_{cm} =   mean value of the compressive strength at the appropriate age in MPa.

The above equation has been used to calculate the E (MPa) value of LWC 35, LWC 50 and NWC 50.

In the above equation f_{cm} is the cylinder strength. In this investigation, however, only cube compressive strength was evaluated as given in Table 2. The CEB-FIP Design Code [16] supplies tables of comparison between strength based on cube specimens and that based on cylindrical ones. Up to the cylinder strength value of 50 MPa, the ratio of the cylinder strength to the cube strength is about 0.8. This ratio increases progressively for strength value above 50 MPa and reaches a value of 0.89 for 80 MPa cylinder strength. Accordingly, cylinder strength was estimated to be ~ 0.8 of cube strength and was used in the calculation of the E value according to the ACI and AS 3600 expression. The calculated values are compared with the corresponding experimental values as shown in Table 5. As can be seen in Table 5, the use of the ACI equation underestimates the E-value in all the concretes of this series of tests. Nevertheless it may still be conservatively used with lightweight concrete.

Table 5 Estimated and experimental E-values

CONCRETE	ρ (kg/m^3)	CUBE STRENGTH (MPa)	EXPERIMENT E (MPa)	CALCULATED E-BASED ON EST. CYL. (MPa)
LWC 35	1795	36	21425	17550
	1795	42	21055	18950
	1795	46	23210	19850
LWC50	1815	48	25400	20650
	1815	61	26855	23230
	1815	65	27710	24000
NWC50	2355	48	32740	30450
	2355	56	34200	32900
	2355	64	36340	35200

Corrosion Potential

The potentials were monitored for nine months and the maximum reading obtained was only −136 mV (see Table 6). These results suggest that at the age of 9 months no active corrosion was taking place in any of the five concretes (readings are more positive than - 200 mV). There is not a marked difference in the half-cell potentials of the differing concrete specimens exposed on seaside. Accordingly, no further discussion is included on half-cell measurements and the results obtained.

Table 6 Half Cell Potentials (mV) measured after 28 and 270 days exposure on seaside

CONCRETE	LWC 35		LWC 50		NWC 50		SLWC 35		SLWC 50	
CURING	28 days	270 days	28 days	270 days	28 days	270 days	28 days	270 days	28 days	270 days
1 SS	-29	-133	-19	-83	-29	-136	-50	-115	-42	-103
3 SS	-21	-95	-17	-69	-21	-107	-44	-89	-39	-82
7 SS	-15	-81	-12	-25	-18	-91	-34	-73	-33	-67

CONCLUSIONS

1. Drying shrinkage of total and sand LWCs tested were found to be more than that of the same strength NWC although the drying shrinkage of SLWC 50 and NWC 50 were practically the same. The indicative value of drying shrinkage of total and sand LWC's tested, in this investigation, were some 20 and 5% higher than that of the corresponding NWC.

2. In spite of the differing values of the compressive strength, modulus of rupture and modulus of elasticity of the total and sand LWCs, and the NWC tested their failure tensile strain capacity (extensibility) was found to be very similar.

3. The results suggest that code equation to estimate E, given cylinder compressive strength and concrete density, is reasonable for LWC as well as NWC.

4. Comparing corrosion potentials within the period of these tests, none of the concretes exhibited half cell potentials that indicated likely corrosion.

REFERENCES

1. SHAH S.P., AHMAD S., High Performance Concrete and Applications, Edward Arnold, 1994

2. BERRA M., FERRARA G., Normal weight and total-lightweight high-strength concretes: A comparative study, ACI- SP 121, 1990, pp. 701-733

3. NILSEN A.U., AITCIN P., Properties of high-strength concrete containing light-, normal-, and heavyweight aggregates. Cement, Concrete, and Aggregates (ASTM), 14, 1992, pp. 8-12

4. PFEIFER D.W., Fly ash aggregate lightweight concrete. ACI Journal, Vol. 68, No.3, 1971, pp. 213-216

5. AL-KHAIAT H., HAQUE N., Strength and durability of lightweight and normal weight concrete. ASCE, Journal of Materials in Civil Engineering, Vol. 11, No. 3, 1999b, pp. 231-235

6. HAQUE N., AL-KHAIAT H., Strength and durability of lightweight concrete in hot marine exposure conditions Materials and Structures, Vol. 32, 1999, pp. 533-538

7. AL-KHAIAT H., HAQUE M.N., Effect of curing on concrete in hot exposure conditions. Magazine of Concrete Research, Vol. 51, No. 4, 1999a, pp. 269-274

8. ASTM, Standard test method for length change of hardened cement, mortar, and concrete. C147-99, Vol. 4.02, 1999, pp. 97-102

9. ASTM, Standard test method for half-cell potentials of uncoated reinforcing steel in concrete, C876-91, Annual book of ASTM Standards, Vol. 4.02, 1999a, pp. 1-6

10. MEHTA P.K., MONTEIRO P.J,M., Concrete-Structure, Properties and Materials, Prentice Hall, New York, 1993, pp. 548

11. NEVILLE A., Properties of concrete, Longman, London, 1995, pp. 844

12. HOUGHTON D.L., Determining tensile strain capacity of mass concrete. ACI Journal, Vol. 73, No. 12, 1976, pp. 691-700

13. ALTOUBAT S.A., LANGE D.A., Creep, shrinkage, and cracking of restrained concrete at early age. ACI Materials Journal, Vol. 98, No. 4, 2001, pp. 323-331

14. ACI, Building Code Requirements for Structural Concrete-ACI 318-95. ACI, Detroit, Farmington Hills, USA, 1995, pp. 369.

15. Standards Australia, "Concrete Structures", AS 3600-2001, Standards Association of Australia. p. 156

16. CEB-FIP, Model Code 1990, Thomas Telford, London, 1993, pp. 437

CONSTRUCTION POTENTIAL OF A MINING BY-PRODUCT

D Ionescu
La Trobe University
Australia

ABSTRACT. Mining by-products are often used as bulk fill in development projects or placed and revegetated at sites specified by the relevant government agencies. However, being very low in organic matter, revegetation of these sites can be difficult and expensive. The findings of an ongoing research at La Trobe University, which focuses on investigating the potential of waste/recycled materials to be used for various civil engineering applications, are presented in this paper. The paper concentrates on the use of a by-product from mining operations under Bendigo as concrete aggregates. Its properties are scrutinised viz standard specifications for aggregates for concrete production. Various mix design are proposed using both crushed spoil rock and crushed good quality rock as aggregates. The behaviour of concrete produced with crushed mining by-product and that produced with good quality crushed aggregates is compared in order to evaluate the advantages and disadvantages of utilizing this spoil rock as concrete aggregate. Results from standard tests suggest that the material is suitable for use as an aggregate in concrete.

Keywords: Aggregates, Concrete, Elastic constants, Mining by-product, Modulus, Recycling, Spoil rock, Waste material.

Ms D Ionescu, Lecturer, Department of Physical Sciences and Engineering, La Trobe University, Australia

INTRODUCTION

Bendigo's rich gold bearing quartz reefs are hosted by tightly forded Ordovician marine sediments. Gold bearing structures were mined by the old timers down to a depth of 1400 m and today over 5000 shafts have been recognised (including 11 over 1000 m deep) on the Bendigo goldfields. Recent drilling and bulk sampling by Bendigo Mining NL have showed that considerable gold resources (12.3 million ounces resource potential) still exist beneath the City of Bendigo. The gold bearing reefs are accessed via a 5.5 m square decline along a 4.5 km traverse. Significant quantities of spoil rock, known as mullock, are generated from tunnelling through the strongly folded and faulted sandstones, shales and slates. The mullock ranges from large boulders around half a metre in diameter down to fine particles passing a 75µm sieve. Some 1000 tonnes of mullock are produced each day, adding to the 250,000 tonnes already stockpiled [1]. Bendigo Mining NL is keen to find useful applications for the mullock. Consequently, the objectives of this preliminary investigation were to determine the engineering properties of the mullock, and establish its suitability for possible use in various civil engineering applications.

MATERIAL REQUIREMENTS

Aggregates are vital ingredients of a concrete mix. They are economical when compared to the cement production cost. Jackson and Dhir [2] summarised the aggregate properties were identified as having considerable effect on the concrete behaviour. These properties include: strength, hardness toughness, durability, porosity, volume change, chemical reactivity and relative density. Preliminary investigation on the possible use of mullock suggested that the by-product satisfy most of the above mentioned properties, therefore they can be used as aggregate in concrete [1].

A good concrete mix shows evidence of equal amounts of pullout (bond) failure and fracturing of the coarse aggregate on the fracture surface [3]. Therefore, besides the previously mentioned properties, it is desirable for the aggregate particles to have a rough surface texture and be well-graded in order to minimize void space. It is generally agreed that particles that are too flat or flaky can decrease concrete strength due to the larger potential fracture planes [4]. Therefore, the content in flat/flaky grains must be strictly controlled. Besides this, aggregates with a high water absorption should be avoided [4].

EXPERIMENTAL PROGRAMME

Representative samples of mullock were obtained from Bendigo Mining's Carshalton Site in Kangaroo Flat. The mullock sample was subjected to crushing in a jaw crusher. The particle size distribution of the crushed material is shown in Figure 1 against the Australian specifications for 20 mm aggregate [5]. It should be observed that the gradation of crushed mullock fits between gradations G2 and G3 [6]. In addition, it contains 41.3 % by weight fines (grains finer than 4.75 mm), which is similar with gradation G3. In order to get a better estimate of the fractions contained by the mullock, the fine aggregates were separated from the crushed material. The gradation of fine and coarse grains of mullock aggregate is presented in Figure 2. The Australian specifications for fine aggregate are also presented in Figure 2 [5, 6].

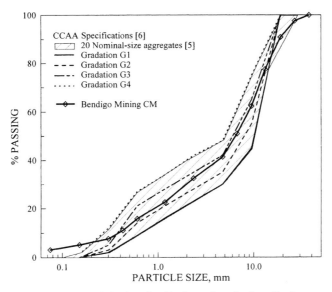

Figure 1 Particle size distribution of crushed mullock

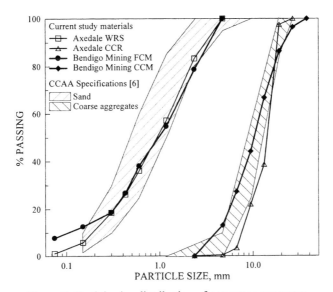

Figure 2 Particle size distribution of concrete aggregates

The currently used concrete aggregates were sourced from Axedale Quarry the main supplier of concrete aggregates for Bendigo. The coarse aggregates, crushed quarry rock – CCR, and fine aggregates, washed river sand - WRS, have the particle size distribution presented in Figure 2. It can be observed that the Axedale WRS fit well within Australian specifications for sand [5, 6]. The characteristics of the concrete aggregates gradations are presented in Table 1. Although the gradation of the fine mullock almost replicated the gradation of river sand, it should be observed that it contained a lager proportion of grains smaller than 0.15 mm than the Australian specifications for sand [5, 6]. When the coarser aggregate from Axedale and Bendigo Mining are compared it should be noted that there are significant differences between the two gradations. While the mullock has an almost proportional content of different fractions of grains, the quarry rock contains about 60 % by weight particles between 13.2 mm and 19 mm.

The coarser grains contained by the crushed mullock can be separated by sieving through a 26.5 mm sieve. Also the finer grains (< 0.15 mm) can be removed by subjecting the crushed mullock to a washing process. However, this would increase both the preparation time and production cost and would be uneconomical when implemented on a large scale. Therefore, in agreement with Bendigo Mining it was decided to use the crushed mullock as was supplied from the crushing plant. The fine aggregate from the two sources were also tested for any clay content . The summary of the consistency tests is presented in Table 2. It should be noted that the consistency test results of finer mullock compare quite well with those of washed sand, both aggregates indicating non plastic fines with insignificant shrinkage on drying.

Table 3 summarises the physical characteristics of the concrete aggregates. The results of the strength tests indicate that the mullock is a tough rock. However, the grains shape tests show that the mining spoil tends to produce flaky particles on crushing, which break along the bedding present in the sedimentary rock. Furthermore, the texture of the mullock particles is relatively smooth. Three concrete mixes were designed for the Axedale CCR and Axedale WRS, Bendigo Mining CM and Bendigo Mining FM, and Bendigo Mining CM and Axedale WRS. The quantities were adjusted for two nominal characteristic compressive strengths 20 MPa and 25 MPa, respectively.

Table 1 Grading characteristics of the concrete aggregates

MATERIAL	D_{10} (mm)	D_{30} (mm)	D_{50} (mm)	D_{60} (mm)	C_u	C_c
Axedale Washed River Sand (AWRS)	0.19	0.48	0.92	1.3	6.84	0.93
Bendigo Mining Fine Mullock (BMFM)	0.10	0.46	1.00	1.5	15.0	1.41
Axedale Coarse Crushed Rock (ACCR)	7.5	11.5	12.5	14.0	1.87	1.26
Bendigo Mining Coarse Mullock (BMCM)	4.0	7.0	10.1	12.0	3.0	1.02

Table 2 Consistency characteristics of the fine aggregates

MATERIAL	SHAPE	TEXTURE	LIQUID LIMIT (%)	PLASTIC LIMIT (%)	PLASTICITY INDEX (%)	LINEAR SHRINKAGE (%)
Axedale Washed River Sand (AWRS)	Sub-angular	Rough	18	14	4	2.0
Bendigo Mining Fine Mullock (BMFM)	Angular	Rough	19	16	3	1.0

Table 3 Physical properties of concrete aggregates

AGGREGATES	MOISTURE COTENT (%)	WATER ABSORBTION (%)	SPECIFIC GRAVITY	FLAKINESS INDEX (%)	LOS ANGELES VALUE (%)
Axedale Washed River Sand (AWRS)	5.0	1.5	2.84	N/A	N/A
Bendigo Mining Fine Mullock (BMFM)	0.2	2.50	2.76	N/A	N/A
Axedale Coarse Crushed Rock (ACCR)	0.8	0.15	2.82	17	14
Bendigo Mine Coarse Mullock (BMCM)	0.1	0.30	2.74	34	18

Based on the Australian specifications [6] it was decided that a concrete batch having a 50 mm slump would be suitable for the range of jobs in which the concrete would be used. Prior to a batch preparation, the aggregates were allowed to soak for 24 hours to bring them to the SSD condition. Care was taken to retain all fines when the excess water was drained. The quantities of materials per m^3 used for each mix considered are detailed in Table 4.

Table 4 Mix design proportions for the selected compressive strength

NOMINAL COMPRESSIVE STRENGTH / AGGREGATES	20 MPa				25 MPa			
	c (kg)	w (kg)	FA (kg)	CA (kg)	c (kg)	w (kg)	FA (kg)	CA (kg)
AWRS & ACCR	355	182	700	970	365	180	700	970
BMFM & BMCM	350	228	690	950	355	224	690	950
AWRS & BMCM	350	188	690	960	360	186	690	960

ANALYSIS OF RESULTS

Concrete Strength

The strength tests were performed in accordance to the Australian specifications [7, 8, 9] and the results are presented in Tables 5 and 6 for the two nominal compression strengths. It should be observed that concrete that was made using the washed river sand and the quarry coarse aggregate shows a very similar strength to the concrete made from washed river sand and coarse mullock aggregate. A close examination of the failure surface showed that specimens containing coarse mullock aggregate displayed a sightly higher amount of pullout aggregate when compared with the specimens prepared with quarry stones. This can be due to two factors a larger percentage of flaky grains present and a smoother texture exhibited by the mullock particle. Poor strength was obtained from specimens that used both fine and coarse mullock aggregate. It was observed that the pullout was the predominant failure mode exhibited by the concrete prepared entirely with mullock aggregates. This can be the result of larger percent of fines present in fine mullock in addition to flaky grains with smooth texture of coarser aggregates discussed above. Nevertheless, a higher w/c ratio required to bring the concrete mix (prepared with mullock aggregate) to the specified slump contributed to lower strength displayed by these specimens.

Table 5 Average properties at 28 days for the nominal 20 MPa mix

AGGREGATES	SLUMP (mm)	COMPRESSIVE STRENGTH (MPa)	INDIRECT TENSILE STRENGTH (MPa)	FLEXURAL STRENGTH (MPa)
Axedale Washed River Sand & Axedale Coarse Crushed Rock	55	37.5	3.2	5.2
Bendigo Mining Fine Mullock & Bendigo Mining Coarse Mullock	50	18.5	2.0	3.8
Axedale Washed River Sand & Bendigo Mining Coarse Mullock	50	37.5	3.3	5.7

Table 6 Average properties at 28 days for the nominal 25 MPa mix

AGGREGATES	SLUMP (mm)	COMPRESSIVE STRENGTH (MPa)	INDIRECT TENSILE STRENGTH (MPa)	FLEXURAL STRENGTH (MPa)
Axedale Washed River Sand & Axedale Coarse Crushed Rock	50	42.5	3.7	5.8
Bendigo Mining Fine Mullock & Bendigo Mining Coarse Mullock	45	23.5	2.1	4.0
Axedale Washed River Sand & Bendigo Mining Coarse Mullock	50	42.5	3.5	6.0

Elastic Properties of Concrete

Results obtained from the elasticity tests [10] are summarized in Table 7. The values of modulus of elasticity employing the empirical relationship in Australian Standards [11] are described below with E_{cj} as the mean value of modulus of elasticity of concrete at a certain age, ρ is density of concrete measured in accordance with the relevant specifications [12] and f_{cm} is the mean value of the compressive strength of concrete at the relevant age. The computed and the measured values of the modulus of elasticity compare quite well, with the former being slightly ($< 20\%$) lower.

$$E_{cj} = \rho^{1.5} \times \left(0.043 \times \sqrt{f_{cm}} \right) \qquad (1)$$

In addition, it should be noted that the trends observed from the strength tests were replicated by the elastic properties of the three concrete mixes. A higher strength is correlated with a higher modulus of elasticity and a lower Poisson's ratio. Specimens produced from concrete with mullock aggregates exhibited the lowest magnitude for the modulus of elasticity and the highest values of Poisson's ratio. Nevertheless, a larger amount of water required during preparation of this concrete mix is associated with the poor performance of this concrete.

The elastic properties of concrete made with coarse mullock aggregates and washed river sand were comparable with those of concrete produced with currently used aggregates, although a slight (~ 10 %) reduction of the modulus of elasticity and a small (< 21 %) increase in the Poisson's ratio values were monitored. Certainly, the slightly lower performance of concrete produced with BMCA and WRS aggregates is due solely to excess in flaky grains and smooth texture of mullock grains.

Table 7 Elastic properties of concrete at 28 days

NOMINAL COMPRESSIVE STRENGTH / AGGREGATES	20 MPa			25 MPa		
	Modulus of elasticity (MPa)		Poisson's ratio	Modulus of elasticity (MPa)		Poisson's ratio
	Measured	Computed		Measured	Computed	
Washed river sand & Coarse crushed rock	40056	33755	0.15	41861	35476	0.13
Mullock fines & Coarse mullock	23706	22018	0.25	24700	22833	0.24
Washed river sand & Coarse mullock	34751	31113	0.17	37268	32546	0.16

CONCLUSIONS

This study confirms that the concrete prepared with mining spoil aggregates exhibits a slightly different behaviour to the concrete prepared with washed rivers sand and crushed quarry rock. This difference is clearly related to the particle size distribution and the grains shape and texture. The excess of fines, flaky grains and smooth texture of the coarser aggregates caused the compressive strength to be below the nominal value for which the mix was designed. In addition, lower values were obtained for both the tensile and the flexural strength. The corresponding modulus of elasticity and Poisson's ratio showed similar poor performance for the concrete prepared only with mullock aggregates.

In contrast, when the finer mullock aggregates were replaced by washed river sand, the concrete strength was not affected. The compressive, flexural and tensile strength of WRS and BMCA concrete was similar to that obtained for concrete prepared with currently used aggregates (WRS and ACCR). However, a slight reduction of the modulus of elasticity and a small increase of the Poisson's ratio were recorded.

Overall, it can be concluded that the use of both fine and coarse aggregate obtained by crushing the mining spoil (mullock) it is not a good option for concrete production, because the concrete performed poorly in both the strength and elastic properties tests. However, when the mullock fines were replaced with washed river sand, the strength and elastic properties of concrete were satisfactory. Furthermore, the cost of the crushed mullock appears to be competitive with the supply of currently used aggregates. Nevertheless, further investigations on the behaviour of concrete produced with coarse mullock aggregates is required to establish the durability performance of this concrete.

The author gratefully acknowledges the financial support received from the La Trobe University Bendigo Faculty Research Grant scheme and support of Bendigo Mining NL. Also the help of Mr Tayne Evans and Mr Joel Miller in preparing the specimens and performing the tests is acknowledged.

REFERENCES

1. TAYNE, E, Geotechnical Investigation of the By-Products from Bendigo Mining NL, La Trobe University, Bendigo, 2002, 67pp

2. JACKSON, N AND DHIR, R K, Civil Engineering Materials, Macmillan, 1996, 534pp

3. NEVILLE, A M, Properties of Concrete, Longman, London, 1995, 779pp

4. GANI, M S J, Cement and Concrete, Chapman & Hall, London, 1997, 212pp

5. STANDARDS AUSTRALIA, Aggregates and rock for engineering purposes - Concrete aggregates, AS 2758.1-1998, Sydney, 1998, 12pp.

6. CEMENT AND CONCRETE ASSOCIATION OF AUSTRALIA, Guide to concrete construction, C&CAA T41, Sydney, 2002, 125pp.

7. STANDARDS AUSTRALIA, Methods of testing concrete - Determination of the compressive strength of concrete specimens, AS 1012.9-1999, Sydney, 1999, 9pp.

8. STANDARDS AUSTRALIA, Methods of testing concrete - Indirect tensile strength of concrete cylinders (Brazil /splitting test), AS 1012.10-2000, Sydney, 2000, 8pp.

9. STANDARDS AUSTRALIA, Methods of testing concrete - Determination of the modulus of rupture, AS 1012.11-2000, Sydney, 2000, 8pp.

10. STANDARDS AUSTRALIA, Methods of testing concrete - Determination of the static chord modulus of elasticity and Poisson's ratio, AS 1012.17-1997, Sydney, 1997, 16pp.

11. STANDARDS AUSTRALIA, Concrete structures, AS 3600-2001, Standards Australia, Sydney, 2001, 176pp.

12. STANDARDS AUSTRALIA, Methods of testing concrete - Determination of mass per unit volume of hardened concrete - Water displacement method, AS 1012.12.2-1998, Sydney, 1998, 4pp.

REFRACTORY PROPERTIES OF INSULATING MATERIALS FROM SECONDARY CEMENTITIOUS MATERIALS (SCMs)

A Jonker

Tshwane University of Technology

J H Potgieter

University of Witwatersrand

South Africa

ABSTRACT. Many industrial processes generate large amounts of wastes. Typical examples include the fertiliser industry (phosphogypsum), steel producers (slag) and the power generating industry (fly ash). Although certain of these wastes are currently used to a limited extend (e.g. fly ash in cement), there is always a need to find more uses and new applications for them. This investigation will describe work done to determine the refractory properties of potential insulating materials. These properties indicate that this refractory material could potentially be used for low temperature insulating applications (1050°C) e.g. lining of preheaters in cement plants.

Keywords: Insulating materials, Secondary cementitious materials, Refractory properties.

Mrs A Jonker, is a lecturer in the Department of Chemistry & Physics at the Tshwane University of Technology in Pretoria, South Africa.

Prof J H Potgieter, is a Professor at the School of Process & Materials Engineering at the University of Witwatersrand in South Africa.

INTRODUCTION

Typical examples of industrial waste include materials from the ferro-alloy and steel industries (slag or iron-rich waste) and the power generating industry (fly ash). These waste products are currently used in limited amounts. It is well known that coal is the main fuel of South Africa [1]. Coal is not only used to generate steam and electricity, but also to produce liquid fuels and chemicals. From an environmental viewpoint, such applications generate large quantities of ash [2] hich mostly end up as large dumps near the points of use.

This investigation was launched to find alternative applications for these materials. One such possible application is to manufacture insulating refractories. The materials used in this investigation were therefore employed in various combinations to try and achieve as many of the desirable properties of refractories as possible. Properties of importance for insulating refractories are the low density of the material and a low thermal conductivity of the product to result in good insulation properties.

Several waste products have been identified for possible incorporation with clays in novel ceramic bodies intended for an insulating refractory. These include fly ash, ball clay, bentonite, phosphogypsum and iron-rich waste.

Insulating Refractories

High-temperature processes require a considerable amount of energy. Often the energy consumption for high-temperature processes is used only partially for the actual technical process. An essential part of the energy escapes through the kiln walls into the atmosphere and is consequently lost for the process. In the case of kilns for ceramics, this loss escaping through walls can amount to 15 to 30% of the total energy consumption required for the sintering process [2]. To keep thermal energy inside the processing room of a thermal plant and prevent its escape into the atmosphere, high-temperature insulating materials are necessary.

Insulating bricks are made from a variety of oxides, most commonly fireclay (alumino silicate) or silica. The high porosity of the brick is created during manufacturing by adding a fine organic material to the mix, such as sawdust or polystyrene. During firing, the organic addition burns out, creating internal porosity. Another way to accomplish high porosity involves the addition of a foaming agent to a slip. Additions of lightweight aggregates like diatomite, is another approach. Because of their high porosity, insulating bricks inherently have lower thermal conductivity and lower heat capacity than other refractory materials [3]. Insulating refractories are used as working linings of furnaces where abrasion and wear by aggressive slag and molten metal are not a concern.

Insulating bricks have a maximum service limit of $1650^{\circ}C$ and are, for example, used in the crowns of glass furnaces and tunnel kilns. Insulating brick based on fireclay aggregate are also available with a combination of high strength and low thermal conductivity and these bricks offer a maximum service limit in the range of $1150^{\circ}C - 1261^{\circ}C$. The internal heat transportation and with it the heat insulation in high-temperature insulating materials are decisively influenced by the structural composition and the temperature. The effectiveness of the temperature influence is also controlled by the structure. Consequently, the structural composition plays a dominating part.

The shrinkage behaviour of an insulating material is important for evaluating its maximum possible temperature of application. For this reason the non-reversible length modification is measured over a long period of time at constant temperatures, the material being heated up on one or all sides in an oxidizing atmosphere without corrosive influences. The classification temperature and the limit of application temperatures (both expressions are used) correspond to the temperature which allows a maximum admissible amount of linear shrinkage. Most countries have established different shrinkage standards. For refractory lightweight bricks and concretes there are shrinkages of 1 to 2% and for refractory fibres 2 to 5%, sometimes even up to 7% [2].

The raw materials used in this study include fly ash, bentonite, ball clay, phosphor-gypsum and iron rich waste.

Fly ash

Fly ash (or pulverised fuel ash) is a material whose characteristics reflect its origin from incombustible mineral matter in powdered coal burned in large electrical power plant boilers. Shrinkage in a body can be lowered by fly ash addition [4]. The fly ash used has a chemical composition as indicated in Table 1.

Ball clay

The ball clay used was W.P. Ball clay (from Kraaifontein), which consists essentially of kaolinite with small amounts of mica and quartz and with a high organic content. Ball clays are used extensively in the whiteware ceramic industry as an ingredient in clay bodies composed largely of non-plastic materials, in order to impart to them the required plasticity and green and dry strengths [5,6].

Bentonite

Bentonite refers to clay of volcanic origin, and it consists mainly of montmorillonite. Bentonite is used in bodies, glazes, enamels, improving plasticity and increasing the thixotropy of the bodies [7]. The reactivity of bentonite depends on the crystal structure and the fine-grained particle size [7,8]. The working moisture of bentonite is high and so is the drying shrinkage, thus amounts of not more than five percent should be used.

Phosphogypsum

Phosphogypsum [9] is a by-product resulting from the phosphoric acid process for manufacturing fertilizers. It consists mainly of $CaSO_4.2H_2O$ and contains some impurities such as P_2O_3, F⁻, and organic substances. Only 15% of the phosphogypsum is utilized by the cement and gypsum industries as a setting moderator for cement and for making gypsum plaster. The remaining 85% of phosphogypsum is not used, causing an environmental problem and creating a need for large areas for disposal. Therefore, attempts have been made to use phosphogypsum in applications such as road and rail works fills, stabilization of base course, and construction.

The particle size factions of this gypsum lie between 0 and 200 µm. Gypsum has little strength and poor adhesive properties, but is added to assist in setting [10].

Iron-rich waste (Fe-rich waste)

There is little, if any, literature recorded on this material so far. Since this is an iron-rich material, it is used to act as a flux. A chemical analysis, given in Table 1, was performed on this material.

METHODOLOGY

Three bodies were formulated from the materials with the following compositions:

- 80% fly ash; 15% ball clay and 5% bentonite, named FBO
- 80% fly ash; 15% ball clay and 5% iron-rich waste named FBI
- 80% fly ash; 5% gypsum and 10% iron-rich waste and 5% over-burden bentonite named FGI.

Each mix was mixed with 30-40% water and two grades of polystyrene, 80% small beads and 60% large beads. The materials were weighed out in 1 kg batches and were hand mixed (to avoid the beads being squashed) with the water. The mixes were separately cast into steel moulds that had been greased (for easier removal of set samples). The moulds were vibrated by hand (it was found that when an electric shaker was used the beads moved upwards in the moulds and were not evenly distributed throughout the cube). The samples were allowed to set overnight (24hours) in the moulds, samples were then removed from the mould and left to air dry overnight (24 hours). The samples were then dried in a drier at 110°C for another 24 hours. The dried samples were fired in a laboratory electric furnace at 1100°C. The heating cycle used was:

- 80°C per hour to the required temperature with a soak of 1 hour
- Allow furnace cool to room temperature.

The following tests were conducted to determine the properties of the mixes employing the following methods:

- Chemical analysis in the form of X-ray fluorescence (XRF) to determine the elements, expressed as oxides, present in each fired sample of the FBI (large and small beads) and FBO (large and small beads) mixes.

- X-ray diffraction (XRD) as the method for mineralogical analysis to determine the minerals present in each fired samples of the FBI (large and small beads) and FBO (large and small beads) mixes.

- Ash fusion temperature test in oxidizing atmosphere to 1550°C on the FBI, FBO and FGI green mixes were done to determine the melting temperature of each mix according to ASTM D 1857 method [11].

- Apparent porosity's, of the FBI (large and small beads) and FBO (large and small beads) mixes were determined by the ISO 5016 [12] and ISO 5017 [13] method.

- Bulk densities of the FBI (large and small beads) and FBO (large and small beads) mixes were determined by the ISO 5016 [12] and ISO 5017 [13] method.

- Apparent specific gravities of the FBI (large and small beads) and FBO (large and small beads) mixes were determined by the ISO 5016 [12] and ISO 5017 [13] method.

- Strength (CCS) was determined by the ASTM C133-84 [14] method on green and fired samples.

RESULTS

Chemical and Mineralogical Composition

The chemical composition of the sample is presented in Table 1. The mineralogical analysis of the samples is presented in Tables 2a to 2d.

Table 1   Chemical composition

ELEMENTS EXPRESSED AS OXIDES	FBI SMALL, %	FBI LARGE, %	FBO SMALL, %	FBO LARGE, %
SiO_2	54.34	52.78	57.56	57.66
TiO_2	2.04	5.56	1.46	1.47
Al_2O_3	30.96	31.51	31.61	31.61
Fe_2O_3	6.81	3.63	3.34	3.38
MnO	0.06	0.09	0.03	0.03
MgO	1.01	0.87	1.03	1.05
CaO	3.86	3.25	3.88	3.92
Na_2O	0.34	0.25	0.24	0.24
K_2O	0.73	0.74	0.87	0.87
P_2O_5	0.41	0.34	0.43	0.43
Cr_2O_3	0.06	0.04	0.03	0.03
V_2O_5	0.05	0.06	0.03	0.03
ZrO_2	0.05	0.07	0.06	0.06
H_2O-	0.0	0.0	0.0	0.0
L.O.I	0.15	0.11	0.18	0.17
Total	100	100	100	100

Table 2a   Mineralogical analysis of FBI small

COLOUR	MINERAL	FORMULA	QUALITATIVE
Green	Albite	$NaAlSi_3O_8$	Major
Purple	Cristobalite	SiO_2	Major
Navy	Mullite	$Al_6Si_2O_{13}$	Major
Red	Quartz	SiO_2	Major
Olive	Perovskite	$CaTiO_3$	Minor
Orange	Rutile	TiO_2	Minor
Blue	Calcite magnesian	Mg.064 Ca .936)(CO_3)	Trace
Pink	Chalcopyrite	$CuFeS_2$	Trace
Yellow	Melanterite	$FeSO_4(H_2O)_7$	Trace

Table 2b   Mineralogical analysis of FBI large

COLOUR	MINERAL	FORMULA	QUALITATIVE
Navy	Mullite	$Al_6Si_2O_{13}$	Major
Red	Quartz	SiO_2	Major
Green	Albite	$NaAlSi_3O_8$	Major
Purple	Cristobalite	SiO_2	Major
Orange	Rutile	TiO_2	Minor
Olive	Perovskite	$CaTiO_3$	Minor
Yellow	Melanterite	$FeSO_4(H_2O)_7$	Trace
Blue	Calcite magnesian	$Mg.064\ Ca\ .936)(CO_3)$	Trace
Pink	Chalcopyrite	$CuFeS_2$	Trace

Table 2c   Mineralogical analysis of FBO small

COLOUR	MINERAL	FORMULA	QUALITATIVE
Navy	Mullite	$Al_6Si_2O_{13}$	Major
Red	Quartz	SiO_2	Major
Green	Albite	$NaAlSi_3O_8$	Major
Pink	Cristobalite	SiO_2	Major
Orange	Rutile	TiO_2	Minor
Yellow	Melanterite	$FeSO_4(H_2O)_7$	Trace
Olive	Perovskite	$CaTiO_3$	Trace

Table 2d   Mineralogical analysis of FBO large

COLOUR	MINERAL	FORMULA	QUALITATIVE
Navy	Mullite	$Al_6Si_2O_{13}$	Major
Red	Quartz	SiO_2	Major
Green	Albite	$NaAlSi_3O_8$	Major
Purple	Cristobalite	SiO_2	Major
Orange	Rutile	TiO_2	Minor
Yellow	Melanterite	$FeSO_4(H_2O)_7$	Trace
Olive	Perovskite	$CaTiO_3$	Trace

Ash Fusion Temperature Test

Table 3   Ash fusion temperature test (oxidizing atmosphere)

SAMPLE IDENTIFICATION	FBI	FBO	FGI
Initial deformation (°C)	1520	>1550	1371
Softening temperature (°C)	1540	>1550	1413
Hemisphere temperature (°C)	>1550	>1550	1439
Fluid temperature (°C)	>1550	>1550	1464

Apparent Porosity, Bulk Density and Apparent Specific Gravity

Table 4   Apparent porosity, bulk density and apparent specific gravity

SAMPLE NAME	APPARENT POROSITY, %	BULK DENSITY, g/cm^3	APPARENT SPECIFIC GRAVITY, %
FBI large	53.03	0.94	2.00
FBI small	60.52	0.83	2.11
FBO large	59.56	0.89	2.21
FBO small	53.67	0.87	1.88

Cold Crushing Strength

Table 5   Cold crushing strength

SAMPLE NAME	COLD CRUSHING STRENGTH, MPa
FBI large	7.89
FBI small	7.19
FBO large	3.62
FBO small	6.39

DISCUSSION

Chemical and Mineralogical Composition

From the chemical and mineralogical analysis it is clear that the materials are mainly alumino silicates which are indicated by the mullite and crystoballite in the mineralogical analysis. The iron content of the mixes containing iron rich waste is higher than the other mixes as is the titanium content. Both titanium and iron would reduce the melting point of the refractory material, rendering it less refractory in nature.

Ash Fusion Temperature Test

Comparing the initial temperature and the hemisphere temperature, it appears that all the mixes have short firing ranges, which indicates that great care must be taken, as deviations from the firing range may cause problems such as low creep resistance and low hot strengths. As suspected, the bodies with the iron-rich waste have lower fusion temperatures than the ones without the iron-rich waste. The lower fusion temperatures are attributed to the higher iron and titanium content of these bodies.

Apparent Porosity, Bulk Density And Apparent Specific Gravity

The bulk densities of the mixes are very similar. The apparent specific gravity of the mixes is also very similar because the amount of fly ash is equal. The porosity of the FBI mix indicates that the small beads increase the porosity. In the FBO mix the large beads give a larger porosity, which can be attributed to defects or cracks in the body.

Cold Crushing Strength

As expected from the AP, BD ASG results, along with the chemical and mineralogical results, the FBI mixture is the stronger material. More sintering takes place because of the higher iron content which lowers the melting temperature of the mixture and therefore increases the bonding, and thus the cold strength. The strength of the FBO large mixture is much lower than expected which may be due to the presence of cracks or defects in the body.

CONCLUSIONS

The FBI small beads give the acceptable results with regard to all the properties. The strength of this mixture is above 7 MPa, indicating a material strong enough to be handled. The porosity is above 60% making the material very porous which should result in a low thermal conductivity. With these properties in place it can be said that the FBI small beads mixture can be used as an insulating refractory material up to temperatures of 1400°C.

ACKNOWLEDGEMENTS

The authors would like to express their gratitude to the NRF for funding the project, Whitney Perrins who helped with the experimental work and the Tshwane University of Technology for the facilities and time with regard to the project.

REFERENCES

1. POHL M. Coal in Crisis Conference, Johanesburg, 1986, p. 6

2. SCHULLE W. and SCHLEGEL E. Ceramic monographs-handbook of ceramics. Germany, Verslag Schmid, 1991, pp. 1,2,4-6.

3. NYIKOS F. and KING E. Refractory materials II. Pretoria: Technikon Pretoria, 2000, pp. 2-4,6-8,10,16-23,25-27.

4. ANDERSON M. Fly ash brick process, Halfwayhouse, Coal Ash Association, 2001, pp. 7-8

5. JONKER A., MAREE D.B.G. and VAN DER MERWE M.J. Guidelines for Ceramic techniques, Pretoria, Technikon Pretoria, 1996.

6. WORRALL W.E. Clay and ceramic raw materials, London, Elsevier Applied Science, 1986, p. 56.

7. SALMANG. Physical & Chemical Fundamentals, London, Butterworth, 1961.

8. WILSON M.G.C. and ANHAEUSSER C.R. The mineralogical resources of South Africa, Pretoria, Council of Geosiences, 1998.

9. SMADI M., HADDAD R. and AKOUR A. Potential use of phospho-gypsum in concrete, Cement and concrete research, 1999, pp. 14-19.

10. S.N. Cement technology II, Pretoria, Technikon Pretoria, 2001, pp. 15-17,404,407.

11. ASTM D 1857. Fusibility of coke and ash, 1987.

12. ISO 5016. Dense shaped refractory products- determination of bulk density, apparent porosity and true porosity, 1988.

13. ISO 5017. Dense shaped refractory products- determination of bulk density, apparent porosity and true porosity, 1988.

14. ASTM C 133-84. Standard test method for Cold Crushing Strength and Modulus of Rupture of refractory brick and shapes, 1984.

MAXIMUM DOSAGE OF GLASS CULLET
AS FINE AGGREGATE IN MORTAR

K Yamada

S Ishiyama

Akita Prefecture University

Japan

ABSTRACT. ASR expansion in mortar specimens made with graded glass cullet originating from waste containers was examined employing both the mortar-bar test method (JIS A 1146) and the rapid test method (JIS A 1804). The pessimum size was 0.6mm regardless of the test method, except for green cullet. The ASR expansion increased according to the glass cullet content. However, expansion was not linearly proportional to glass content. At the same time, mixed size cullet caused larger expansion than that estimated from the expansion of each single size cullet multiplied by the dosage. Extrapolating the result, authors suggest that the maximum dosage where ASR expansion remains below 0.1 % is about 10 % even for mortar specimen made of mixed-size clear glass cullet.

Keywords: Glass cullet, Alkali-silica reaction, ASR, Alkali-aggregate reaction, AAR, Pessimum size, Sustainable construction, Concrete aggregate.

Dr K Yamada is a Certificated Architectural Engineer and professor at the Department of Architecture and Environmental Engineering at the Akita Prefecture University. He has been in charge of the Materials Laboratory that includes research and development related to fracture and ductility of fibre reinforced concrete, durability of concrete and other building materials, wooden structures, life-cycle assessment and sustainable construction. His research interests cover every aspect of the design, assessment and use of concrete and wood in an environmentally friendly way.

Mr S Ishiyama is a research associate at the Department of Architecture and Environmental Engineering at the Akita Prefecture University. He received his MS in engineering from the Tohoku Institute of Technology. His research interests include durability of fibre reinforced concrete, permeability of concrete and recycling of waste materials in concrete.

INTRODUCTION

Glass is commonly used for various containers, window panes, video displays and light bulbs. Although recycled glass cullet comprises the major raw material for glass, nearly one million tons of glass is not recycled within the limited scope of container glass alone. Though this remaining glass cullet is used for roadbed and aggregate for asphalt concrete at the moment, wider application preferably with functional advantages is required. One possible way of using glass cullet in that regard is for aggregate in concrete and mortar.

There is much literature revealing the problems and advantages of glass cullet for concrete aggregate [1, 2]. It is commonly accepted that because glass cullet has high strength, low water absorption and adequate hardness, it is suitable for concrete aggregate if alkali-silica reaction (ASR) is avoided.

There are several ways of avoiding ASR expansion for glass cullet without modifying glass chemistry or cement chemistry. One is the use of ground powder glass [3]. The other is the combined use of mineral admixtures such as metakaolin [4], fly ash [5], slag [6] and silica fume [7]. Along with the use of some admixtures, knowledge of the maximum level of glass cullet in concrete is a useful and easy way to control ASR expansion.

In this research, authors examined ASR expansion of mortar specimens with graded glass cullet adapting rapid test method (JIS A 1804) and mortar-bar test method (JIS A 1146) for the purpose of revealing the maximum level of glass cullet in concrete.

EXPERIMENTS

Table 1 shows the mix proportions for mortar bar test method (JIS A 1146), in which the glass content is 100 % and the aggregate-cement ratio is 2.25. Table 2 shows the mix proportions for rapid test method (JIS A 1804), which was conducted to obtain the pessimum size. Each NaOH solution was prepared to meet the requirement of total alkali in specimens, which is 1.2 % Na_2O equivalent for the mortar bar test or 2.5 % for the rapid test, making the amount of solution 100g. Mixed particle size specimens were also made for the rapid test method, observing the same mix proportions as the mortar bar test method. The number of specimens tested was 3 for each mixture.

Procedures from manufacturing specimens to measuring expansion are strictly stipulated in JIS standards, which the authors observed. Though measurement of weight increase is not required in the JIS methods, the authors measured the weight of specimens immediately before the length of specimens was measured. The principal test conditions of the JIS standards are listed in Table 3. There is a major difference between rapid tests (JIS A 1804 and ASTM C1260).

The existing test methods mentioned above are used for evaluation of whether the aggregate possesses possible ASR reactivity or not. There should be a test method to examine the quantitative expansion of the reactive aggregate in concrete, which may be different from them, because glass cullet has obvious ASR reactivity. Even so, the tendency to expand and the threshold value for detrimental expansion (0.1 %) can be verified by these methods, as for natural aggregate.

Table 1 Mix proportions of specimens evaluated by mortar bar test method

	GLASS CULLET (g)					SAND (g)	CEMENT (g)	NaOH (g)
	Size (mm)							
	2.36	1.18	0.60	0.30	0.15			
Mixed Size	45.0	112.5	112.5	112.5	67.5	0.0	200.0	100.0
Single Size	450 g for each single size					0.0	200.0	100.0

Table 2 Mix proportions of specimens evaluated by rapid test method

	GLASS CULLET (g)					SAND (g)	CEMENT (g)	NaOH (g)
	Size (mm)							
	2.36	1.18	0.60	0.30	0.15			
Single Size	from 0 to 400 g for each single size from 400 to 0 g Glass Cullet + Sand = 400 g						200.0	100.0

Table 3 Comparison of ASR test method including ASTM standard for reference

METHOD	DIMENSIONS	TEST ENVIRONMENT			MIX PROPORTION		
	Specimen (mm)	Cure after demoulding	Test condition	Alkali dosage	Aggregate / Cement	Water / Cement
Rapid test JIS A 1804	40×40 L160	20°C 24 hours	Water at 127°C 4 hours	2.50 % vs. cement	2	0.5
Mortar bar test JIS A 1146	40×40 L160	- -	Saturated humidity at 40°C 6 months	1.20 % vs. cement	2.25	0.5
Rapid test ΛSTM C 1260	25.4×25.4 L285.8	80°C 24 hours	Solution of NaOH (1 mol/l) at 80°C 14 days	none	2.25	0.47

ASR EXPANSION BY MORTAR BAR TEST METHOD

Figure 1 represents the expansion of specimens after 6 months of exposure, which shows large expansion (above 0.8 %) for clear cullet at the pessimum size (0.6mm). The tendency and the value are almost the same for brown cullet. But green cullet shows a different tendency displaying a different pessimum size (1.18mm) and indicating half the value of clear cullet. Jin and Meyer [3] indicated that green cullet has no ASR expansion with a 10 % dosage under the ASTM rapid test, but our result is for a 100 % dosage. This result shows that green cullet also shows ASR expansion in different ways, possibly due to minerals added for the colour.

The dotted line in Figure 1 represents the result from clear cullet by rapid test. Although the mix proportions are slightly different in respect to the aggregate-cement ratio, the tendency of expansion with respect to particle size can be adequately noted. The pessimum size (0.6mm) is the same for clear cullet despite the different test method. It is suggested that the rapid test has a less severe effect compared to the mortar bar test in terms of the value of expansion.

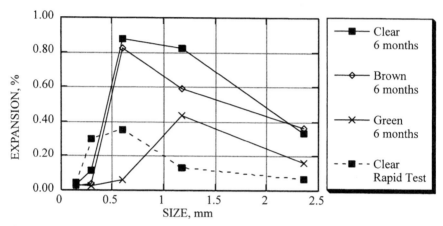

Figure 1 Expansion of specimens evaluated by mortar bar test with a reference by rapid

ASR EXPANSION BY RAPID TEST METHOD

Expansion of Clear Glass Cullet in Single Use

Figure 2 represents the result of the rapid test method, which indicates that the pessimum size is between 0.3mm and 0.6mm rather than changing from 0.3mm (low dosage) to 0.6mm (high dosage). From Figure 2, it can be observed that expansion increases in accordance with an increase in dosage. Though natural aggregate is known to have a maximum expansion at a certain combination of reactive and non-reactive aggregate, such phenomenon was not clearly found for glass cullet. Cullet of size 0.15mm produces no harmful expansion. Though the cullet of size 2.36mm apparently shows no harmful expansion in Figure 2, the result that indicates detrimental expansion in Figure 1 should also be considered.

Increase of Weight of Specimen Due to ASR

Figure 3 represents expansion against mass increase of specimens containing a 100 % dosage of clear cullet. The mass increase is mainly due to the weight of water imbibed ASR gel, although the weight increase due to the progress of hydration and spalling of the specimen (or squeezing of the gel) would affect the weight change to some extent. Though conceding that the mass increase was not totally reliable, it is noted that the mass increase by rapid test is smaller than that of mortar bar test, suggesting the reaction is less sufficient compared to that of the mortar bar test. As the expansion for the rapid test is about the half of the mortar bar test, so also is the mass increase.

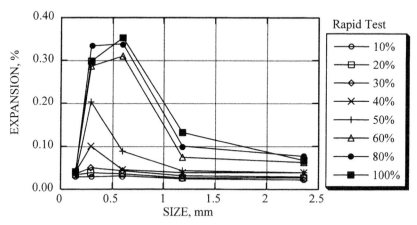

Figure 2 Pessimum size of clear glass cullet evaluated by rapid test method

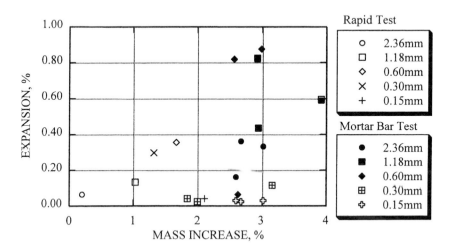

Figure 3 Relationship between expansion and mass increase of specimens that were made of clear glass cullet with the dosage of 100 % and evaluated by rapid test or mortar bar test

Bending Strength of Specimen after ASR

Figure 4 plots bending strength as a function of cullet particle size. The bending strength of the reference specimen made of only natural sand aggregate was 4.94MPa. Thus, the high dosage of glass cullet has the effect of decreasing bending strength. The remarkable phenomenon is that the bending strength decreases in accordance with the expansion of the specimen. This phenomenon is consistent with a theoretical prediction made by Bazant [8, 9].

Air Inclusion by Glass Cullet

Figure 5 shows the normalized weight of specimens immediately after demoulding and before exposure to an accelerated ASR testing environment. It shows the tendency for weight to decrease in accordance with the dosage of glass cullet. One reason for this is that the density of glass is 2.48 while the density of sand is 2.64 ,which makes the specimen light relative to the theoretical mass as depicted by the dotted line. Air inclusion is another reason for this, reflecting the well-known fact that glass cullet has an air entraining effect [1]. It is observed that this effect becomes stronger as the size of cullet becomes smaller.

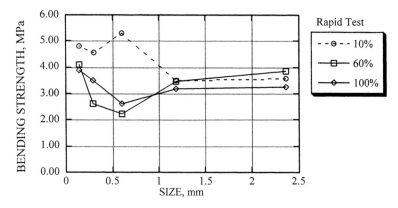

Figure 4 Pessimum size appeared in bending strength

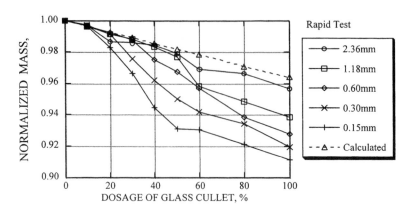

Figure 5 Normalized mass as a function of dosage of glass cullet

MAXIMUM DOSAGE TO ACHIEVE EXPANSION BELOW 0.1 %

Though Figure 2 indicates that expansion is related to dosage, the relationship between expansion and dosage is not simple. Table 4 represents some results from Figures 1 and 2 as well as other experimental ones in the case of pessimum size, which shows that the relationship between dosage and expansion is a nonlinear one. Little is known about numerical model for prediction of ASR expansion at this moment as Bazant [10] pointed out, empirical means of prediction may be the only way.

Expansion of a specimen of mixed size cullet can be estimated from the expansion of single size cullet multiplying the content ratio, if a simple mixture rule is applied. Table 5 composes the calculated expansion and weight increase of mixed size cullet with the experimental values. Although the calculated weight increase is close to the experimental one, the result of expansion is not consistent suggesting the existence of a pessimum size combination.

From Table 4, the expansion of 10 % dosage for pessimum size cullet can be extrapolated to be 0.061 % under the mortar bar test. At the same time, from Table 5, the magnification coefficient for mixed size is expected to be about 1.5 for clear cullet (pessimum colour). So, the predicted expansion may be 0.092 in the case of mixed size for pessimum colour. Though Figure 2 indicates that the maximum dosage where expansion becomes 0.1 % is under 30 % by rapid test, the authors suggest the maximum dosage is 10 %. Further, little influence to strength and air inclusion is observed in Figures 4 and 5 for a 10 % dosage.

Table 4   Nonlinearity of expansion with respect to dosage of glass cullet

TEST METHOD	COLOUR	SIZE	EXPANSION (%)			RATIO	
			Dosage			30 % vs. 50 %	50 % vs. 100 %
			30 %	50 %	100 %		
Rapid Test	Clear	0.6 mm	0.043	0.087	0.353	0.487	0.248
Mortar Bar Test	Clear	0.6 mm	-	0.307	0.875	-	0.351

Table 5 Nonlinearity of expansion with respect to mixture of glass cullet

TEST METHOD		MORTAR BAR TEST			RAPID TEST		
		Experimental (%)	Predicted (%)	Ratio	Experimental (%)	Predicted (%)	Ratio
Mass Increase	Clear	2.744	2.927	0.937	1.721	1.336	1.288
	Brown	2.274	2.783	0.817			
	Green	1.995	2.454	0.813			
Expansion	Clear	0.705	0.489	1.441	0.304	0.185	1.643
	Brown	0.514	0.402	1.279			
	Green	0.335	0.151	2.219			

CONCLUSIONS

ASR expansion in mortar specimens made of sieved glass cullet particles originating from waste bottle glass was examined employing both the mortar bar test method (JIS A 1146) and the rapid test method (JIS A 1804). The expansion rate under the mortar bar method for 6 months of exposure was substantially higher than that by using the rapid test method. This fact suggests the necessity of new methods for evaluation of ASR expansion due to glass cullet.

The pessimum size was 0.6mm regardless of test method except for green cullet which displayed a larger size (1.18mm). ASR expansion increased according to the dosage of glass cullet, which is different compared to ASR expansion of certain kinds of reactive natural aggregate in concrete. However, expansion was not linearly proportional to dosage. At the same time, mixed size cullet caused larger expansion than that estimated from the expansion of each single size cullet multiplied by the dosage, which suggests possible existence of a pessimum size combination.

The maximum dosage for every single particle size that induced little expansion was 30 %, if evaluated by the rapid test method. However, since expansion was larger using the mortar bar test method, the authors suggest that the maximum dosage is 10 % where expansion falls below 0.1 % even for mortar specimen made of mixed-size clear glass cullet and also bending strength does not substantially decrease.

REFERENCES

1. DHIR, R.K., LIMBACHIYA, M.C., DYER, T.D. eds.: Recycling and Reuse of Glass cullet, Thomas Telford, 2001, 3, 292 pp.

2. LIU, T.C., MEYER, C. eds.: Recycling Concrete and Other Materials for Sustainable Development, ACI, SP-219, 2004, 164 pp.

3. JIN, W., MEYER, C., BAXTER, S., Glascrete-Concrete with Glass Aggregate, ACI Materials Journal, March-April, 2000, pp 208-213.

4. SABIR, W.B., Metakaolin and Calcined Clays as Pozzolans for Concrete - A Review-, Cement & Concrete Composites, No. 23, 2001, pp 441-454.

5. ACI COMMITTEE 232, Use of Fly Ash in Concrete, Technical Report 232.2R-96, ACI, 1996.

6. ACI COMMITTEE 233, Ground Granulated Blast-Furnace Slag as a Cementitious Constituent in Concrete, Technical Report 233R-95, ACI, 1995.

7. ACI COMMITTEE 234, Guide for the Use of Silica Fume in Concrete, Technical Report 234R-96, ACI, 1996.

8. BAZANT, Z.P., ZI, G., MEYER, C., Fracture Mechanics of ASR in Concretes with Waste Glass Particles of Different Sizes, Journal of Engineering Mechanics, Vol. 126, No. 3, March, 2000, pp 226-232.

9. BAZANT, Z.P., JIN, W., MEYER, C., Microfracturing Caused by Alkali-Silica Reaction of Waste Glass in Concrete, Proceedings of FRAMCOS-3, AEDIFICATIO, 1998, pp 1687-1693.

10. BAZANT, Z.P., STEFFENS, A., Mathematical Model for Kinetics of Alkali-Silica Reaction in Concrete, Cement and Concrete Research, No. 30, 2000, pp 419-428.

DURABLITY PERFORMANCE OF RECYCLED AGGREGATE CONCRETE

A R M Ridzuan   A Ibrahim
A M M Ismail
Universiti Teknologi Mara
A B M Diah
Universiti Sains Malaysia
Malaysia

ABSTRACT. The effects of using crushed waste concrete as course aggregates upon compressive strength and durability against to carbonation and sulfate attack were investigated. Waste concrete cubes, which had been tested for compressive strength in compliance with construction specifications, were crushed and utilized as coarse recycled concrete aggregates (RCA) in new concrete. It is important to mention that, in order to simulate the real life conditions, waste concrete with very minimal information about its origin was used in its natural moisture condition. Tests on the aggregates showed that the RCA had lower specific gravity and bulk density, but a higher water absorption capacity than the natural aggregates. The resistance to mechanical actions such as impact and crushing for RCA is also lower. Concrete mixes with water cement ratios ranging from 0.47 to 0.70 were prepared using this RCA as coarse aggregate and tested. From the strength point of view the RCA concrete compared well with natural aggregate concrete. Therefore it could be considered for various potential applications. With respect to resistances to carbonation and sulfate attack the RCA concrete showed comparable performance.

Keywords: Recycled aggregate concrete, Durability, Carbonation, Sulfate attack, Strength.

Dr Ahmad Ruslan Mohd. Ridzuan, is a Senior Lecturer, Faculty Civil Engineering, Universiti Teknologi MARA, Malaysia. He present interests include recycle aggregate concrete, utilization of waste in concrete and self-compacting concrete. He published papers in this area.

Dr Azmi Ibrahim, is an Associate Professor of Faculty of Civil Engineering, Universiti Teknologi MARA, Malaysia. His research interests include waste utilization of reinforced concrete and structural.

Abdul Manaff Mohd Ismail is a Lecturer at Faculty of Civil Engineering, Universiti Teknologi MARA, Malaysia. Interest in concrete materials and structures.

Dr Abu Bakar Mohamad Diah, Associate Professor at Science University of Malaysia. His research interest includes Blended Cement Concrete and Durability of Concrete.

INTRODUCTION

Concrete, being the most widely used construction material, has seen its consumption rise with the growth in population and urbanization. However, along with an increased consumption, there is also an increased generation of wastes. These waste concrete are derived from a number of sources. The most common are from demolished concrete structures, fresh batch leftovers, rejected precast elements and waste concrete cubes that have been tested for compliance with building specifications. This debris is usually thrown away as landfill or dumped, causing environmental pollution [1]. Furthermore with the depletion of quality aggregates in the future and greater awareness of environmental protection, the need to use recycled aggregates more effectively has never been greater. This has led to the use of recycled aggregate in new concrete production, which is deemed to be a more effective utilization of concrete waste. However, the information on concrete using recycled concrete aggregate is still insufficient in this region and contradictory results have sometimes been obtained [2-12] but in general they are in agreement regarding the properties of the recycled aggregate. Hence, it is advisable to obtain more detailed information about the characteristics of concrete using recycled concrete aggregate (RCA).

In Malaysia, very little is known about the use of recycled aggregate in the manufacture of new concrete. Therefore, the main thrust of this investigation was to evaluate recycled aggregate concrete from waste concrete cubes (with granite as coarse aggregate) that have been exposed to local climatic conditions for use as coarse aggregate in new concrete, as such material is likely to provide both environmental and economic advantages. In this study, an investigation has been carried out to quantify the properties of concrete made by fully and partially replacing natural coarse aggregate with RCA. The properties investigated were aggregate characteristics, workability, compressive strength development and resistance to carbonation and sulfate attack.

MATERIALS AND TECHNIQUES

Materials

The materials used in this study were ordinary Portland cement in compliance with MS 522 [13], river sand and coarse natural aggregate, which were of granite and comply with BS 882 [15]. The waste concrete used in this study was concrete cubes brought from neighboring construction sites, which sent concrete cubes for compressive strength testing at 7 and 28 days in compliance with construction specifications. The cubes were crushed at an existing plant for the production of crushed rock aggregate comprising of a jaw crusher to produce recycled concrete aggregate of nominal maximum size of 20mm. The crushed products were then sieved into single size fractions of 20 mm and 10 mm. The 20 mm and 10 mm coarse aggregate were then combined in the ratio of 2:1 respectively for both the natural aggregate and RCA concrete mixes.

Physical Properties of Aggregates

The grading of both the natural aggregates and RCA are shown in Figure.1 and 2 in relation to the upper and lower limits set by BS 812 [14] and BS 882 [15]. The physical properties of all the aggregates in terms of specific gravity, bulk density, aggregate impact and crushing values, and water absorption are presented in Table 1, determined in accordance with BS 1881:part 3 [16].

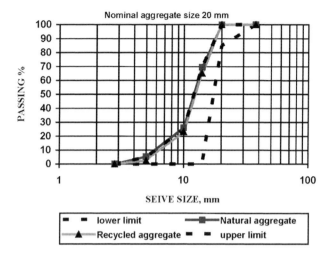

Figure 1 Gradation limits (BS 882: 1992) nominal size 20 mm of
Recycled Aggregate and Natural Aggregate

The gradation of the aggregates shows that it is possible to obtain RCA within the limit of the
available standard without much difficulty. From the results obtained it seems that the RCA
material had lower resistance to impact and crushing load and to have higher water
absorption. This is mainly due to the low density and more porous old mortar attached to the
recycled aggregates, which also contributed to the lower specific gravity and bulk density.

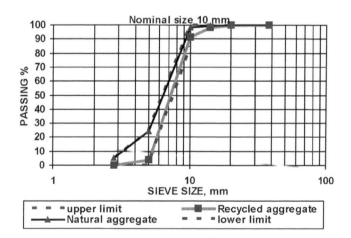

Figure 2 Gradation limits (BS 882: 1992) nominal size 10 mm of
Recycled Aggregate and Natural aggregate

Table 1 Physical Properties of Aggregate

PROPERTIES	AGGREGATE_TYPE		
	Natural (NA)	Recycled (RA)	Ratio (Na/Ra)
Specific gravity			
(i) SSD	2.55	2.31	1.10
(ii) Oven dried	2.52	2.23	1.13
(iii)Relative	2.61	2.41	1.08
Water absorption (%)	1.35	3.30	0.41
Bulk density (kg/m^3)	1390	1255	1.11
Aggregate impact value (%)	16.5	31.4	0.53
Aggregate crushing value (%)	16.0	31.0	0.52

Mix Design of Recycled Aggregate Concrete (RAC) and Natural Aggregate Concrete (NAC)

The designed water-cement ratios of the mixes range from 0.47 to 0.70. Since there is no existing standard method of designing concrete mixes incorporating recycled aggregates the recycled aggregate concrete mixes were derived simply by replacing the natural coarse aggregates proportion in the natural aggregate concrete mix design developed using a conventional mix design method [17], with recycled coarse aggregate (air dried). The mix proportions are shown in Table 2.

Compressive Strength Test

For each concrete mix, 18 100mm cubes were made in standard steel moulds. After 24 hours, the cubes were removed from the mould and placed in water at 23°C until testing. The Cube specimens were tested for compressive strength in accordance with BS 1881:Part 116 [18]. Reported observations are average of three measurements of compressive strength at each age.

Measurements of Depth of Carbonation

The sample for carbonation investigation were left in the laboratory ambient environment after curing under the respective curing environment. The depth of carbonation was determined by spraying the surface of the broken concrete, which has been split perpendicular to the exposed faces with a solution of phenolthalein.

Sulfate Attack Test

The specimens for the sulfate test were prepared and cast in cylindrical moulds as shown in Fig.5. Specimens were demoulded after 24 hours and air cured in laboratory ambient conditions at 29°C with high humidity for 7 days before immersion in sulfate solutions. Test solutions of 0.3 molar, sodium sulfate (5%) and magnesium sulfate (5%) were prepared and stored in plastic containers and covered with polythene sheet to minimize evaporation. The volume of the solution and its level were adjusted so that the specimens were always immersed in the solutions. To maintain the pH in the solutions within the limit, pH value were monitored every three days and adjusted by titration with sulphuric acid. Physical deterioration due to sulfate attack on the hardened concrete cylindrical specimens were evaluated in terms of dimensional and weight changes every 7 days. At the scheduled intervals, the specimens were retrieved, air-dried for one day in the laboratory environment and measurements taken.

Table 2 Mixes proportion for RAC and NAC

WATER CEMENT RATIO	MIXES	CEMENT kg/m^3	WATER kg/m^3	COARSE AGGREGATE, kg/m^3				SAND kg/m^3
				20 mm		10mm		
				RA	NA	RA	NA	
0.70	NAC M1	295	205	-	516	-	260	1115
	RAC M1 – 50	295	205	233	258	116	129	1115
	RAC M1 – 75	295	205	348	129	175	65	1115
	RAC M1 – 100	295	205	465	-	233	-	1115
0.64	NAC M2	320	205	-	537	-	268	1065
	RAC M2 – 50	320	205	242	268	121	134	1065
	RAC M2 – 75	320	205	363	133	182	67	1065
	RAC M2 – 100	320	205	484	-	242	-	1065
0.58	NAC M3	355	205	-	550	-	275	1010
	RAC M3 – 50	355	205	248	275	124	137	1010
	RAC M3 – 75	355	205	372	137	186	69	1010
	RAC M3 – 100	355	205	496	-	248	-	1010
0.55	NAC M4	375	205	-	570	-	285	960
	RAC M4 – 50	375	205	257	285	129	143	960
	RAC M4 – 75	375	205	385	143	193	72	960
	RAC M4 – 100	375	205	514	-	257	-	960
0.52	NAC M5	395	205	-	585	-	292	915
	RAC M5 – 50	395	205	263	293	131	146	915
	RAC M5 – 75	395	205	395	146	198	73	915
	RAC M5 – 100	395	205	526	-	263	-	915
0.47	NAC M6	435	205	-	595	-	295	860
	RAC M6 – 50	435	205	268	298	134	149	860
	RAC M6 – 75	435	205	401	149	201	74	860
	RAC M6 – 100	435	205	535	-	267	-	860

N.B. 50, 75, 100 denotes percentage replacement of coarse natural aggregate with coarse recycled aggregate. M1, M2, M3, M4, M5 and M6 are series.

RESULTS AND DISCUSSIONS

Fresh Concrete

The fresh properties of the RCA concrete mixes were determined by measuring the workability (slump) and (compacting factor). In general, the results showed a reduction in slump and compacting factor value with increasing RCA content in the mix, but this remained essentially within specified tolerances (± 25 mm).

Strength Development

Figure 3 shows the relationship between the 28 day compressive strength and water cement ratio for the mixes. Generally the trend is similar for the NAC and RAC mixes, that is the higher the water cement ratio the lower the compressive strength. The relationship also exhibits a good correlation. From the results it can be seen that all the mixes managed to attain the 28 day target strength. For the same design strength the RAC mixes show higher compressive strength as compared to the corresponding NAC mixes. At 28 days the compressive strength of the RAC mixes was 3 – 16 % higher than the corresponding NAC mixes. The higher compressive strength of the RAC may be attributed to high water absorption of recycled concrete aggregates which reduce the effective water-cement ratio of the RAC mixes, thus resulting in higher compressive strength than the corresponding NAC mixes. Also the shape of the recycled aggregate was more angular with a rough texture, which provided better bonding to the concrete matrix compared to natural aggregates. However the difference in compressive strength between the RAC and NAC become smaller as the water-cement ratio is reduced. This may be due to the fact that at higher design strengths as the recycled aggregate fails earlier than the cement paste due to the limited strength of the RCA which has less resistance to crushing loads.

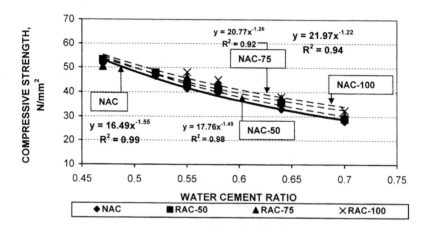

Figure 3 Relationship between 28 days compressive strength
and water cement ratio of RAC and NAC

Durability Performance of RAC and NAC with Respect to Carbonation

The results of the depth of carbonation measurements after 1.5 year exposure to laboratory ambient conditions for the RAC and NAC mixes are shown in Figure 4. The depth of carbonation of the recycled aggregate concrete mixes RAC-100, RAC-75 and RAC-50 are comparable to the corresponding natural aggregate concrete NAC mix with only very small differences between them being observed. The results also followed expected behaviour with improvements in performance with a decrease water-cement ratio.

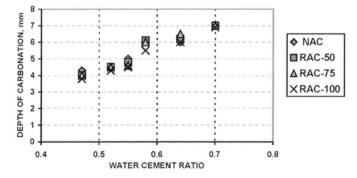

Figure 4 Relationship between depth of carbonation after 1.5 years exposed in laboratory ambient condition and water cement ratio of RAC and NAC.

Influence of RCA Content on Carbonation

Figure 5 illustrates the effect of percentage replacement of coarse recycled aggregate on the carbonation of RAC. The greater the replacement level the greater the tendency for the depth of carbonation to reduce. The reason for this is that the higher water absorption of the recycled aggregate reduces the effective water-cement ratio, hence improving its strength and hence the quality of the cement gel of the resulting concrete. Hence it can be deduced that the utilization of RCA does not have a detrimental effect on the durability of the RAC concrete with respect to resistance to carbonation.

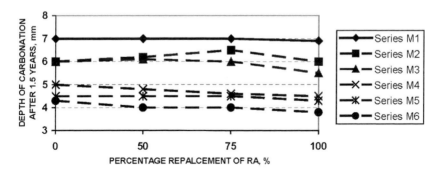

Figure 5   Effect of recycled aggregate content on the depth of carbonation of RAC.

Relationship between Compressive Strength and Durability to Carbonation

Figure 6 illustrates the relationship between the depth of carbonation and compressive strength at 28 days for the RAC and NAC. From regression analysis the depth of carbonation and compressive strength values exhibit good correlation and followed a logarithmic function. The relationship also followed expected behavior with improvement of resistance to carbonation with increasing compressive strength.

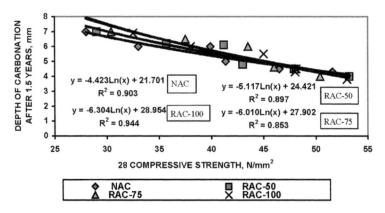

Figure 6   Relationship of depth of carbonation after 1.5 years in laboratory ambient condition and 28 days compressive strength of RAC and NAC

Resistance To Sulfate Attack

The results of the sulfate attack measurements are shown in Table 3. The results also show that the RAC mixes possess durability characteristics similar to those of the corresponding NAC mixes.

Table 3   Comparison of resistance to sulfate attack of RAC and NAC after 32 weeks exposure

MIXES	WATER CEMENT RATIO					
	0.47	0.52	0.55	0.58	0.64	0.70
	INCREASE IN WEIGHT, %					
	Sodium Sulfate (5%)					
NAC	1.46	1.39	2.45	3.50	3.83	3.09
RAC – 50	1.62	1.43	2.37	3.45	3.88	3.11
RAC – 75	1.41	1.43	1.24	1.18	4.31	2.93
RAC –100	1.67	1.60	1.38	3.61	4.44	3.34
	Magnesium Sulfate (5%)					
NAC	1.43	1.86	2.15	1.89	2.05	1.14
RAC – 50	1.54	1.29	2.06	1.86	1.69	1.18
RAC – 75	1.21	1.31	1.15	1.93	2.18	1.21
RAC –100	1.55	1.28	1.05	2.06	2.02	1.22
	INCREASE IN DIAMETER, %					
	Sodium Sulfate (5%)					
NAC	0.24	0.58	0.03	1.88	1.48	0.95
RAC – 50	0.19	0.05	0.07	1.22	2.62	0.96
RAC – 75	0.33	0.25	0.55	1.28	0.87	0.60
RAC –100	0.13	0.60	0.63	1.19	1.08	0.99
	Magnesium Sulfate (5%)					
NAC	0.23	0.06	0.13	0.20	1.45	0.42
RAC – 50	0.06	0.05	0.11	0.15	1.57	0.58
RAC – 75	0.10	0.18	0.08	0.12	1.59	0.42
RAC –100	0.18	0.09	0.13	0.37	1.44	0.46

CONCLUSIONS

The study shows that quality recycled aggregate can be produced using a similar crushing process to that used for the production of crushed rock aggregates. Tests also showed that it is possible to obtained RCA that complied to the standard. It must be mentioned here that the crushed RCA was kept in a sealed container and used when required. No adjustment of mix water content was made to account for the higher water absorption of the RCA. From a strength point of view, the RAC compared well with the corresponding NAC and could be considered for various potential applications. The durability performance of the RAC and NAC were comparable providing the mixes were designed to have equal water-cement ratios and design strength.

The benefits resulting from the study are not purely economic and environmental, but also provide the construction industry with technical information on a valuable resource that has a key role in meeting the challenges of sustainable construction.

REFERENCES

1. BUCK, A D. Recycled Concrete as Source of Aggregate, American Concrete Institute Journal, 212, 1977.

2. FRONDISTOU, Y. Waste Concrete as a n aggregate for new concrete, Journal of the American Concrete Institute, Proceedings, 74: 8, pp 373, 1977.

3. DIAH, A B M. Utilisation of Waste Concrete As Aggregate for Environmental Protection, Malaysian Science & Technology Congress '98 Symposium D: Biodiversity & Environmental Sciences, 1998.

4. HANSEN, T C AND NARUD, H. Strength of Recycled Concrete made from Crushed Concrete Coarse Aggregate Concrete International – design and Construction, 5: 1 pp.79-83, 1983.

5. RAVINDARAJAH, R S AND TAM, C T. Properties of concrete with Crushed Concrete as Coarse Aggregate. Magazine of Concrete Research. 37 130, 1985.

6. RAVINDARAJAH, R S LOO, Y H AND TAM, C.T. Recycled Concrete as Fine and Coarse Aggregates. Magazine of Concrete Research 39, 141, 214, 1987.

7. HANSEN, T C. Recycled Aggregates and Recycled Aggregate Concrete, Third State – of – the Art Report 1949 – 1989, RILEM Report 6, Recycling of Demolished Concrete and Masonry; Spoon, 1992.

8. RIDZUAN A R M., DIAH A B M. Utilization Of Recycled Aggregate In New Concrete – The Malaysian Technologist, Vol. 1 / 2, 2000.

9. RIDZUAN A R M., DIAH A B M.The Influence Of Recycled Aggregate On The Early Compressive Strength And Drying Shrinkage Of Concrete – Proceeding Of Structural Engineering Mechanics And Computation SEMC 2001, Elservier, South Africa, 2001

10. RIDZUAN A R M, DIAH A B M AND JELANI M A. – Influence of Recycled Aggregate on the Performance and Durability of OPC Concrete, Proceeding 2nd World Engineering Congress 2002. Kuching Sarawak Malaysia.pp 109-113, 2002.

11. RIDZUAN, A R M., DIAH, A B M, KAMARUZAMAN, K B AND HAMIR, R. The Influence of Recycled Aggregate On The Early Compressive Strength Of OPC Concrete, Journal of Physical Science, Universiti Sains Malaysia,Vol. 11, pp 57-65, 2000.

12. YODA, K, YOSHIKANE, T, NAKASHIMA, Y, AND SOSHIRODA, T. Proceedings of the Second International RILEM Symposium on Demolition and Reuse of Concrete and Masonry, (Chapman and Hall, London, 1985). P 527

13. MS 522: PART 1, Specification for Portland cement, SIRIM Malaysia. 1989.

14. BRITISH STANDARD INSTITUTION BS 812: PART 3, Method for Determination of Particle Size Distribution, BSI London, 1985.

15. BRITISH STANDARD INSTITUTION BS 882, Specification for Aggregates from Natural Sources for Concrete, BSI London, 1992.

16. BRITISH STANDARD INSTITUTION BS 1881: PART 3, Mechanical Properties of Aggregates, BSI London, 1975.

17. TEYCHENNE, D.C., FRANKLIN, R.E AND ERNTROY, H.C. Design of Normal Concrete Mixes, 1988.

18. BRITISH STANDARD INSTITUTION BS 1881: PART 116, Method of Determination of Compressive Strength of Concrete Cubes, BSI London, 1983.

PERFORMANCE OF CONCRETE CONTAINING FLY ASH AT EARLY AGES

J Wang
P Yan
Tsinghua University
China

ABSTRACT. The adiabatic temperature increase of concrete containing various percentage of fly ash was measured to understand heat emission during hydration. A temperature-matching curing (TMC) schedule in accordance with adiabatic temperature increase is adopted to simulate the situation in real massive concrete. The performance of concrete cured both in TMC and standard conditions were investigated. The possibility of prediction of strength in real structures using an equivalent age approach is discussed. The hydration of high volume fly ash concrete is delayed slightly. Its final degree of hydration is high enough under the condition of low water-binder (W/B) ratio to make the adiabatic temperature increase higher than the concrete containing a lower percentage percent of fly ash. Elevated temperatures enhance the hydration of binder to benefit the gain of compressive and flexural strength of the concrete containing complex binder.

Keywords: Temperature-matching curing, Equivalent age, Strength, Fly ash.

Wang Jiachun, Doctoral student in Department of Civil Engineering, Tsinghua University, Beijing, China. His research interest is cracking performance of concrete materials in early age.

Dr Yan Peiyu, Professor in Department of Civil Engineering, Tsinghua University, Beijing, China. His research interests include durability of concrete and hydration of cementitious materials.

INTRODUCTION

Much ready-mix concrete contains mineral admixtures, such as fly ash or ground granulated blast-furnace slag, at a range of quantities. Mineral admixtures improve the workability and durability of concrete [1]. A hindrance baffling the wide use of mineral admixtures is that the strength developing rate of concrete with high volumes of mineral admixture is slower than that containing only Portland cement. It is found that the performance of concrete in real structures is quite different from concrete prepared in laboratory and cured under standard conditions. The heat of hydration of Portland cement makes the temperature in the core of concrete structure rise gradually, which should enhance greatly the hydration of composite binders containing high volumes of mineral admixture. Therefore, the strength of concrete containing high volumes of mineral admixture increases faster in real structures than in standard condition. A maturity method is used to predict the strength development of concrete in real structures.

A temperature-matching curing (TMC) schedule in accordance with the adiabatic temperature increase of concrete was adopted to simulate the situation in real massive concrete. The performance of concrete containing different percentages of fly ash and cured both in TMC and standard conditions was investigated. Additionally, the concept of equivalent age is discussed.

EXPERIMENTAL

Materials

PO 42.5 ordinary Portland cement complying with the Chinese National Standard GB 175-1999 and a kind of fly ash qualified as first class according to the Chinese National Standard GB 1596-91 were used. Their chemical composition and physical properties are given in Table 1. A polycarboxylic superplasticizer was used to prepare concrete.

Table 1 Physical properties and chemical compositions of cement and fly ash

PHYSICAL PROPERTIES	CEMENT	FLY ASH
Density, g/cm^3	3.15	2.20
Fineness (remaining on 45μm sieve), %	-	9.0
Specific surface, Blaine, m^2/kg	350	290
Median particle size, μm	17.0	35.4
Water requirement, %	-	95

CHEMICAL COMPOSITION, %

	SiO_2	Al_2O_3	Fe_2O_3	CaO	MgO	Na_2O	K_2O	Na_2O_{eq}	SO_3	LOI
Cement	22.80	4.55	2.82	65.34	2.74	-	-	0.55	2.92	3.97
Fly ash	57.6	21.9	7.7	3.87	1.68	2.51	1.54	-	-	2.9

Crushed limestone with a size range of 5~20 mm and natural river sand with a fineness modulus of 3.0 were used as coarse and fine aggregate.

Proportions and Properties Measuring of Concrete

The proportions of the concrete mixes are summarized in Table 2. All mixes were prepared in a 30 l capacity rotary pan mixer. The dry materials were premixed for 1 min. Then water was added and the mixing was continued for an additional 3 min. Superplasticizer was used to enhance the workability of mix C3.

Table 2 Proportions of the concrete mixtures

MIX NO.	W/(C+FA)	WATER	CEMENT	FLY ASH	FINE AGGREGATE	COARSE AGGREGATE	SP	SLUMP
		kg/m^3	kg/m^3	kg/m^3	kg/m^3	kg/m^3	kg/m^3	mm
C1	0.53	195	368	-	681	1111	-	82
C2	0.50	184	294	74	682	1111	-	75
C3	0.26	96	147	221	791	1187	4.78	220

The adiabatic temperature increase of fresh concrete was determined using a computer controlled measuring system. The peripheral temperature of a concrete specimen was controlled at 0.1°C lower than the central temperature. Data were recorded once every 5min until the temperature ceased to change.

Cubic specimens of dimensions 100×100×100 mm for compressive strength testing, prismatic specimens of dimensions 100×100×400 mm for flexural strength testing were prepared. The specimens were covered with plastic film and water-saturated burlap at 20±2°C for 12 h before demoulding. After demoulding, specimens were kept over water in a temperature-matching container. The temperature in the container rose from ambient temperature according to the adiabatic temperature profile of each concrete mixture. The temperature in the container was not controlled and fell to ambient temperature in 3 h after the curing time of 60 h for the C1 mixture, 112 h for the C2 mixture and 140 h for the C3 mixture. Then all specimens were cured in standard condition. At predetermined intervals of 1, 2, 3, 4, 5, 7 and 28 days specimens were removed from their container and cooled to ambient temperature. Then their strength was measured.

RESULTS AND DISCUSSION

Adiabatic temperature rising of fresh concrete containing different percentage of fly ash

Adiabatic temperature curves of fresh concrete containing different percentages of fly ash are shown in Figure 1. Concrete C1 whose binder is composed of only pure Portland cement shows the highest temperature increase and largest temperature increase rate among the three concrete mixes. Concrete C2 whose binder is composed of 20% fly ash shows a similar temperature increase rate in the initial hydration period. Its temperature increase rate declines gradually with hydration time. Its temperature increase is lowest amongst the three concrete mixes. The temperature increase of concrete C3, whose binder is composed of 60% fly ash, is delayed slightly due its small amount of cement. Its final temperature rising is higher than that of concrete C2. It suggests that the degree of hydrating of Portland cement in concrete C3 under the conditions of low W/B ratio and much dilution by fly ash is much higher than that in concrete C1 and C2.

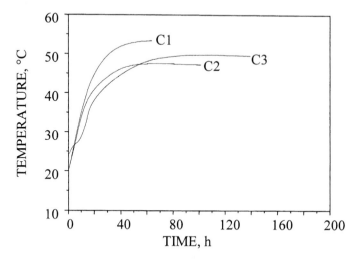

Figure 1   Adiabatic temperature curves of concretes

Mechanical properties of concrete cured under various conditions

The compressive strengths of concretes cured under various conditions are shown in Figure 2. Concrete cured under TMC achieve higher strengths than that cured in standard conditions at the same ages, except concrete C1 containing Portland cement. The strength of concrete C1 increases little in later age when it is cured under TMC. It achieves the lowest 28 days strength. Concrete C2 shows a similar strength development rate with C1 in early age but a higher rate than C1 at later ages. High volume fly ash concrete C3 with a low W/B ratio achieves high strength at a high rate both in early and later ages.

The elevated temperature enhances the hydration of the binder. Coarse hydration products are likely to form and a non-compact paste structure is yielded when hydration goes too fast. As a result, the strength development rate of concrete C1 declines. The pozzolanic reaction of fly ash is accelerated by elevated temperatures.

It consumes the coarse $Ca(OH)_2$ crystals formed during the hydration of Portland cement and forms a fine dense gelatinous hydration product, which strengthens the paste structure greatly. Therefore, concrete containing fly ash shows much better performance in a real structure than when cured under standard conditions.

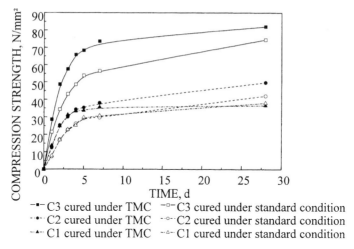

Figure 2   Compressive strength of concrete cured under the various conditions

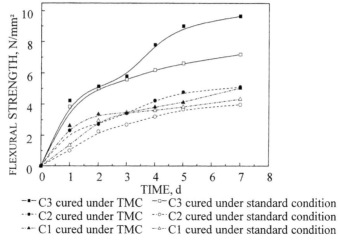

Figure 3   Flexural strength of concrete cured under the various conditions

The development of flexural strength of concrete cured under various conditions is similar to their compressive strength (Figure 3). W/B ratio is still the most important factor in determining the strength of concrete. Concrete C3 achieved much higher flexural strength than the other two concretes, although it contains less Portland cement and a high volume of fly ash. The TMC process enhances the gain of strength. The enhancing effect becomes apparent only after some time has elapsed. Concrete C3, cured both under TMC and standard conditions, achieves nearly same flexural strength in the first 3 days. It displays a large increase of strength after 4 curing days for the sample cured under TMC conditions.

Equivalent age

The concept of equivalent age is an alternative to the temperature-time factor to account for the combined effects of temperature and time on strength development [2]. Equivalent age represents the age at a reference curing temperature that would result in the same fraction of the limiting strength that would occur from curing at other temperatures. Freiesleben Hansen and Pedersen (FHP) [3] developed the equivalent age function, shown in Eq (1):

$$t_e(T_r) = \sum_0^t \exp(\frac{E}{R}(\frac{1}{273+T_r} - \frac{1}{273+T}))\Delta t \qquad (1)$$

where $t_e(T_r)$: equivalent age at reference curing temperature, h;
Δt: chronological time interval, h;
T: average concrete temperature during time interval Δt , °C;
T_r: reference temperature, °C;
E: activation energy, J/mol;
R: universal gas constant, 8.3144 J/mol/K.

It has been known that the activation energy is not a constant. It is temperature-dependant and sensitive to the composition and physical characteristics of binder. Due to the much higher reactivity of modern Portland cement and the extensive use of mineral admixture to substitute Portland cement at different percentages, Jonasson, Groth, and Hedlund [4] proposed a new temperature-dependant formulation of activation energy of complex binder in Eq (2):

$$E(T) = 44066(\frac{30}{10+T})^{0.45} \qquad (2)$$

where T is concrete curing temperature.

It is proposed by R. C. Tank and N. J. Carino [5] that relative strength can be represent as a function of the equivalent age as follows:

$$R = \frac{S}{S_u} = \frac{K_r(t_e - t_{or})}{1 + K_r(t_e - t_{or})} \qquad (3)$$

where R: relative strength;
S: compressive strength;
S_u: compressive strength at reference age;
K_r: rate constant at the reference temperature;
t_e: equivalent age at the reference temperature;
t_{or}: age at the start of strength development at the reference temperature.

In this study, T_r=20°C, S_u is compressive strength on 7 days of TMC curing. With equations (1), (2) and (3), K_r and t_{or} is calculated by least squares fit of compressive strength of concretes cured both under standard and TMC conditions (Table 3). The ages of concrete cured in TMC process are converted into the equivalent age. It can be seen that K_r gets small and t_{or} prolongs with increasing substitution of fly ash. The relative strengths were plotted versus equivalent age and a regression curve was obtained in Figure 4. The departure of each curve from the testing results is satisfactorily small. It can be used to predict the strength development of concrete in real structures.

Table 3 Parameters of the rate constant function

	C1	C2	C3
K_r (h^{-1})	0.037	0.028	0.025
t_{or} (h)	12.8	15.0	16.75

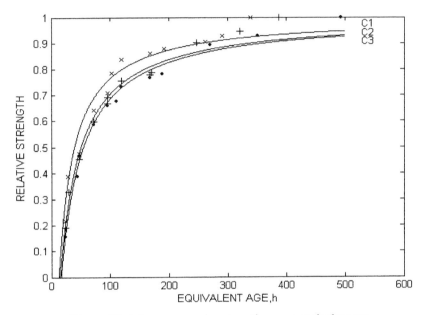

Figure 4 Relative compressive strength versus equivalent age

CONCLUSIONS

The hydration of high volume fly ash concrete is delayed slightly. Its final hydration degree is high enough under the condition of low W/B ratio to make the adiabatic temperature rising higher than the concrete containing a lower percentage of fly ash.

Elevated temperature enhances the hydration of binder to benefit the gain of compressive and flexural strength for concrete containing complex binders. Elevated temperature is not beneficial for the late age strength development of concrete containing only Portland cement. The equivalent age concept can be used to predict the strength of concrete containing fly ash in real structures.

REFERENCES

1. MALHOTRA, V.M., MEHTA, P.K., High-performance, high volume fly ash concrete, Supplementary cementing materials for sustainable development, Inc., Ottawa, Canada, 2002, pp 101.

2. SCHINDLER, A.K., Effect of temperature on hydration of cementitious materials, ACI materials Journal, Vol. 101, 2004, No. 1, pp 72-81.

3. HANSEN, P.F., PEDERSEN, J., Maturity computer of controlled curing and hardening of concrete, Nordisk Betong, Vol. 1, 1977, pp 19-34.

4. JONASSON, J.E., GROTH, P., HEDLUND, H., Modeling of temperature and moisture field in concrete to study early age movements as a basis for stress analysis, Proceedings of the international RILEM symposium on thermal cracking in concrete at early ages, R. Springenschmid, ed., E&FN Spon, London,1995, pp 45-52.

5. TANK, R.C., CARINO, J.N., Rate constant functions for strength development of concrete. ACI materials Journal, Vol. 88, 1991, No. 1, pp 74-83.

USE OF RECYCLED TYRE RUBBER AS AGGREGATES IN SILICA FUME CONCRETE

M Gesoğlu

E Güneyisi

T Özturan

Boğaziçi University

Turkey

ABSTRACT. In this paper, some physical and mechanical properties of rubberized concretes with and without silica fume were studied. Two types of tyre rubber, namely crumb rubber and tyre chips, were utilized as fine and coarse aggregate, respectively. The rubberized concrete mixtures were produced by partially replacing the aggregate with rubber at different rubber contents ranging from 2.5 to 50% by total aggregate volume. Silica fume was included by partial substitution of cement at varying amounts of 5 to 20 %. A total of 35 concrete mixtures with a water-cementitious material ratio of 0.60 were cast and tested for compressive strength, static modulus of elasticity, and split-tensile strength according to relevant ASTM standards. It was found that there was a marked reduction in the slump, unit weight, strength and elastic modulus with increasing rubber content. The use of the silica fume improved the mechanical properties of both non-rubberized and especially rubberized concretes.

Keywords: Elastic modulus, Rubberized concrete, Silica fume, Strength, Tyre rubber.

Dr M Gesoğlu, is a Research Assistant in the Department of Civil Engineering at Boğaziçi University, Istanbul, Turkey.

Dr E Güneyisi, is a Research Assistant in the Department of Civil Engineering at Boğaziçi University, Istanbul, Turkey.

Dr T Özturan, is a Professor of concrete technology in the Department of Civil Engineering, Boğaziçi University, Istanbul, Turkey. His research interests include the use of fly ash and silica fume in concrete, durability of concrete, high performance concretes.

INTRODUCTION

The increasing sensitivity on environmental and energy saving issues makes the possibility of a widespread utilization of waste materials very interesting. It is well known that the disposal of the waste tyres is a growing problem due to the fact that the waste rubber is not easily biodegradable even after a long period landfill treatment. However, the usability of waste rubber as a material and energy source in the construction industry may be an efficient way to the disposal of this waste material. Especially, the use of the waste tyres in portland cement concrete mixtures, particularly for highway engineering, may also be considered. However, very limited work has been done by investigators on the potential use of rubber tyres in conventional concrete mixtures [1-9].

The objective of this study is to utilize the waste tyre rubber as aggregates in the production of concrete mixtures with and without silica fume. Two types of rubber, namely crumb rubber and tyre chips, were used as fine and coarse aggregates, respectively. The initial compressive strength of the non-rubberized concrete mixtures was about 54 MPa. The concrete mixtures made with tyre rubber were tested for the compressive and split-tensile strengths and the static modulus of elasticity.

EXPERIMENTAL STUDY

Fine and coarse aggregates, tyre rubber, ASTM Type I portland cement, silica fume (91.0% SiO_2 content), and a high range water-reducing admixture were used in the production of non-rubberized and rubberized concrete mixtures. Natural sand having a maximum particle size of 4 mm was used as fine aggregate while crushed limestone having 20 mm maximum size was used as coarse aggregate. Crumb rubber and tyre chips were utilized as fine and coarse material with specific gravities as 0.83 and 1.02, respectively. The gradation curves of the aggregates and the crumb rubber are shown in Figure 1. The gradation curve for the tyre chips could not be determined as for normal aggregates, since they were elongated particles between 10 and 40 mm.

Control mixtures were produced at a water-cementitious material ratio (w/cm) of 0.60 with cement content of 350 kg/m^3. Silica fume was used as a partial replacement at 5, 10, 15, and 20% by weight of cement. The mix proportions of control mixtures with and without silica fume are given in Table 1. To develop the rubberized concrete mixtures, all mix design parameters were kept constant except for the aggregate constituents. Tyre rubber was used as a replacement for an equal part of aggregate by volume. The rubberized concretes contained both types of rubbers (crumb and chip) with six designated rubber contents of 2.5, 5, 10, 15, 25, and 50% by total aggregate volume. The rubber content was divided equally between the crumb and the chip. That is, for a 50% rubber content, crumb rubber replaced 50% of the sand volume and tyre chips replaced 50% of the coarse aggregate volume. The control mixtures were designed to have a slump of 180 ± 20 mm, which was realized by using a high range water-reducing admixture. The mixing process was conducted in a power-driven revolving pan mixer as per ASTM C192. The fresh concrete properties such as slump and unit weight were measured, and then three 150 mm cube and three 150x300 mm cylinder specimens were cast from each batch. Test specimens were demoulded 24 hr after casting and then water cured for 7 days. Thereafter, the specimens were kept in a curing room at a temperature of 21 ± 1 °C and at a relative humidity of 60 ± 5% until the time of testing. All specimens were tested at the age of 90 days.

For hardened concrete, all mixes were tested for compressive strength, static elastic modulus, and split-tensile strength. The compressive strength tests were conducted on the cube specimens according to ASTM C39 while the split-tensile strength and the modulus of elasticity were performed on the cylinders as per ASTM C496 and ASTM C469, respectively, by means of a 3000 kN capacity testing machine. Static elastic modulus of concrete was determined by fitting the specimen with a compressometer containing a dial gage capable of measuring deformation to 0.002 mm and then loading and unloading three times to 40% of the ultimate load of the companion cube specimen. The first set of readings of each cylinder was discarded and the modulus was calculated from the average of the second and third readings. The same specimens were later tested for the split-tensile strength. Three specimens were tested for each property.

Table 1 Specific gravity of materials and mix proportioning of control concrete

MATERIAL (kg/m³)	SPECIFIC GRAVITY	SILICA FUME (%)				
		0	5	10	15	20
Cement	3.15	350	333	315	298	280
Silica fume	2.33	0.0	17.5	35.0	52.5	70.0
Water	1.00	210	210	210	210	210
Superplastizer	1.18	5.3	5.3	5.3	5.3	5.3
Crushed stone	2.70	1076	1073	1070	1066	1063
Sand	2.62	697	695	693	691	688

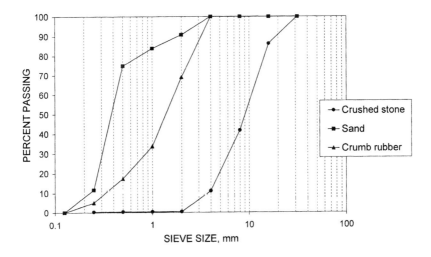

Figure 1 Grading curves of the aggregates

TEST RESULTS AND DISCUSSION

The variation in the slump and unit weight of the non-rubberized and rubberized concretes with and without silica fume were shown in Figures 2 and 3, respectively. Both slump and unit weight of the concretes continuously decreased with increasing rubber content. At a rubber content of 50%, the slump nearly reduced to zero so that the mix was not workable at all and additional effort was required for compaction. Similarly, for the concrete with 50% rubber, there was a 25% reduction in the unit weight, irrespective of the silica fume content.

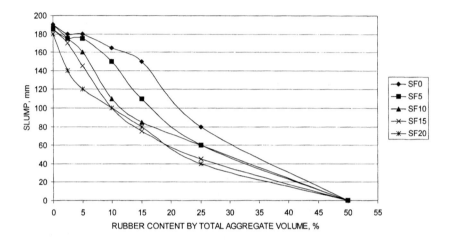

Figure 2   Slump of the concretes containing tyre rubber and
silica fume at varying proportions

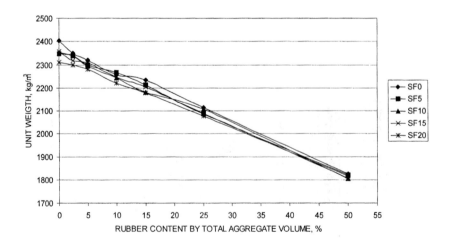

Figure 3   Unit weight of the concretes containing tyre rubber
and silica fume at varying proportions

Test results regarding the compressive strength, static modulus of elasticity, and split-tensile strength of concretes are shown in Figures 4, 5, and 6, respectively. Moreover, Figure 7 demonstrates the percent enhancement of the mentioned concrete properties with the increase in silica fume content.

As seen in Figure 4, the concretes had compressive strengths varying from 60 to 7 MPa depending on the silica fume and rubber used. It was observed that there was a systematic reduction in compressive strength with the increase in rubber content for the concretes with and without silica fume. When 50% of the total aggregate volume was replaced by rubber, irrespective of the silica fume content, the strength loss was about 85%. The effect of silica fume on the compressive strength of the concretes at varying rubber contents is well observed in Figure 7. The percent increase in compressive strength ranged from 6 to 12% for the non-rubberized concretes but from 2 to 43% for the rubberized concrete, depending on the silica fume and rubber contents. This was expected because of the filling effect of the silica fume in the interfacial zone, thus providing a good adherence between the rubber particles and the cement matrix.

Figure 5 indicated that the static elastic modulus of the non-rubberized and rubberized concretes with and without silica fume ranged from about 37 to 6 GPa. Similar to the reduction in compressive strength, the static elastic modulus also decreased with increasing rubber content. For instance, the modulus of elasticity reduced to about 6 GPa, indicating an 82% reduction, when 50% rubber content by the total aggregate volume was used. As illustrated in Figure 7, with the use of silica fume, the increase in the elastic modulus of the concretes with different rubber contents were not consistent with the increase in the silica fume content as opposed to the case in the compressive and tensile strengths. Moreover, it was also observed that the increase in the elastic modulus was much smaller as compared to that in the compressive strength. Generally, the non-rubberized and rubberized concretes showed an increase in the modulus values up to 12 and 17%, respectively, depending on the amount of silica fume used.

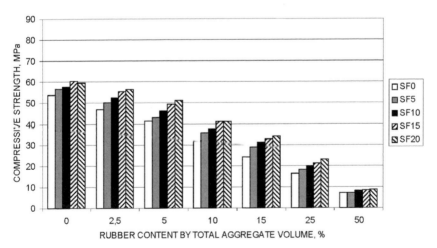

Figure 4   Compressive strength of the concretes containing tyre rubber and silica fume at varying proportions

As shown in Figure 6, the concretes had split-tensile strengths ranging from 3.7 to 0.7 MPa. It was found that the split-tensile strength of the concretes also reduced systematically with the inclusion of the rubber material, irrespective of the silica fume content. At 50% rubber content, the strength loss was as high as 80%. However, as seen in Figure 7, with the use of silica fume the tensile stregths were enhanced remarkably, but the percent increase depended mainly on the amounts of the rubber and the silica fume used.

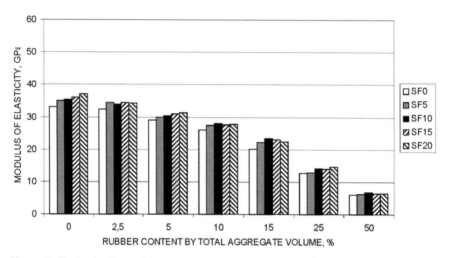

Figure 5   Static elastic modulus of the concretes containing tyre rubber and silica fume at varying proportions

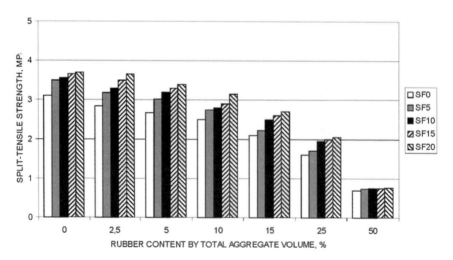

Figure 6   Split-tensile strength of the concretes containing tyre rubber and silica fume at varying proportions

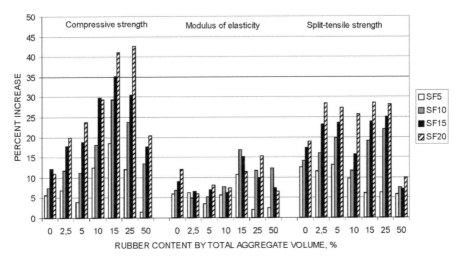

Figure 7   Effect of silica fume on the hardened properties of the concretes at different rubber contents

CONCLUSIONS

The following conclusions may be drawn on the basis of this experimental study.

1. The results indicated that there was a systematic reduction in the hardened properties of the concretes with increasing rubber contents. The loss in compressive strength, elastic modulus, and split-tensile strength reached to 85, 82, and 80%, respectively, when the rubber content was increased to 50% by total aggregate volume.

2. Inclusion of silica fume into the matrix resulted in remarkable improvement in both non-rubberized and especially rubberized concretes. The compressive strength increased by about 12 and 43% for the former and the latter, respectively. However, this increment was less for both modulus of elasticity and split-tensile strength.

3. This study also revealed that workable (80–150 mm slump) concretes with different silica fume contents could be obtained having a compressive strength of 25-35 MPa, split-tensile strength of 2-2.8 MPa and elastic modulus of 20-23 GPa with the inclusion of 15% rubber by volume of total aggregate, providing also about 10% reduction in the unit weight.

REFERENCES

1. ELDIN N.N. and PIEKARSKI J.A. Scrap tyres: Management and economics. J. Environ. Eng. Vol. 119, No. 6, 1993. pp. 1217-1232.

2. ELDIN N.N. and SENOUCI A.B. Use of scrap tyres in road construction. J. Const. Eng. M. Vol. 118, No. 3, 1992. pp. 561-576.

3. WANG Y., ZUREICK A.H. and CHO B.S. Properties of fiber reinforced concrete using recycled fibers from carpet industrial waste. J. Mater. Sci. Vol. 29, No. 16, 1994. pp. 4191-4199.

4. SEGRE N. and JOEKES I. Use of tyre rubber particles as addition to cement paste, Cement and Concrete Research. Vol. 30, 2000. pp. 1421-1425.

5. OLIVARES F.H., BARLUENGA G., BOLLATI M. and WITOSZEK B. Static and dynamic behavior of recycled tyre rubber filled concrete. Cement and Concrete Research. Vol. 32, 2002. pp. 1587-1596.

6. KHATIP Z.K. and BAYOMY F.M. Rubberized Portland cement concrete. ASCE-Journal of Materials in Civil Engineering. Vol. 11, No. 3, 1999. pp. 206-213.

7. TOPÇU I.B. and AVCULAR N. Collusion behaviors of rubberized concrete. Cement and Concrete Research. Vol. 27, No. 12, 1997. pp. 1893-1898.

8. TOUTANJI H.A. The use of rubber tyre particles in concrete to replace mineral aggregates. Cement and Concrete Composites. Vol. 18, 1996. pp. 135-139.

9. GUNEYISI E., GESOGLU M. and OZTURAN T. Properties of rubberized concretes containing silica fume. Cement and Concrete Research. In press, 2004.

ENGINEERING AND DURABILITY PROPERTIES OF CONCRETE CONTAINING WASTE GLASS

M C Tang

L Wang   R K Dhir

University of Dundee

United Kingdom

ABSTRACT. Glass constitutes a major component of solid waste in many of the developed and developing countries. It is therefore vital to examine and find various routes to exploit the inherent properties of glass in order to recycle this material in high-value and large volume applications. Recent interest in using this material in the construction as concrete components provides such means of doing so. However, it is vital to understand how this material will perform when used as an aggregate material, as filler, or as a cement component, in order for engineers to gain confidence in using this material in concrete. This paper examines some of the engineering and durability properties of concrete containing waste glass.

Keywords: Waste glass, High value, Large volume, Engineering properties, Durability properties, Concrete

M C Tang is a Research Fellow in the Concrete Technology Unit in the Division of Civil Engineering, School of Engineering, at the University of Dundee, UK. His research interest focuses mainly on the sustainable use of construction and industrial by-products as cement and aggregate components with special interest to alkali-aggregate reactions in concrete.

L Wang is a research student currently doing his MSc with the Concrete Technology Unit at the University of Dundee

Professor Ravindra K Dhir is Director and Professor of the Concrete Technology Unit at the University of Dundee. He is a member of numerous national and international technical committees and has published extensively on many aspects of concrete technology, cement science, durability and construction methods.

INTRODUCTION

The use of glass as a construction material in the past has been mainly for architectural and decorative purposes such as window panelling, surface glazing or etchings and many other creative utilisations of this material. However, recent research has also discovered that recovered or recycled glass can also be incorporated into concrete as a cement or aggregate component.

Studies have shown that glass was not suitable to be used as coarse aggregate due to its poorly shaped particles causing the increased bleeding, segregation and marked strength

regression of concrete [1]. This was attributed to the high brittleness of glass when used as a coarse fraction, leading to cracks in the interfacial zones between the aggregates and cement paste [2]. It was then suggested that waste glass was more suitable to be used as fine aggregate or as a cement component rather than as coarse aggregate, as pozzolanically reactive fine waste glass could be utilised to further enhance the properties of concrete.

Clearly, it is vital to fundamentally understand how this granular (when crushed), inorganic material with similar characteristics of natural sand will perform when used as an aggregate material, or how when crushed to powder form, the high silica content of glass might proved to be pozzolanic. Studies into using waste glass as fine aggregate and as a cement component have shown promising results where the inclusion of glass seem to be improving the particle packing of the mix and enhancing the strength of the concrete via pozzolanic activity[3, 4, 5].

This paper reports on a selection of engineering and durability properties from a study into the performance of concrete containing glass cullet, as either fine aggregate, filler, or as a cement component. The results presented in this Paper are selected from a larger study examining a range of issues relating to developing methodologies of optimising glass concrete by exploring replacement levels and particle sizes to fully highlight the inherent properties of waste which may enhance the properties of concrete.

EXPERIMENTAL METHODS

MATERIALS

The materials used in the study are described below. The particle size range and chemical composition of glass used are given in Table 1 and Table 2, respectively.

Portland Cement (PC)

A single source of CEM I 42.5N cement conforming to BS EN 197-1[6] was used in the project. It was supplied in 25kg moisture sealed bags.

Waste Glass

All the waste glass used in this study is green colour soda lime container glass which are cleaned and graded according to the various applications.

Glass used as Fine Aggregate

The waste glass obtained for use as fine aggregate in concrete was requested from the supplier, crushed and graded to the requirements under BS EN 12620 [7].

Cullet Used as a Filler Aggregate

The grading used for the filler aggregate corresponds to the coarsest filler aggregate grading given in BS EN 12620 [7].

Cullet Used as a Cement Component

Cullet for use as a cement component in the concrete study was prepared by grinding the waste glass to form a powdered material which passes through a 63μm sieve. The particle size distribution of the resulting material was almost similar when compared to those of Portland cement.

Table 1   The particle size range of the waste glass used in study

MATERIAL	PARTICLE SIZE RANGE	DESIGNATION
Fine glass aggregate	0 – 8 mm	GS
Filler glass aggregate	0 – 2 mm	GF
Powdered glass as cement component	< 63 μm	GC

Table 2   Chemical composition of cement and glass

COMPONENTS	MAJOR OXIDE ANALYSIS, % by mass	
	PC	Glass
SiO_2	20.7	73.3
Al_2O_3	4.3	2.0
Fe_2O_3	3.4	0.2
CaO	63.9	9.0
MgO	2.3	0.8
P_2O_5	0.1	0.0
TiO_2	0.3	0.1
SO_3	3.2	0.1
K_2O	1.1	0.4
Na_2O	0.9	18.9
MnO	0.1	0.1

Experimental Programme

In order to establish the ideal amount of waste glass to be included as concrete components, a series of mixes were carried out and tested to determine the effects of different amounts of glass on concrete performance. It should be noted however, the glass content for this study are selected based on acceptable performances in compressive strength and ASR expansion. The optimum levels of replacement for glass for this study are as follows:

- Glass replacements of fine aggregate 50% by mass

- Glass was employed as filler at a single fine aggregate replacement level of 5% by mass

- Glass as cement component at levels of 15% by mass of the total cement

Based on the glass content established from the on-going project, three sets of glass concrete mixes are designed and tested. The details of the mix proportions are presented in Table 3.

Table 3   Mix proportions compressive strength and of concrete made with glass

MIX	W/C RATIO	MIX PROPORTIONS, kg/mm^3							DENSITY kg/m^3	SLUMP mm
		PC	Water	Aggregate				Glass		
				Sand	10mm	20mm	overall			
GS	0.61	295	180	305	430	860	1595	295	2350	45
RS	0.61	295	180	610	430	860	1900	0	2400	50
GF	0.65	275	180	605	430	855	1890	30	2360	50
RF	0.65	275	180	635	430	855	1920	0	2350	60
GC	0.55	275	180	580	430	860	1870	40	2360	30
RC	0.55	325	180	580	430	860	1870	0	2370	35

he first letter "G" in the mix name describes Glass cullet, and "R" describes Reference concrete. The second letters in mix indicates method of replacement, namely Sand, Filler and Cement.

TEST RESULTS AND DISCUSSIONS

Discussions on test results on engineering properties of concrete

Tests on fresh concrete produced using PC 42.5N and waste glass of various proportions are given in Table 3. The unit weight of concrete was slightly reduced when large amounts of glass are replaced into the mix (i.e. 50% sand replacement). The 2% reduction in unit weight could be attributed to the fact that the specific gravity of waste glass is slightly lower than that of natural sand. It was also discovered a slight decrease of slump (5mm) with concrete containing waste glass due to the poor geometry of the glass particles. However, it should be noted that these differences are very small.

The incorporation of glass cullet in concrete seems have significant effect on the rate of the strength development especially when the glass was used as cementitious material in the concrete. The test results for compressive strengths of the glass concretes are presented in Figure 1 and 2. The compressive strength of concrete specimens containing waste glass achieved comparable if not higher strength than corresponding reference mixes at 28 days, with also greater strength developing potential beyond 28 days. For example, the concrete containing glass as sand produced 7-day strength of about 70% of that attained at 28 days, while its reference mixes initially gained strength more rapidly with 7-day strength well exceeding 75% of the 28-day strength. However, the concrete containing glass sand managed to achieve 125% of its 28 day strength at the age of 180 days, when compared with the 110% increase of the reference mix. The increase in later age strength was even more significant when examining the concrete containing waste glass as cement component, with a difference of almost 30% of relative gain in compressive strength at 180 days. The increase in later age strength could be attributed to finely-divided waste glass undergoing pozzolanic activity and contributing to the strength development of the concrete, thus the differences being more apparent in concrete mixes containing waste glass as cement component as they contain a larger amount of powdered glass.

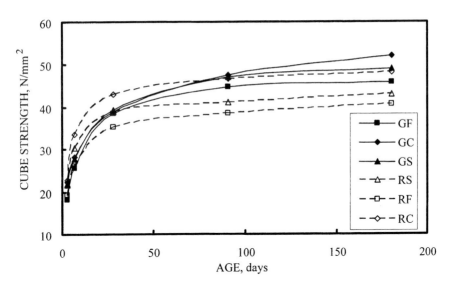

Figure 1   Strength development of concrete containing waste glass

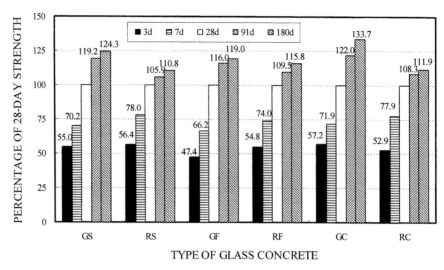

Figure 2   Relative rate of strength gain in concrete containing waste glass

The addition of waste glass into concrete does not significantly alter the usual relationship between compressive strength and flexural strength or between the compressive strength and the modulus of elasticity. Flexural strength was observed to change inconsistently with the various replacement of waste glass where no noticeable effect could be seen with the inclusion of waste glass. Data presented in Table 4 showed the pozzolanic reactivity of the glass in concrete affect the compressive strength, flexural strength and modulus of elasticity of concrete in nearly the same manner.

Table 4   Flexural strength and modulus of elasticity of concrete made with glass

MIX	W/C	CUBE STRENGTH N/mm^2		FLEXURAL STRENGTH N/mm^2		F/C STRENGTH RATIO %		MODULUS OF ELASTICITY kN/mm^2			
								Tested		Calculated	
		28day	180day	28day	180day	28day	180day	28day	180day	28day	180day
GS	0.61	39.5	49.0	4.4	4.6	11.1	9.5	26.0	28.5	26.6	29.6
RS	0.61	39.0	43.0	4.9	5.1	12.6	11.8	26.0	27.0	26.4	27.7
GF	0.65	38.5	46.0	5.0	5.3	13.1	11.5	27.5	29.0	26.3	28.7
RF	0.65	35.0	40.5	4.1	4.7	11.6	11.7	24.0	25.5	25.0	26.9
GC	0.55	39.0	52.0	4.3	4.7	10.9	9.1	26.0	29.5	26.4	30.5
RC	0.55	46.0	48.0	5.3	5.5	11.6	11.4	28.0	29.0	28.7	29.3

The calculated modulus of elasticity is from the expression $E_c = 4.23(f_c)^{0.5}$ recommended by ACI 318-89 (Revised 1992)

Drying shrinkage curves in Figure 3 showed that, when fine aggregate was replaced by waste glass up to 50% by volume in concrete, shrinkage was initially reduced up to 15% after 20 weeks for drying curing. Similar improvement can be found in the concrete containing glass as filler aggregate. Concrete containing waste glass as a cement component also displayed a decrease in shrinkage after 10 weeks, even though it initially exhibited higher shrinkage.

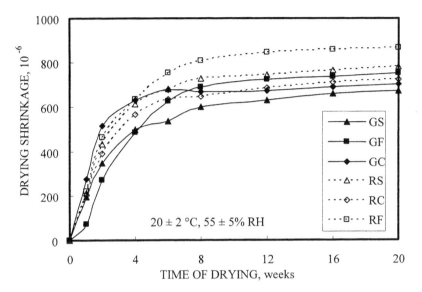

Figure 3   Drying shrinkage of concrete containing waste glass

Test results on durability properties of concrete

The test results for carbonation (method in CEN/TC51/WG12/TG5 (1995) and modified in Dundee University (1998)), freeze-thaw resistance (CEN/TC 51-draft pr EN (BSI 1994)), sulphate resistance (BRE Digest 363, Parts 1 and 2) and abrasion resistance are presented in Table 5. The carbonation test results were inconsistent where when waste glass was used in the concrete. The inclusion of waste glass as sand and cement does appear to increase the carbonation depth as seen in Figure 4, as the higher amount of fine glass is reducing the alkalinity of the concrete via pozzolanic reaction.

The freeze-thaw resistance of concrete containing waste glass is presented in Figure 5 where better resistance was discovered with concrete containing 5% filler glass while the concrete containing 50% fine glass aggregate displayed a decrease in freeze-thaw resistance. This could be attributed to the geometry of the glass particle, whilst improving the particle packing of the concrete when using as a filler material, would similarly decrease the packing when using it as fine aggregate.

226 Tang, Wang, Dhir

Table 5   Durability properties of glass concrete and reference concrete

MIX	W/C	PC kg/mm³	GLASS kg/mm³	CUBE STRENGTH N/mm²		CA. DEPTH mm	FREEZE-THAW kg/m²	SULFATE ATTACK %	ABRASION DEPTH mm
				28day	180day	9week	56cycle	20week	28day
GS	0.61	295	295	39.5	49.0	15.0	0.045	-0.002	1.2
RS	0.61	295	0	39.0	43.0	7.5	0.020	-0.003	0.9
GF	0.65	275	30	38.5	46.0	8.5	0.006	-0.017	0.4
RF	0.65	275	0	35.0	40.5	16.0	0.030	-0.001	1.2
GC	0.55	275	40	39.0	52.0	14.0	0.024	-0.008	0.8
RC	0.55	325	0	46.0	48.0	4.0	0.020	-0.003	0.7

CA.=Carbonation

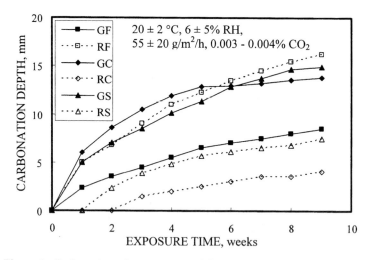

Figure 4   Carbonation of concrete containing waste glass up to 9 weeks

Sulphate resistance of concrete containing waste glass are increased especially when used as a filler material as seen in Figure 6. This is mainly attributed to the filler glass improving the particle packing of the concrete, making the concrete more impermeable to sulphates. The pozzolanic reactivity of glass was also seen to increase sulphate resistance as seen in the mixes with waste glass as cement component, as glass binds the amount of free alkalis in cementitious compounds, rendering it unavailable for sulphate reaction.

The inclusions of glass as sand or cementitious material are harmful with respect to the abrasion resistance of concrete. Test results shown in Table 5 show the abrasion depth increased by 33.3% with 50% by volume of sand in concrete. However, the abrasion depth in the specimen containing glass filler aggregate was at least 3 times smaller than the reference concrete. No apparent effect could be seen when glass was used as cement component. This is again attributed to the geometry of waste glass and the improved particle packing when the waste glass is used as a filler material.

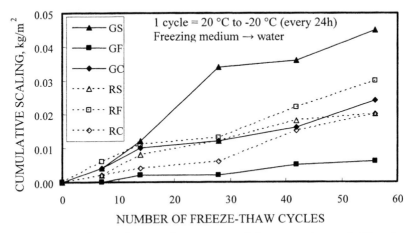

Figure 5 Freeze-Thaw resistance of concrete containing waste glass up to 56 cycles

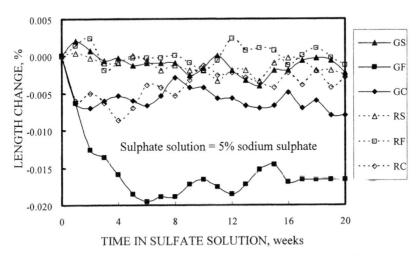

Figure 6 Sulphate resistance of concrete containing waste glass up to 20 weeks

CONCLUSIONS

The engineering and durability properties of concretes containing waste glass at levels which were selected based on acceptable performances in compressive strength and ASR expansion are carefully examined and presented in this paper.

The resulting data show that there is a great potential for the utilisation of waste glass in concrete; as fine aggregate, filler material or as cement component. It was found that strength gain of concrete containing glass is enhanced due to the pozzolanic reactivity of glass. Other engineering properties of concrete with waste glass are also shown to be satisfactory. The test results for the durability properties are more complex as they showed that whilst the inclusion of waste glass in one form could improve a certain kind of deterioration, it would proved to be detrimental in another. However, it can be demonstrated that careful and judicious selection of particle size and grading of waste glass could mitigate the various forms of deterioration.

REFERENCES

1. POLLEY C., CRAMER S.M. , DE LA CRUZ R.V., Potential for using waste glass in Portland cement concrete, Journal of Materials in Civil Engineering, Vol.10, 1998, pp210-211

2. TOPCU I.B., MEHMET C., Properties of concrete containing waste glass, Cement and Concrete Research, Volume 34, Issue 8, 2004, pp 1307-1312

3. SHAO Y., THIBAUT L., SHYLESH M., RODRIGUEZ D., Studies on concrete containing ground waste glass, Cement and Concrete Research 30, 2000, pp91-100

4. DHIR R.K., DYER T.D., TANG M.C., CUI Y., Towards maximizing the value and sustainable use of glass in concrete, CONCRETE, Jan 2004, Vol. 38, No.1, pp38-40

5. SHAYAN A., XU AIMIN., Value-added utilisation of waste glass in concrete, Cement and Concrete Research, Vol. 40, 2004, pp81-89

6. BRITISH STANDARDS INSTITUTE, BS EN 197-1:2000, Cement. Composition - Part 1: Specifications and conformity criteria for common cements, 2000, 46pp

7. BS EN12620 BRITISH STANDARDS INSTITUTE, BS EN 12620:2002, Aggregates for Concrete, 2002, 47pp

POTENTIAL FOR RECYCLING DEMOLISHED CONCRETE AND BUILDING RUBBLE IN KUWAIT

S Al-Otaibi

M El-Hawary

Kuwait Institute for Scientific Research

Kuwait

ABSTRACT. Demolished concrete and building rubble waste presents an environmental problem in terms of ways of disposal and use. The amount of structural waste in Kuwait for 1996 was around 2 million tons, out of which 30% is concrete. The recyclable concrete reached 764 thousand tons in 2002 and is expected to exceed 1210 thousand tons in 2020 as many old buildings will be demolished. Dumping was the only measure taken previously, but it is no longer a valid option as it occupies land which is in great demand for urban development. Recycling that form of solid waste into construction is a better solution. In Kuwait attempts are being made in adopting the recycling approach in the last few years. This paper states out the problem and presents a review of these attempts and research carried out to asses the suitability of crushed concrete and masonry for use in concrete and in sand lime brick manufacture. Recycled concrete can be crushed and utilized as aggregates in the production of new concrete. The results also show that fine powder, obtained from crushed concrete and masonry blocks contains adequate amounts of lime ($Ca(OH)_2$) resulting from the hydration process which may react with silica under high temperature and pressure to produce lime-silca bricks. The production process including autoclaving time and temperature is included along with the properties of the resulting bricks. The reported properties include specific gravity, compressive strength, and absorption. The properties of the resulting bricks are compared with the specifications requirements.

Keywords: Demolishing, Rubble, Recycling, Concrete.

Dr S Al-Otaibi, Associate Research Scientist, Kuwait Institute for Scientific Research, Division of Environmental and Urban Development, Building and Energy Technologies Department.

Dr M El-Hawary, Associate Research Scientist, Kuwait Institute for Scientific Research, Division of Environmental and Urban Development, Building and Energy Technologies Department.

INTRODUCTION

Waste management is one of the most complex and challenging problems in the world which has a great impact on environment. Due to the high rate of construction development that Kuwait has witnessed during the past forty years, increased demolition activity to replace older structures, and maintenance and restoration works required for the preservation of old buildings, tremendous quantities of building debris have been accumulated in Kuwait, to the point where the situation has become increasingly worrying for the government authorities.

Construction and rubble waste need to be disposed of in an appropriate way, in view of the pollution caused by debris to the environment. In Kuwait, dumping has been used extensively to dispose of such debris. Debris is usually dumped in sites, which have been previously used as sand and gravel quarries. In some cases, such debris is used on shores as breakers and in some kind of land reclamation. Waste disposal practices such as these have contributed to the loss of land areas that might otherwise have been available for construction development projects. Preservation of the environment and natural resources, as well as control of land exhaustion and minimization of dumping activities have become matters of priority for Kuwait Municipality in the last few years.

Recycling of construction and rubble waste presents a very viable option and it is reported around the world that demolished concrete and blocks and asphalt can be recycled into construction applications successfully provided that certain guides and specification requirements are met. More buildings in Kuwait are being demolished and the resulting rubble waste is in huge quantities. A report by Kuwait municipality estimates the quantities in the years 1995 and 1996 around 2 million tons (2,087,300 and 1,980,700 respectively) [1].

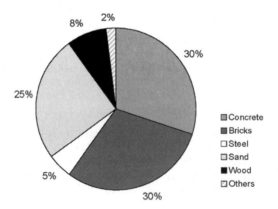

Figure 1 Components of Structural Waste in Kuwait

The structural waste is classified to different components as shown in Figure 1. The main components are concrete, bricks and sand [1]. In a study by Al-Meshaan and Ahmed [2], the amount of waste from demolished buildings that need to be buried was estimated as shown in Table 2, along with the required size of landfill.

Table 2. Amount of Building Waste and Required Landfill in Kuwait

YEAR	BURIED QUANTITY (1000 TON)	LANDFILL (1000 M^2)
1995	741	49
2000	843	56
2005	957	64
2010	1134	76
2015	1262	84
2020	1424	95

Most of this waste has been dumped in dumping sites around the country (Figure 2) These sites lie very near to urban areas; hence present an environmental problem and large areas of land which is in great demand for development and hinder the growth of the cities around them.

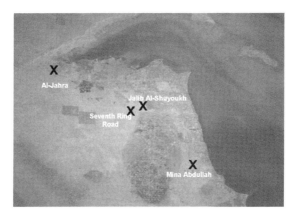

Figure 2 Location of Rubble Dumping Sites

Kuwait Municipality embarked on an ambitious program to solve the problem of structural solid waste in 1999 by offering investment opportunities in the field of recycling of structural waste through a BOT program. Two local companies were granted permission to establish recycling plants in 2002. One plant (Environment Preservation Industrial Company) started in early 2004 the production of recycled concrete and asphalt products. The plant consists of designated areas for:

- Building waste and debris reception
- Sorting and classification
- Crushing and Screening (Figure 3)
- Storage and Shipping

The main products that result from the crushing process are recycled aggregates of different sizes (Figure 4).

Figure 3 Crushing /screening of waste concrete   Figure 4 Crushed concrete aggregates

Building demolition waste was used in the manufacture of concrete and concrete products such as bricks. In a study by Zakaria and Cabera on the performance and durability of concrete made with demolition waste, compressive strength and permeability were determined and the results showed promising results [3]. Other researchers reported that aggregates used in concrete products can be replaced up to 100% by demolition waste[4,5]. Al-Mutairi and Haque found that concrete produced using coarse aggregate obtained from crushed concrete resulted in a good quality concrete [6].

Extensive research work was conducted on the use of crushed old concrete as aggregate for new concrete (Brown, 1998), the strength of the resulting concrete was found to be relatively lower than that produced using conventional aggregates, depending on the properties of the parent concrete. The strength, however, may improve if concrete was crushed and reused in a short period of time (Ramamurthy and Gumaste, 1998; Tavakoli and Soroushian, 1996 and Rashwan and Abourizk, 1997). Some work was also conducted on the performance (Shayan and Xu, 2003 and Nagataki et al., 2000) of the resulting concrete. In the highway industry the crushed concrete aggregates may be used as road base material or in producing new concrete pavements (Chini et al., 1996 and Cuttell et al., 1996). Under natural conditions sand in a sand-lime mixture is an inert material incapable of interacting with lime. However, in a steam-saturated atmosphere (100% humidity) at a temperature of 170°C and above, sand silica becomes chemically active and quickly combines with lime according to the reaction.

$$Ca(OH)_2 + SiO_2 +(n-1)H_2O = CaO.SiO_2.nH_2$$

Which yields calcium hydrosilicates of great strength and high water resisting properties. The crushed recycled concrete contains calcium hydroxide, lime, from hydrated cement and siliceous particles from the aggregates. The grind recycled concrete may be autoclaved to get strong lime-silica bricks. The concrete may be grind and used to produce bricks or it may be crushed to the required size and used as aggregates and the remaining crusher fines may be autoclaved to produce lime-silica bricks.

TEST METHODOLOGY

Use of Recycled Aggregates in Concrete

Materials

Cement: Ordinary Portland cement, conforming to the requirements of BS EN 197-1 was used in this investigation. Aggregates: Coarse aggregates were both normal aggregates and recycled aggregates. The same sand was used for all the mixes. All aggregates were in SSD condition. Water: Potable water was used.

The concrete mixes used were:

CM: A grade 40 MPa control mix using normal aggregates and having w/c=0.45.

RAC1: A recycled aggregate concrete with the same proportions as the control mix replacing the coarse aggregates by crushed concrete aggregates and having w/c=0.45.

RAC2: A recycled aggregate concrete with the same proportions as the control mix replacing 50% of the coarse aggregates by crushed concrete aggregates and having w/c=0.45.

RAC3: A recycled aggregate concrete with the same proportions as the control mix replacing the coarse aggregates by crushed concrete aggregates and having w/c=0.4.

The details of the mixes are shown in Table 3.

Table 3 Concrete Mix Proportions

MIX No.	OPC kg/m^3	WATER kg/m^3	FINE AGG. kg/m^3	COARSE AGG. 10 mm kg/m^3	COARSE AGG. 20 mm kg/m^3	w/b %
CM	450	203	600	380	770	0.45
RAC1	450	203	600	380	770	0.45
RAC2	450	203	600	380	770	0.45
RAC3	450	180	600	380	770	0.40

Testing Procedure

Properties of the Coarse Aggregates

The specific gravity of coarse aggregate and water absorption were determined following the ASTM C-127 procedure. The sieve analysis is in accordance with ASTM C-136

Compressive Strength

100mm^3 cubes were prepared to determine compressive strength according to BS 1881:1983.

Sand-Lime Bricks from Crushed Concrete Powder

Old concrete cubes with a compressive strength of 37 N/mm^2, are crushed with a jaw crusher, sieved then ground to have a very fine powder. Then the specimens were prepared as follows:

The powder was mixed with sand and water added with water w/p = 0.144, the mixture is placed in brick cube moulds and compacted. A compressive strength of 5 N/mm^2 was applied and the mould kept for 24 hours. Specimens were autoclaved for 15 hrs, cooled then tested for compressive strength.

RESULTS AND DISCUSSION

Recycled Aggregate for Concrete

The specific gravity of coarse aggregates and water absorption results shown in Table 4 indicate that recycled aggregate have lower density and higher water absorption.

Table 4 Specific gravity and water absorption of coarse aggregates

Sample	Specific Gravity		Water Absorption
	Bulk	Apparent	(%)
Normal Aggregate			
Agg. 19.0 mm	2.70	2.69	0.39
Agg. 9.5 mm	2.68	2.66	0.51
Recycled Aggregate			
Agg. 19.0 mm	2.44	2.63	5.07
Agg. 9.5 mm	2.34	2.63	8.09

The sieve analysis results for the normal and recycled coarse aggregates with 20 mm nominal size shown in Figure 5 indicate that the coarse aggregate grading achieved in the crushing of concrete is somewhat close to the results for the normal aggregates. These results cannot be generalized as there is a lot of variation in the recycled aggregates depending on the source

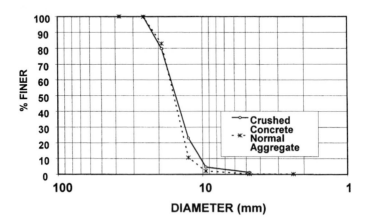

Figure 5 Grain size distribution of 20 mm coarse aggregates

Compressive Strength

The compressive strength development of all the four mixes are shown in Figure 6. Results indicate control mix CM at the desired design strength while the RAC1 (recycled aggregates) achieves 20% lower strength. RAC2, at 50% recycled coarse aggregates, has a slightly increased strength but lower than the control. RAC3, less water at w/c=0.4, has strength close (slightly higher) to the control. Thus desired compressive strengths with recycled aggregates and proper mix proportions allow acceptable results.

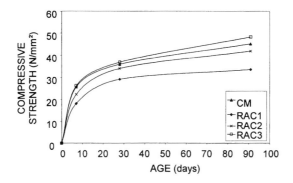

Figure 6 Compressive Strength Development

Brick Production

The average compressive strength of the resulting bricks was found to be 15.8 N/mm^2. ASTM C-90 (Table 5) requires that compressive strength of bricks to be above 6.9 N/mm^2 for weather exposed bricks and above 4.8 N/mm^2 for internal or weather-protected bricks. This illustrates the excellent potential finely ground concrete waste in brick production.

Table 5 ASTM C90 Requirements for Bricks

GRADE	COMPRESSIVE STRENGTH, Min., psi (N/mm^2)		WATER ABSORPTION, MAX, lb/ft^3 (kg/m^3) (average of 3 units) with oven-dry weight of concrete, lb/ft^3 (kg/m^3)			
	Average Gross Area		Weight Classification			
	Average of 3 Units	Individual Unit	Light Weight		Medium Weight	Normal Weight
			Less than 85 (1362)	Less than 105 (1682)	125 to 105 (2002 to 1682)	125(2002) or more
N-I N-II	1000 (6.9)	800 (5.5)	...	18 (288)	15 (240)	13 (208)
S-I^A S-II^A	700 (4.8)	600 (4.1)	20 (320)	...	...	...

^{A} Limited to use above grade in exterior walls with weather-protective coating and in walls not exposed to the weather.

CONCLUSIONS

The main conclusions are:

1. Recycled aggregate has good potential to be used in construction in Kuwait.
2. Direct replacement of normal aggregates with recycled aggregates gives lower strengths.
3. Concrete with good strength using recycled aggregates from crushed concrete is possible.
4. Attention to the properties of recycled aggregates is needed before use in concrete mixes.
5. Recycled concrete crushed powder could be used in brick production.
6. Further research on different applications to utilise recycled concrete products is required.

REFERENCES

1. KUWAIT MUNICIPALITY, Specifications of Projects: use of structural waste, 1999. (*Arabic*)

2. AL-MESHAAN M. AND AHMED F. Environmental Strategies for Solid Waste Management, 1997, Public Authority for Environment, 1997, Kuwait, (*in Arabic*)

3. ZAKARIA, M.: CABERA, J.G. Performance and Durability of Concrete Made With Demolition waste and Artificial Fly Ash Aggregates, Waste Management, 16, 1-3, 1996, pp. 151-158

4. DE VERIES, P., Recycled Materials for Concrete, Quarry Management, 22, 12, Dec. 1995.

5. POON, C. S.; KON, S.C.; LAN, L. Use of Recycled Aggregates in Molded Concrete Bricks and Blocks, Construction and Building Materials, v16, n5, July 2002, pp. 281-289.

6. AL-MUTAIRI, N. AND HAQUE, M. Strength and Durability of Concrete, Proceedings: Int. Symp.Advances in Waste Management, CTU, Univ. of Dundee, UK, Sept. '03, pp. 10-18.

7. AL-WAHEEL, E.I AND KISHAR, E. 1996. "Hydrothermal Characteristics of Sand-Lime Brick Production", ZKG International, No. 5.

8. BROWN, B., 1998. "Aggregates for Concrete", Concrete (London), Vol.32, No. 5, May.

9. CHINI, S.A.; SERGENIAN, J.J. AND ARMAGHANI, J.M. 1996. "Use of Recycled Aggregates for Pavement", Proceedings of the 1996 4th Materials Eng. Con., Washington DC, USA.

10. CROSS, S.A.; et al. 1996. "Long Term Performance of Recycled Portland Cement Concrete Pavement", Transportation Research Record, No. 1525, September.

11. CUTTEL, G.D.; SNYDER, M.B.; Vandenbossche, J.M. and Wade, M.J. 1996. "Performance of Rigid Pavements: Recycled Concrete Aggregates",Transportation Research Record, 1574,

12. RAMAMURTHY, K. AND GUMASTE, K.S. 1998." Proberties of Recycled Aggregate Concrete", Indian Concrete Journal, Vol. 72, No.1, January.

13. RASHWAN, M. AND ABURIZK, S. 1997. "Properties of Recycled Concrete", Concrete International, Vol. 19, No.7, July.

14. RAVINDRARAJAH, R.; LOO, Y.H. AND TRAN, C.T. 1987. "Recycled Concrete as Fine and Coarse Aggregates in Concrete", Magazine of Concrete Research, 39, 141, December.

15. SHAYAN, AHMED AND XU, A. 2003. "Performance and Properties of Structural Concrete Made with Recycled Concrete", ACI Materials Journal, Vol.100, No. 5, September.

16. TAVAKOLI, M. AND SOROUSHIAN, P. 1996."Strengths of Recycled Aggregate Concrete Made Using Field-Demolished Concrete as Aggregates", ACI Materials, vol. 93, March/April.

THE MANUFACTURE OF PRECAST BUILDING BLOCKS UTILISING RECYCLED CONSTRUCTION AND DEMOLITION WASTE

N Jones   **M N Soutsos**   **S G Millard**   **J H Bungey**   **R G Tickell**

University of Liverpool

J Gradwell

Enviros Ltd

United Kingdom

ABSTRACT. The potential for using recycled construction and demolition waste as a replacement for natural crushed limestone in the manufacture of precast concrete building blocks is being investigated at the University of Liverpool. Both the economic case and technical capabilities are being considered. Laboratory studies reported here have examined the use of both concrete and masonry derived waste materials, as a replacement for coarse aggregate and fine aggregate in block manufacture. Initial results include the development of a procedure for laboratory replication of industrial block manufacturing processes, together with trials with representative locally obtained materials to meet typical block specification requirements.

Keywords: Precast concrete, Blocks, Construction and demolition waste, Recycled aggregates, Sustainable construction.

Dr N Jones is a Research Associate in the Department of Civil Engineering, University of Liverpool.

Dr M N Soutsos is a Lecturer in the Department of Civil Engineering, University of Liverpool and leads work on Recycled Aggregates.

Dr S G Millard is a Senior Lecturer in the Department of Civil Engineering at the University of Liverpool.

Prof J H Bungey is Professor of Civil Engineering at the University of Liverpool and Head of the Materials and Infrastructure Group.

Mr R G Tickell is a Senior Lecturer in the Department of Civil Engineering, University of Liverpool.

Ms J Gradwell is an Environmental Consultant with Enviros Ltd.

INTRODUCTION

The European Commission initiated in 1991 the Priority Waste Streams Programme for six waste streams, of which one is construction and demolition waste (C & DW) [1]. Whilst this reportedly accounts for about 17% of the estimated annual UK waste in recent years [2] only about 4 % is recycled for high specification applications. A further 30 % of the 70 million tonnes involved has been reused in low grade applications such as bulk fill, where some crushing and separation of metals and wood is required [3]. Disposal charges for the remainder have been traditionally low. Development of higher level uses has been hampered by the perceived risks involved [4], and ever-increasing legislation relating to handling of waste materials.

Set against this, around 200 million tonnes of new construction aggregates are extracted annually in the UK [5], with associated environmental costs leading to the introduction of an aggregates levy in 2002 [4, 5, 6]. Landfill charges are increasing substantially, whilst a recent Quality Protocol published by WRAP [7] seeks to address quality issues in waste-derived aggregates for the benefit of producers and purchasers of such materials.

It is against this background that recent work at the University of Liverpool has considered the economic and practical issues, as well as technical aspects, of use of recycled concrete and masonry C & DW in the manufacture of concrete building blocks. The project, in collaboration with environmental consultants Enviros, has established a network of appropriate stakeholders to guide and assist the work. Building blocks have been selected as a relatively high volume commodity product, for which the potential effects of contaminants on reinforcing steel are avoided.

Technical objectives are to develop definitive mix design guidance and specifications for the use of such materials for high quality building products. The results presented in this paper focus on the laboratory replication of industrial manufacturing processes to produce comparable specimens for testing purposes, and initial results with both masonry and concrete waste obtained and processed locally.

ECONOMIC CASE

This will vary regionally in light of the demands and availability of both primary and potential recycled materials with transport costs being a significant factor. A cost model is being developed [8] and a preliminary assessment of the North West region indicates appropriate potential taking account of current and anticipated regeneration, provided that laboratory studies can provide appropriate technical solutions.

USE OF RECYCLED AGGREGATES IN CONCRETE

Some previous studies have indicated promising results when using recycled materials from tightly controlled sources into fresh concrete [9, 10], whilst others have demonstrated the successful re-use of crushed recycled material from precast products into new products [11, 12]. The current situation has recently been summarised by Dhir et al [13] including compliance with British and European standards.

The work at Liverpool addresses a different market level, and focuses on C & DW arising from the demolition of building structures (about which few details may be known regarding materials quality). Such materials, even if nominally designated as concrete or as masonry, are unlikely to be pure. Masonry is usually found to contain some plaster, tile and glass as well as metal, paper and wood whilst concrete typically contains some small quantities of masonry.

PRECAST CONCRETE BUILDING BLOCKS

Standard precast concrete building blocks are $440 \times 215 \times 100$ mm in dimension with a weight of less than 20 kg and are predominantly manufactured and sold regionally. Such blocks conform to BS6073 [14] and typically contain 6 mm naturally quarried limestone aggregate to achieve a minimum compressive strength of 7 N/mm^2 at 28 days. Factory manufacturing procedures vary, but typically involve a combination of vibration and static pressure applied to a relatively dry mix containing approximately 100 kg/m^3 of cement, sometimes followed by some form of heat treatment over a 24 hour period.

LABORATORY BLOCK MAKING PROCEDURES

Establishing a consistent and efficient procedure for block fabrication in the laboratory was an essential precursor to the programme to determine the characteristics of blocks made with the range of materials under consideration. Following visits to, and supply of typical materials by a major block manufacturer, efforts were thus initially concentrated on replicating the properties of standard blocks. 6 mm and 4 mm-to-dust fractions of limestone were combined with Rapid Hardening Portland Cement at 100 kg/m^3 and sufficient water added to achieve a dry consistency that will just bind when squeezed by hand.

Initial trials utilised standard 150 mm cube moulds and static pressure alone from a compression testing machine as indicated by Poon [12]. This proved to be unsatisfactory, possibly due to material angularity. Compaction was then attempted in 300×150 mm cylinder moulds using a hand-held vibrating hammer with a circular foot attachment and a vertical measuring device. This was more successful but trials indicated that the additional use of a vibrating table had no beneficial effect on the compaction or the strength of blocks produced. It did however prove very difficult to achieve correct specimen dimensions particularly in terms of squareness of the top surface.

In the light of these results, it was decided to utilise half-length blocks ($220 \times 215 \times 100$ mm) cast on their long edges as in industrial manufacturing processes. This economised the amounts of materials required and also reduced repetitive handling Health and Safety issues in the laboratory. Comparative crushing trials between full-size and sawn half-length blocks undertaken by the manufacturer indicated no difference in measured compressive strengths. A reusable, quick-release steel mould was designed and fabricated comprising a baseplate and two side elements as shown in Figure 1(a).

A total of 18 baseplates and 6 sets of sides were made. The height of the moulds was oversized to enable the correct amount of fresh material to be added necessary to provide a compacted block of dry mass just under 10 kg. Hand operation of the vibro-compaction hammer fitted with a rectangular foot again caused difficulties in creating square top surfaces to blocks. A rig comprising a steel stand and slide rods was thus also fabricated to allow the vibro-compaction hammer-drill to slide down to a predetermined height to give correct block dimensions as shown in Figure 1(b).

(a) Moulds for half-length block specimens   (b) Compaction alignment rig

Figure 1 Block fabrication equipment

This apparatus has been found to work well and provides a uniformly compacted block, with square sides as required. Blocks are demoulded immediately after compaction, enabling rapid re-use of mould sides after cleaning for the same mix, following compaction, blocks are then left on the base plate until an age of 16-24 hours.

Initial trials used a curing procedure which involved wrapping the blocks in wet hessian and polythene sheeting and placing in an oven at 40 – 45°C for 24 hours followed by storage in water up to 7 days in an attempt to replicate industrial processes. Subsequent tests however indicated that comparable results could be achieved by simply curing at room temperature under wet hessian and polythene for 24 hours, followed by storage on open laboratory shelving until testing.

Compressive strength testing was carried out on air-dry specimens at 7 days using a standard fibreboard capping as in the factory and outlined in BS6073 [14]. A conversion factor of × 1.06 was then established experimentally to convert these values to equivalent 28-day strengths of wet-tested, mortar-capped, specimens. It was reassuring to note that this factor is identical to that commonly used by the block manufacturers. It should be noted that care is needed with fibreboard capping to ensure appropriate operation of the compressive testing machine to avoid indication of a false premature failure due to limited crushing of the fibreboard material at a stress of around 7 N/mm^2.

TRIALS WITH LIMESTONE AGGREGATES

These procedures were then used to establish relationships between compressive strength and density for blocks made with the limestone materials supplied. It was noted that although an acceptable range of dry block density of 1850 to 2050 kg/m^3 is specified, all of the factory-supplied blocks were found to be at the upper end of this range. For these materials, a difference approaching 50 kg/m^3 was found between the fresh and hardened materials and some typical results are shown in Figure 2.

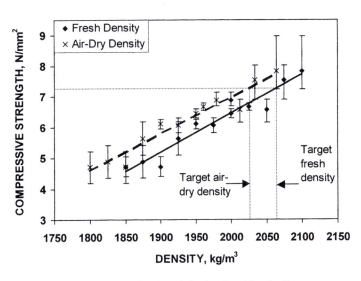

Figure 2 28-day strength versus fresh and air-dry densities for limestone aggregates

From Figure 2, it can be seen that a target fresh density of 2065 kg/m³ was suitable yielding a typical air-dry block density of 2025 kg/m³. It was also found that the range of water/cement ratios which enabled these materials to be successfully compacted, whilst retaining block integrity after demoulding was quite large (0.3 to 0.9). An optimum in the region of 0.45 was established for an air-dry block of density 2025 kg/m³ as shown in Figure 3, with 9.7 kg of fresh concrete being placed in the mould for compaction.

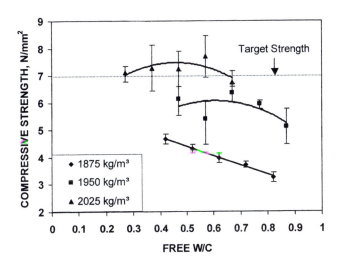

Figure 3 Effect of free water/cement ratio on compressive strength for limestone aggregates

USE OF REPLACEMENT AGGREGATES

Concrete and masonry derived aggregates were obtained and crushed locally and tests undertaken to establish their appropriate physical and chemical properties. Both types were found to have a very high moisture absorption when compared to quarried limestone whilst the density of masonry derived aggregates was significantly less than that of limestone. Typical results of tests for loose bulk density are summarised in Table 1.

Table 1. Loose bulk densities of aggregates

AGGREGATE TYPE	LOOSE BULK DENSITY (OVEN DRY), kg/m^3
Fine Limestone (N)	1355
Coarse Limestone (N)	955
Fine Masonry (R)	664
Coarse Masonry (R)	541
Fine Concrete (R)	1037
Coarse Concrete (R)	1115

(N) Naturally quarried   (R) Recycled

In producing any mix, it is thus crucial that the value of aggregate moisture absorption is accurately estimated. Table 1 shows that the loose bulk densities of the recycled masonry aggregates are within the loose bulk density limits for lightweight aggregates, of 1200 kg/m^3 for fine aggregate and 1000 kg/m^3 for coarse aggregate [15], the loose bulk density of the recycled concrete fine aggregate is also within the limits for a lightweight aggregate. The coarse concrete aggregate was just outside the maximum limit.

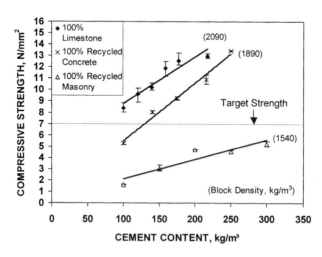

Figure 4 Results found using 100 % recycled materials

In light of these similarities to lightweight aggregates, a mixing procedure was adopted for all mixes of pre-mixing aggregate with one half of the water, followed by the cement and then the remaining water, which is similar to the mixing procedure used with lightweight aggregates. Replacement aggregate quantities were also calculated on a volumetric rather than mass basis.

Preliminary tests were undertaken using 100 % recycled materials and are summarised in Figure 4 for the maximum practically achievable air-dry densities. It can be immediately seen that compressive strengths are significantly lower than achieved with limestone, and substantial increases in cement content are needed to satisfy the block strength requirement. This would have serious implications for economic viability, thus partial replacement of limestone by recycled materials is indicated.

Subsequent tests have shown that partial replacement of coarse limestone by 60 % recycled concrete aggregates showed minimal strength reduction and this is consistent with the findings of Poon [12]. Replacement of the fine material by concrete derived aggregate produces a strength reduction above 20 % replacement. Replacement of limestone aggregate by recycled masonry has a significant effect at lower levels. Current work is focussing on establishing the optimum replacement levels for the relevant combinations of fine and coarse materials.

CONCLUSIONS

It has been demonstrated that a laboratory procedure using vibro-compaction and half-sized specimens with curing at room temperature can produce limestone concrete blocks with comparable properties to those achieved by commercial production processes. The degree of compaction is shown to be more critical than the water/cement ratio in achieving appropriate strength.

Crushed materials from building or demolition sites, although nominally separated as concrete or masonry, contain many impurities and have lower densities and higher water absorptions than limestone. This necessitates moisture content testing and mixing procedures similar to those used for lightweight aggregates.

Results show that it is unlikely that adequate block strengths can be achieved by 100 % replacement by recycled materials without a significant increase in cement content, thus partial replacement is required. Replacement of coarse material has less effect on strength than found with replacement of fine material.

ACKNOWLEDGEMENTS

The authors are grateful to the ONYX Environmental Trust and the Flintshire Community Trust Ltd (AD Waste Ltd) for funding this project. The authors would also like to thank the following industrial collaborators for their assistance with the project: Clean Merseyside Centre, Marshalls Ltd, Forticrete Ltd, Liverpool City Council, Liverpool Housing Action Trust (LHAT), RMC Readymix Ltd, WF Doyle & Co Ltd, DSM Demolition Ltd, and Environmental Advisory Service – Merseyside.

REFERENCES

1.   AGGREGATES ADVISORY SERVICE, Construction and Demolition waste – The European Union priority waste streams programme, http:www.planning.detr.gov.uk/aas/, Digest No 011 (3:1/99), p 4.

2.   ENVIRONMENT AGENCY, Waste Production, http:www.environment-agency.gov.uk/commondata/105385/swmiwprind.pdf, p 14.

3.   WILLIAMS, P.T., Waste Treatment and Disposal, John Wiley & Sons Limited, 1998.

4.   HM TREASURY, Consultation on the Objectives of Sustainability Fund under the Aggregates Levy Package, Sustainability Fund Consultation, Environmental Tax Team, HM Treasury, London, http://www.hm-treasury.gov.uk/docs/2000/sfcons2108.html.

5.   DEPARTMENT OF THE ENVIRONMENT TRANSPORT AND THE REGIONS, Planning for the Supply of Aggregates in England – A Draft Consultation Paper, http://www.detr.gov.uk/, October 2000, p 70.

6.   DEPARTMENT OF THE ENVIRONMENT, TRANSPORT AND THE REGIONS, Building a Better Quality of Life – A strategy for more Sustainable Construction, DETR free literature, Wetherby, http://www.detr.gov.uk, April 2000, p 33.

7.   WRAP, The Quality Protocol for the production of aggregates from inert waste, Banbury, 2004.

8.   GRADWELL, J., TICKELL, R.G., MILLARD, S.G., BUNGEY, J.H., JONES, N., Determining the economic viability of precast concrete products made with recycled construction and demolition waste, Proc. Int. Conf. Sustainable Waste Management and Recycling: Challenges and Opportunities, Kingston University, London, 2004.

9.   JOHNSTON, A.G., DAWSON, A.R., BROWN, G.J., Recycled Concrete Aggregate – an Alternative Concept in Sustainable Construction: Use of Recycled Aggregate, University of Dundee, 1998, pp 31-44.

10.   DHIR, R.K., LIMBACHIYA, M.C., LEELAWAT, T., Suitability of Recycled Concrete Aggregate for Use in BS5328 Designated Mixes, Proceedings of the Institution of Civil Engineers: Structures & Buildings, August 1999, pp 257-274.

11.   COLLINS, R., Recycled Concrete, Advanced Concrete Technology – Processes, ed. J Newman and B.S. Choo, Butterworth-Heinemann, 2003.

12.   POON, C.S., Use of Recycled Aggregates in Moulded Concrete Bricks and Blocks, Construction and Building Materials, Vol. 16, 2002, pp 281-289.

13.   DHIR, R.K., PAINE, K.A., DYER, T.D., Recycling construction and demolition wastes in concrete, Concrete, Vol. 38, No. 3, 2004, pp 25-28.

14.   BRITISH STANDARDS INSTITUTION, BS 6073-1: Precast Concrete Masonry Units – Specification for Precast Concrete Masonry Units, 1981.

15.   BRITISH STANDARDS INSTITUTION, BS 3797: Lightweight Aggregates for Masonry Units and Structural Concrete, 1990.

BEST PRACTICABLE ENVIRONMENTAL OPTIONS (BPEOs) FOR RECYCLING DEMOLITION WASTE

A Whyte

T D Dyer   R D Dhir

University of Dundee

United Kingdom

Abstract. Development of a new accessible and user-friendly guide able to assist in the determination of Best Practicable Environmental Options (BPEOs) for demolition waste is discussed. The paper explores the key aspects of this new practitioner's guide to the management of the waste-stream of construction demolition arisings. The guide is divided into several constituent material categorisations which are, in turn, examined in terms of their specific disposal alternatives. Each disposal alternative is assessed in terms of its environmental impact, measured in eco-points. Development of this new BPEO shall allow quick comparisons to be made by those charged with disposal responsibilities and, the opportunity to cross-reference preferred options with a detailed environmental justification.

Keywords: Best practicable environmental options, Demolition waste, Life cycle assessment

Dr A Whyte is a Research Fellow in the Concrete Technology Unit, University of Dundee. He is involved in research projects examining issues related to life-cycle analysis and, technology and knowledge transfer in the concrete construction industry.

Dr T D Dyer is a Research Lecturer in the Concrete Technology Unit, University of Dundee. He is involved in a number of research projects examining issues related to sustainability and recycling in concrete construction.

Professor R K Dhir is director of the Concrete Technology Unit, University of Dundee, Scotland UK. He specialises in binder technology, permeation, durability and protection of concrete. His interests also include the use of construction and industrial wastes in concrete to meet the challenges of sustainable construction. He had published and travelled widely and serves on many Technical Committees.

INTRODUCTION

Developing an environmentally sound strategy for recycling construction and demolition arisings is vital to address a waste stream that makes up 17% of the UK's total annual waste arisings. *Legislation-push* from the aggregates levy and landfill restrictions, together with *technology-pull* from modern recycling techniques able to process more than 95% of demolished material, will continue to exert pressure on recycling rates for demolition materials. To allow construction practitioners to recognise the full potential of construction demolition as a building resource, improved guidance is needed.

Given the considerable volume of research already carried out into the environmental impact of various building materials processing and manufacturing techniques, empirical comparisons are now possible to allow stakeholders to make informed decisions regarding recycling and disposal options for demolition and construction waste arisings. A new guide able to assist in determining best practicable environmental options (BPEOs) for demolition waste is currently being developed to assist practitioners and legislators in the construction industry to make objective decisions regarding the management of demolition waste. The potential benefits of guidance based on best practicable environmental options includes: improved environmental credentials, savings in disposal and transportation costs, revenue generation from reuse beyond current recycling rates of 35-37% and, reductions in the demand for increasingly scarce virgin materials [1].

This paper discusses the constituents and categorisation of construction and demolition waste in terms of European Waste Catalogue definitions. The need to establish the full range of alternative waste disposal options is then outlined. The paper then charts the development of respective waste disposal schematics and the calculation of environmental impact (Ecopoints). Using cable as an example, guidance in choosing the best practical environmental option from a range of disposal alternatives is highlighted.

WASTE CATEGORISATION AND ENVIRONMENTAL IMPACT

The UK's devolved Waste Plans increasingly use best practicable environmental options to underpin environmental legislation and, where appropriate, dictate waste management solutions that are enforceable by law [2]. Methodologies for establishing BPEOs for waste seek a 'balance between economic, environmental and social costs and benefits, in a deliverable and affordable manner and in compliance with the law'. Current guidance concentrates on *municipal* waste management. No BPEOs deal specifically with the arisings generated by construction and demolition [3].

Construction and demolition (C&D) waste is one of the Environmental Protection Agency's thirteen specific classifications of waste (alongside arisings from agriculture, mining, dredged spoils, municipal, commercial, and industrial wastes) and is regarded as requiring priority waste management attention. Regulators acknowledge that much work is required not only to devise and implement appropriate C&D waste management schemes but also, to establish accurate regional measurement data.

Procedures to establish a BPEO for waste generally requires stakeholders to establish a baseline of material generated, ascertain decision criteria for disposal, and then define, appraise, shortlist and consequently choose optimum waste management options [4].

Waste Constituents and Categorisation

Demolition waste is composed chiefly of concrete rubble and masonry and, to a lesser extent, asphalt, with the remainder a mix of ceramics, wood, roof tiles, metals and glass. The waste stream is said to constitute a quarter of the European community's controlled waste arisings and has been divided, by the European Waste Catalogue (EWC), into more than two dozen construction and demolition waste coding sub-categories. These sub-divisions indicate the level of sorting possible and should, in theory, assist segregation of arisings by contractors to increase the potential for reuse and recycling [5]. A broad categorisation of construction and demolition waste may be described as:

Concrete:	reinforced & mass, rubble, with bricks, blocks
Ceramics:	bricks, tiles, drainage pipes, sanitary ware
Inerts:	topsoil, sub-base, sand, clay, gravel, natural stone
Gypsums:	gypsum-based, mortar, cement, screed, external cladding
Pavements:	bitumens based materials
Metals:	aluminium, copper, galvaniseds, tin, cables, zinc, brass, lead, iron
Insulation:	mineral wool, glass fibre, polystyrene, styrofoam
Timber:	general, plywood, MDFs, chipboard
Plastics:	UPVC windows, doors, DPCs
Other:	Packaging (cardboard, paper), office/canteen waste

Waste disposal options and feasibility

Managing the constituents of construction and demolition waste falls under the two main removal options of landfill and, re-use and recycling. Choices include [6]:

- Land-filling to non-hazardous sites, inert material sites and hazardous waste sites.
  - Exempt-site disposal at locations authorised by local Environmental Regulators.
  - Unlicensed disposal and illegal fly-tipping sites particularly in rural areas
- Low-level processing of waste to road-fill, bunds and temporary works.
- High-level re-processing as reconstituted building materials and component salvage.

Whilst the technical expertise exists to utilise waste materials as new building products transportation issues, relating to the location of both feedstock source and recycling facility, as well as consumer resistance towards secondary materials, persist. Costs and revenue-generating potential influence how waste is managed by industry, with low specification incorporation of waste preferable to heavily-processed high-grade products [7].

A range of measures such as command regulations, fiscal controls and encouragement via so-called 'green' procurement processes (contract conditions and design specifications that are sympathetic towards recyclates) can best encourage C&D waste re-use and recycling [8]. Greater industry awareness, linked with the fact that recycled aggregates are now permitted constituents for most applications in the Specification for Highway Works, increases uptake of recyclable materials [9].

Waste management methodologies, seeking to identify best practical environmental options, require consideration of ease of use, legislation to safeguard user-groups, financial cost and environmental impact.

Life-Cycle Analysis

It has been argued that to achieve sustainability in construction, social issues be given a 20% weighting of importance, economics be given an importance weighting of 30% but that environmental issues, with a weighting of 40%, are most important [10]. The BPEO for the management of demolition waste similarly demands environmental impact comparison of disposal options, particularly *life-cycle analysis* to quantify and assess the inputs and outputs affecting the performance of particular disposal alternatives. *Life-cycle assessment* (LCA) takes analysis to the stage of environmental impact examination and provides a means of quantifying materials and energy used and pollutants generated.

Using life cycle assessment, the environmental impacts of the life-cycle stages of construction materials and recyclates can be compared [11]. Given difficulties in making judgements about the benefits of different materials and their effect on environmental impact comparison categories, ranking and a simplified system of scoring or weighting environmental impacts, allows data to be more accessible. Weightings are required for the environmental impact categories of: Climate change, Ozone depletion, Low-level ozone creation, Ecotoxicity, Fossil fuel depletion, Water extraction, Acid deposition, Human toxicity to air, Human toxicity to water, Eutrophication, Minerals extraction and, Waste disposal.

Environmental estimating tools, are beginning to allow users to calculate the environmental impacts in a simplified points basis. One such system, that of Ecopoints, attributes marks across cumulative elements. As a single unit of measurement an Ecopoint measures the total environmental impact of a product or process, derived across the defined range of weighted environmental impact categories and, normalised by taking the total number of Ecopoints for all impacts that arise per UK citizen amounting to 100. Higher Ecopoint values indicate greater environmental impact so that a simple comparison of two alternatives identifies the lowest Ecopoint value as a preferred option [10].

USER GUIDE

An examination, by the authors, of each of the waste categories identified above has been conducted in terms of the constituent resources and energy associated with the input feedstock, the pollution and waste of the processed material and building product output. The spectrum of alternatives for each of the categories has been prepared in the form of Waste Disposal Schematics depicting the nature of the demolition material, disposal options and, the range of processes that contribute to the overall environmental impact of material specific alternatives.

The assessment procedure to establish a means to predict and subsequently compare the environmental impact of the range of waste disposal options builds upon previous research conducted by the CTU, University of Dundee and others in the application of environmental life-cycle analysis techniques for plant utilisation, material production and emissions. Environmental assessment data have subsequently been calculated in the form of Ecopoints to assist comparison.

Waste Disposal Schematics

Disposal option schematics have been prepared across the range of demolition waste materials under the headings of: Cable, Plastic, Insulation, Plasterboards/ Gypsums, Wood, Concrete/ Masonry/ Ceramics, Soil, Pavement/ Bitumens, Glass, and Ferrous and Non-ferrous Metals. An example of a waste disposal schematic is presented below (Figure 1) for the waste *Category* of Cable where landfill, recycling and incineration present themselves as the *Disposal Options* and analyses of respective processes determine *Environmental Impact*.

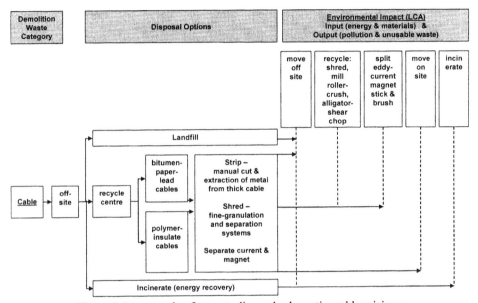

Figure 1 An example of a waste disposal schematic: cable arisings

Environmental Impact

Life-cycle assessment data for specific waste disposal options allow comparison of alternatives in terms of environmental impact. In the example of cable (Figure 1) both bitumen-paper-lead/metal cables and, polymer-insulated cables (PVC and PU) are considered, where waste generally consists of a metal conductor (copper) with a semi-conductive sheet (carbon black) and an insulator, bedded and sheathed in plastic. Disposal options may be summarised in the following terms

- *Landfilling* waste cable breaks down only 3% of the original polymer material during the first 100 years, although up to 80% of the plasticisers in PVC cable are broken down. Limestone filler in the plastic materials are relatively stable in landfill conditions.
- *Recycling* requires liberation and separation by: stripping –cutting and the extraction of metal from cable elements, shredding –fine-granulation and, separation –corona wires sprayed with a charge to allow metallic particle conduction to a magnet and resultant polymeric materials to be crushed off.
- *Incineration* of waste cable may be used as a means of energy recovery.

Secondary research provides input, output and emission data for cable waste such that:

- Input materials are made up of approximately 70% plastic and 30% (copper) conductor.
- Input energy of up to 1,600 kWh of electrical power together with over 500 kWh of power from oil is used to *recycle* the constituent plastic and conductor of 1t of cable.
- Output energy during *Landfill* energy collection is estimated at between 10 and 200 kWh per 1t of cable. *Incineration* of 1t of waste cable generates 3,000 to 5,000 kWh of power.

Environmental impact figures based on energy/ pollution values are carried forward towards a normalised impact-value of total Ecopoints for 1 tonne of waste cable associated with landfill, incineration and recycling procedures respectively. Table 1 summarises environmental impact data.

Table 1 An example of impact values for landfill, incineration and recycling: cable calculated using weighted environmental profiles (Ecopoints) valuation system

IMPACT CATEGORY	UNITS	ENVIRONMENTAL IMPACT PER TONNE OF WASTE CABLE (plastic 70% and copper 30%)				
		Landfill		Incineration		Recycling thermoplastic
		PVCs	PUs	PVCs	PUs	
Climate change	kgCO$_2$eq-100yrs	2.7300	0.4468	2.4533	3.3947	0.4655
Acid deposition	kg SO$_2$eq	0.0000	0.0000	0.3938	0.0071	0.0365
Human toxicity	kg tox	0.0056	0.0004	0.0067	0.0032	0.0085
Low-level ozone	kg ethane eq	0.0301	0.0052	0.0000	0.0000	0.0005
Eutrophication	kgPO$_4$eq	0.0495	0.0064	0.0059	0.0078	0.0308
Waste disposal	tonnes	0.4850	0.5583	0.1963	0.2333	0.0000
Fossil fuel depletion	toe	-0.0802	-0.0049	-1.3600	-1.8936	0.4919
Minerals extraction	tonnes	0.1842	0.1895	-0.1841	-0.1895	-0.1869
	Ecopoint	3.4062	1.2018	1.5129	1.5631	0.8470

Given that high Ecopoint values indicate a greater environmental impact, comparison of the disposal options for waste (PVC) cable suggest that landfill is the least attractive option, with recycling the most attractive, best practicable environmental option, notwithstanding the variables of cost and legislation compliance.

CONCLUSIONS

Building upon a category-by-category exploration of waste disposal options, a new Guide to the BPEO for recycling demolition waste is being developed to allow stakeholders to consider the environmental impact of the constituent resources and energy requirements in obtaining the input feedstock and output pollution and residue in the preparation of recycled building products. Work to date concentrates on environmental profile Ecopoint valuations for the disposal options. This shall complement an examination of available arrangements for waste management in terms of performance, rated across decision criteria made up of, not only environmental issues but also, economics, compliance with legislation that affects society in general and, practicality of option implementation in local authorities throughout the UK.

ACKNOWLEDGMENTS

The authors would like to express their gratitude to the RMC Environmental Fund for the funding of this project and the various industrial partners, namely AMEC Capital Projects Ltd., British Cement Association, Greater Manchester Waste Ltd., John Doyle Construction Ltd., Quarry Products Association, Scottish Environment Protection Agency, Tayside Contracts and, WS Atkins Consultants Ltd for their valuable involvement.

REFERENCES

1. HOBBS,G., Management of construction & demolition waste', BRE IP1/96, CI/SfB-A5-T6, Watford, 1996

2. RTPI / CORDY, T., Planning and environmental protection: an introductory guide, Thomas Telford, London, 2002

3. SCOTTISH EXECUTIVE / SCOTTISH ENVIRONMENT PROTECTION AGENCY (SEPA), 'The National Waste Plan 2003', Charlesworth, UK, 2003

4. SCOTTISH ENVIRONMENT PROTECTION Agency (SEPA), National Waste Strategy, Best Practicable Environmental Option Decision Making Guidance', UK, 2000

5. EUROPEAN WASTE CATALOGUE, CONSLEG: 2000 DO 532 – 01/01/2002, Publications for the European Communities, 2000

6. DEPT. OF ENVIRONMENT/ HOWARD HUMPHRIES, Managing demolition and construction wastes: report of the study on the recycling of demolition and construction wastes in the UK' HMSO, 1994

7. GUTHRIE,P., COVENTRY,S. & WOOLVERIDGE,A., Waste minimisation and recycling in construction – technical review', CIRIA Report 28, CIRIA London, 1999

8. COLLINS, R., 'Fit for Purpose Specifications: using recycled and secondary aggregates' BRE, Watford, 2001

9. BARRIT, J., Overcoming barriers to recycling -coarse aggregates for concrete in BS8500:part2(2002), 2nd Int. Conf. on Pavement Engineering Liverpool, 2003

10. DICKIE,I. & HOWARD,N., Assessing environmental impacts of construction: industry consensus, BREEAM and UK Copouts', BRE Digest 446, ISBN1 86081 3984, 2000

11. ANDRESON, J., EDWARDS,S., MUNDY,J. & BONFIELD,P., 'Life cycle impacts of timber: a review of the environmental impacts of wood products in construction', Digest 470, BRE Watford, ISBN 1860815863, 2002

COST-EFFECTIVE AND GOOD-PERFORMANCE CONCRETE FOR SUSTAINABLE CONSTRUCTION THROUGH RECYCLING

G Moriconi

Technical University of Marche

Italy

ABSTRACT. A judicious use of natural resources, achieved by the use of by-products and waste materials, and a lower environmental impact, achieved through reduced carbon dioxide emission and reduced natural aggregate extraction from quarries, represent two main actions that meet sustainable construction development. Recycled-aggregate concrete containing high-volume fly ash is an example of construction material in harmony with this concept, whereby sustainable construction development is feasible with satisfactory performance, in terms of both safety and serviceability of structures, at lower costs and with environmental advantages over ordinary concrete. In this paper, criteria are discussed on the basis of which recycled-aggregate concrete mixture proportions can be optimized as well as the fresh concrete behaviour during placing. Moreover, when using recycled aggregates appropriately, some important properties of the hardened concrete such as ductility and durability can be better engineered, as this paper emphasizes.

Keywords: Sustainable construction, Recycled-aggregate concrete, High volume fly ash concrete.

Professor G Moriconi, is the Head of the Department of Materials and Environment Engineering and Physics at the Technical University of Marche, Ancona, Italy. He is the author or co-author of numerous papers in the field of cement chemistry, concrete technology and building materials. In 2003, he was awarded by ACI for his contribution to the research development in the general field of concrete durability.

INTRODUCTION

Concrete is basically made of aggregates glued by a cement paste which is made of cement and water. Each one of these concrete primary constituents, to a different extent, has an environmental impact and gives rise to different sustainability issues [1-2]. The current concrete construction practice is thought unsustainable because, not only is it consuming enormous quantities of stone, sand, and drinking water, but also one billion tons a year of Portland cement, which is not an environment-friendly material from the standpoint of energy consumption and global warming [3].

However, notwithstanding the energy consumption of cement production and the related remarkable carbon dioxide emissions, concrete can "adsorb" these negative effects and become an environmentally sustainable material. This outstanding effect is mainly attributable to the opportunity of easily incorporating in concrete mineral additions quite different in nature, composition and origin, thanks to concrete technology developments particularly connected to advances in concrete admixtures. In fact, many by-products and solid wastes can be used in concrete mixes as aggregates or cement replacement, depending on their chemical and physical characterization, if adequately treated. The capacity of concrete for incorporating these secondary raw materials is very wide and the main limit is their availability, which has to be comparable with the cement stream, since it is not worthwhile to develop new cementitious materials if their availability on the market cannot be guaranteed.

As an example, environmental issues associated with carbon dioxide emissions from the production of Portland cement demand that supplementary cementing materials in general, and fly ash as well as ground granulated blast-furnace slag in particular, be used in increasing quantities to replace it in concrete. Given the almost unlimited supply of good quality fly ash worldwide, new concrete technology such as high-volume fly ash concrete has been developed, based on the combined use of superplasticizers and supplementary cementing materials, leading to economical high-performance concrete with enhanced durability [4-8].

Therefore, much of the discussion of the sustainability of the concrete industry to date has dealt with materials issues such as the use of Portland cement replacement materials and recycling of concrete removed from existing structures. However, any discussion of the sustainability of the concrete industry must consider industry concerns much broader than those of "greenness" of a given technology. For example, if the public or designers perceive concrete as a non-durable material or as a material that is more difficult to design, the sustainability of the industry is affected [9]. A related comment is that public funding has become a very limited resource with many demands running after limited discretionary funding. As a result, publicly funded infrastructure simply must last longer, since the replacement of these structures before a reasonable life span cannot be allowed. However, in general, there is an increased interest in durability of structures and life-cycle costing. Projects have recently been completed where 1,000 year service life for the concrete has been requested and achieved through high-performance, high-volume fly ash concrete [10]. While these projects are unusual, service life requirements for 100 years for bridges in severe environments are becoming more common [9].

Finally, it must be realized that resources are limited. In particular, the mineral resources that are necessary for cement and concrete production are being stretched or exhausted in some locations. Yet in spite of the growing awareness that resources are being depleted, there is an extremely strong movement against developing new sources [9]. However, since sustainability is becoming an important issue of economic and political debates, the next developments to watch in the concrete industry will not be new types of concrete, manufactured with expensive materials and special methods, but low cost and highly durable concrete mixtures containing the largest possible amounts of industrial and urban wastes that are suitable for partial replacement of Portland cement, virgin aggregate and drinking water. The goal is to transform all concrete into a general-purpose building material that is composed of eco-friendly components, and produces crack-free highly durable structures [3].

Recently published reports [11-12] show that the goal of complete utilization of construction and demolition wastes is attainable. For instance, it has been found [11] that the finely ground fraction from these wastes, when used as a partial replacement for cement, improved the bond strength between mortar and fired-clay brick in masonry units. For use in structural concrete mixtures, it was showed [12] that the strength loss resulting from complete replacement of natural coarse and fine aggregates with recycled-concrete aggregates can be compensated by incorporation of fly ash and water-reducing chemical admixtures into fresh concrete mixture.

This paper reviews research to promote recycling into structural concrete for common use in building construction, with the aim of emphasizing the feasibility, as well as the advisability, of such an action. High-volume fly ash recycled-aggregate concrete is an attainable result of these efforts and it may be regarded as a cost-effective and good-performance material for sustainable construction achieved through recycling.

HIGH-VOLUME FLY ASH RECYCLED-AGGREGATE CONCRETE

Mixture Proportions

Recycled-aggregate concrete (RAC) for structural use can be prepared by completely substituting natural aggregate, with a water/cement ratio of 0.30 in order to achieve the same strength class as the reference concrete, manufactured by using only natural aggregates with a water/cement ratio of 0.60 as commonly used in practice, with the aim of compensating the recycled aggregate weakness by a safety margin [13]. This is obviously a provocation, since so large a stream of recycled aggregates to allow for full substitution of natural aggregates is not available. However, it is useful to prove that to manufacture structural concrete by partly substituting natural with recycled aggregates by up to fifty percent is absolutely feasible. In any case, since the adoption of very low water/cement ratios implies unsustainably high amounts of cement in the concrete mixture, recycled-aggregate concrete may also be manufactured by using a water-reducing admixture in order to lower both water and cement dosage, or even by adding fly ash as a partial fine aggregate replacement and by using a superplasticizer to achieve the required workability, without disregarding in all cases the required strength class [12]. Moreover, high-volume fly ash recycled-aggregate concrete (HVFA-RAC) can be manufactured with a water/cement ratio of 0.60, by simultaneously adding to the mixture as much fly ash as cement, substituting the fine aggregate fraction [14]. Thus, a water to cementitious material (binder) ratio of 0.30 is obtained enabling the concrete to reach the required strength class (see Table 1).

This procedure is essential for designing an environmentally-friendly concrete. All the concretes can be prepared maintaining the same fluid consistency by proper addition of an acrylic-based superplasticizer.

Table 1   Concrete mixture proportions (kg/m^3).

Concrete Mixture		Natural-Aggregate Concrete	RAC	HVFA-RAC
Water		230	230	230
Cement		380	760	380
Fly Ash		-	-	380
Natural Sand		314	-	-
Fine Recycled Fraction		-	-	-
Crushed Aggregate		1338	-	-
Coarse Recycled Fraction		-	1169	1057
Superplasticizer		-	-	6.8
Water/Cement		0.60	0.30	0.60
Water/Binder		0.60	0.30	0.30
Compressive Strength (MPa)	3 days	16	26	20
	28 days	27	31	29
	60 days	32	34	36

Fresh Concrete Behaviour

When concrete shows high fluidity besides good cohesiveness, it is said to be self-compacting. This recent achievement of concrete technology, which has lead to several advantages, is in fact a development of the well-known rheoplastic concrete [15-17], achieved thanks to superplasticizers, in which segregation and bleeding are suppressed thanks to filler addition and the use of a viscosity-modifying agent. However, these additions may not be sufficient, if the maximum volume of coarse aggregate and minimum volume of fine particles (including cement, fly ash, ground limestone, etc.) are not complied with. Furthermore, from rheological tests on cement pastes, it has been observed that, for maximum segregation resistance, the yield stress of the paste should be high [18-21] and the difference in density between the aggregate and the paste should be low. This would mean that segregation will be particularly reduced when lighter aggregate, such as recycled aggregate, is used [22]. Moreover, this behaviour seems to be enhanced when rubble powder, that is the finest fraction produced during the recycling process of rubble to make aggregates, is reused as a filler. In this condition, the segregation resistance appears so high that the coarse recycled aggregate can float on a highly viscous cement paste, and an adjustment could be attempted by adding fly ash which, when used alone as a filler, confers reduced flow-segregation resistance and increased flowability to concrete.

Durability

Some aspects of the durability of recycled-aggregate concretes have already been studied. In particular, attention has been focused on the influence of concrete porosity on drying shrinkage and corrosion of embedded steel bars as well as on concrete carbonation, chloride ion penetration and concrete resistance to freeze-thaw cycles [14, 23-26]. Results showed that, when fly ash is added to recycled-aggregate concrete:

1. Pore structure /macropore is improved, volume is reduced with benefitial mechanical performance, such as compressive, tensile and bond strength [14, 26]. Ordinary concrete with natural aggregate, has reduced stiffness of recycled-aggregate concrete containing fly ash, which should be taken into account during structural design [26].

2. From a serviceability point of view the drying shrinkage of recycled-aggregate concrete does not appear a problem since, due to the reduced stiffness of this concrete, the same risk of crack formation results as for ordinary concrete under restrained conditions [26].

3. Testing of resistance against freezing and thawing cycles shows no difference between natural-aggregate concrete and high-volume fly ash recycled-aggregate concrete [14].

4. The addition of fly ash is very effective in reducing carbonation and chloride ion penetration depths in concrete (Figures 1 and 2), because of pore refinement of the cementitious matrix due to a filler effect and pozzolanic activity of fly ash. Moreover, the strong beneficial effect of the presence of fly ash on chloride penetration depth is evident since the chloride ion diffusion coefficient in high-volume fly ash concrete is one order of magnitude less than that into concrete without a fly ash addition [14, 25].

5. As far as corrosion aspects are concerned, the use of fly ash does not decrease the corrosion resistance of steel reinforcement (Figure 3), as long as the concrete strength is adequate, whilst it appears very effective in protecting galvanized steel reinforcement (in Figure 4 the zinc layer is totally consumed only for natural-aggregate concrete) also in porous concrete, as it can occur when recycled aggregates are used, even in the case of cracked concrete [25-26].

6. In general, it is confirmed that concrete containing high volume of fly ash presents no problem with respect to corrosion of reinforcement, because of the very low permeability of concrete, even when a porous aggregate, such as recycled aggregate, is used. In fact, if on the one hand fly ash addition reduces the concrete pore solution alkalinity by altering the passivity conditions of steel reinforcement, on the other hand it improves significantly the concrete microstructure by making the penetration of aggressive agents and the onset of corrosion increasingly difficult.

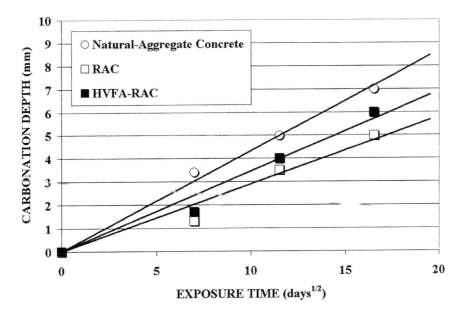

Figure 1   Carbonation depth as a function of the time of exposure to air.

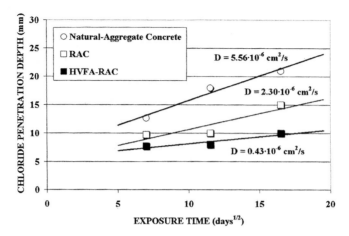

Figure 2 Chloride penetration depth as a function of the time of exposure to a 10% sodium chloride aqueous solution

Figure 3 Visual observation of the corrosive attack at the crack apex on bare steel plates embedded in natural-aggregate concrete (left), RAC (middle) and HVFA concrete (right).

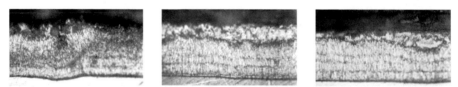

Figure 4 Metallographic cross section of galvanized steel plates embedded in cracked natural-aggregate concrete (left), RAC (middle) and HVFA concrete (right).

Economical Evaluation

As in most common structural applications, if a strength class value of 30 MPa is required, recycled-aggregate concrete without any mineral addition would not perform satisfactorily, whereas recycled-aggregate concrete with high-volume fly ash would have excellent performance. For this reason an economical comparison should be made for comparable performances [12] between natural-aggregate concrete and recycled-aggregate concrete with high-volume fly ash of the same strength class.

On the basis of current costs of the individual constituents, at present in Italy traditional costs evaluation can be carried out leading to the cost of high-volume fly ash recycled-aggregate concrete being slightly higher (about 5%) than natural-aggregate concrete (Table 2). This result is nearly obvious since both types of concrete belong to the same strength class. However, besides the traditional cost of aggregates, it would be important to take into account their environmental cost. The *eco-costs* [27], which are the expenses necessary to eliminate the environmental impact caused by the extraction of natural aggregates from quarries, should be considered as well as the *negative eco-costs*, that are the expenses to eliminate the environmental load if rubble from building demolition, and also fly ash from thermal plants, are not utilized to produce concrete. By considering the environmental costs of aggregates [27], though undeterminable/ changeable with social and political factors, it is predicted that high-volume fly ash recycled-aggregate concrete in the future could be remarkably cheaper than the natural-aggregate concrete, following an ineluctable development of the aggregates market.

Table 2 Traditional (T) and eco-balanced [1] (E-B) costs referred to 1 m^3 of concrete.

Ingredient	Unit cost (€/kg)	Natural-Aggregate Concrete		RAC		HVFA-RAC	
		T	E-B	T	E-B	T	E-B
Water	0.001	0.30	0.30	0.30	0.30	0.30	0.30
Cement	0.121	45.98	45.98	91.96	91.96	45.98	45.98
Fly Ash	0.022	-	-	-	-	8.36	8.36
Fly Ash Disposal	0.250	-	-	-	-	-	-95.00
Natural Sand	0.015	4.55	4.55	-	-	-	-
Fine Recycled Fraction	0.007	-	-	-	-	-	-
Crushed Aggregate	0.013	17.26	17.26	-	-	-	-
Coarse Recycled Fraction	0.006	-	-	7.54	7.54	6.82	6.82
Rubble Disposal	0.050	-	-	-	-58.45	-	-52.85
Superplasticizer	1.435	-	-	-	-	9.76	9.76
Total		68.09	>68.09 [1]	99.80	41.35	71.22	-76.63

[1] Only negative eco-costs, deriving from waste disposal, are taken into account.

CONCLUSIONS

An important conclusion from this paper is that the compressive strength of the recycled-aggregate concrete can be improved to equal or exceed natural-aggregate concrete by adding fly ash to the mixture as a fine aggregate replacement. In this way, a given strength class value, as required for a wide range of common uses, can be reached through both natural-aggregate concrete and recycled-aggregate concrete with high-volume fly ash, by adequately decreasing the water to cement ratio with the aid of a superplasticizer in order to maintain the workability. Concrete manufactured with recycled aggregate and high-volume fly ash shows no deleterious effect on the durability of reinforced concrete, with some improvement in some cases. From an economical point of view, if only traditional costs are taken into account, recycled-aggregate concrete with fly ash is less attractive than natural-aggregate concrete. However, if eco-balanced costs are considered, the exact opposite could be valid.

REFERENCES

1. MEHTA, P K. Reducing the Environmental Impact of Concrete. Concrete International. Vol. 23, No. 10, 2001. pp 61-66.

2. MEHTA, P K. Greening of the Concrete Industry for Sustainable Development. Concrete International. Vol. 24, No. 7, 2002. pp 23-28.

3. MEHTA, P K. The Next Revolution in Materials of Construction. Proceedings of the VII AIMAT Congress, Ancon, Italy, June 29 – July 2, 2004. Keynote Paper 1, 19 pp.

4. MALHOTRA, V M. Superplasticized Fly Ash Concrete for Structural Applications. Concrete International. Vol. 8, No. 12, 1986. pp 28-31.

5. MALHOTRA, V M. Making Concrete 'Greener' With Fly Ash. Concrete International. Vol. 21, No. 5, 1999. pp 61-66.

6. MEHTA, P K. Concrete Technology for Sustainable Development. Concrete International. Vol. 21, No. 11, 1999. pp 47-53.

7. MALHOTRA, V M AND MEHTA, P K, High-Performance, High-Volume Fly Ash Concrete, Supplementary Cementing Materials for Sustainable Development Inc., Ottawa, Canada, ISBN: 0-9731507-0-X, 2002, 101 pp.

8. MALHOTRA, V M. Concrete Technology for Sustainable Development. Proceedings of the Two-Day International Seminar on "Sustainable Development in Cement and Concrete Industries", Milan, Italy, October 17-18, 2003. pp 11-18.

9. HOLLAND, T C. Sustainability of the Concrete Industry – What Should Be ACI's Role?. Concrete International. Vol. 24, No. 7, 2002. pp 35-40.

10. MEHTA, P K AND LANGLEY, W S. Monolith Foundation: Built to Last '1000 Years'. Concrete International. Vol. 22, No. 7, 2000. pp 27-32.

11. MORICONI, G, CORINALDESI, V AND ANTONUCCI, R. Environmental-Friendly Mortars: A Way to Improve Bond Between Mortar and Brick. Materials and Structures. Vol. 36, No. 264, 2003. pp 702-708.

12. CORINALDESI, V AND MORICONI, G. Role of Chemical and Mineral Admixtures on Performance and Economics of Recycled-Concrete Aggregate. American Concrete Institute. SP-199, 2001. pp 869-884.

13. CORINALDESI, V, ISOLANI, L AND MORICONI, G. Use of rubble from building demolition as aggregates for structural concretes. Proceedings of the 2nd National Congress on Valorisation and Recycling of Industrial Wastes, L'Aquila, Italy, 5-8 July 1999. pp 145-153.

14. CORINALDESI, V AND MORICONI, G. Durability of Recycled-Aggregate Concrete Incorporating High Volume of Fly Ash. Proceedings of the 9th International Conference on "Durability of Building Materials and Components", Brisbane, Queensland, Australia, March 17-20, 2002. Paper 71.

15. COLLEPARDI, M. Assessment of the "Rheoplasticity" of Concrete. Cement and Concrete Research. 6, 1976. pp 401-408.

16. COLLEPARDI, M. A Very Close Precursor of Self-Compacting Concrete (SCC). Proceedings of the Third CANMET/ACI International Symposium on "Sustainable Development of Cement and Concrete", Supplementary Papers, San Francisco, U.S.A., 2001. pp 431-450.

17. COLLEPARDI, M. Self-Compacting Concrete: What is New?. American Concrete Institute. SP 217, 2003. pp 1-16.

18. BILLBERG, P. Fine mortar rheology in mix design of SCC. Proceedings of the First International RILEM Symposium on "Self-Compacting Concrete", Eds. A. Skarendahl & O. Petersson, Stockholm, Sweden, 1999. pp 47-58.

19. EMBORG, M. Rheology Tests for Self-Compacting concrete – How Useful are they for the Design of Concrete Mix for Full Scale Production?. Proceedings of the First International RILEM Symposium on "Self-Compacting Concrete", Eds. A. Skarendahl & O. Petersson, Stockholm, Sweden, 1999. pp 95-105.

20. SAAK, A W, JENNINGS, H M AND SHAH, S P. Characterization of the Rheological Properties of Cement Paste for Use in Self-Compacting Concrete. Proceedings of the First International RILEM Symposium on "Self-Compacting Concrete", Eds. A. Skarendahl & O. Petersson, Stockholm, Sweden, 1999. pp 83-93.

21. SAAK, A W, JENNINGS, H M AND SHAH, S P. New Methodology for Designing Self-Compacting Concrete. ACI Materials Journal. Vol. 98, No. 6, 2001. pp 429-439.

22. CORINALDESI, V AND MORICONI, G. The Role of Recycled Aggregates in Self-Compacting Concrete, American Concrete Institute, SP 221, 2004. SP-221-57.

23. TITTARELLI, F AND MORICONI, G. The Effect of Fly Ash and Recycled Aggregate on the Corrosion Resistance of Steel in Cracked Reinforced Concrete. Proceedings of the 9th International Conference on "Durability of Building Materials and Components", Brisbane, Queensland, Australia, March 17-20, 2002. Paper 70.

24. CORINALDESI, V, TITTARELLI, F, COPPOLA, L AND MORICONI, G. Feasibility and Performance of Recycled Aggregate in Concrete Containing Fly Ash for Sustainable Buildings. American Concrete Institute. SP 202, 2001. pp 161-180.

25. CORINALDESI, V, MORICONI, G AND TITTARELLI, F. Durability of Reinforced Concrete for Sustainable Construction. American Concrete Institute. SP 209, 2002. pp 169-186.

26. MORICONI, G. Third Millennium Concrete: A Sustainable and Durable Material. L'Industria Italiana del Cemento. 787, 2003. pp 430-441.

27. TAZAWA, E. Engineering Scheme to Expedite Effective Use of Resources. Proceedings of the International Symposium on "Concrete Technology for Sustainable Development in the Twenty-First Century" (Ed. P.K.Metha), Radha Press, New Delhi, India, 1999. pp 23-42.

ROOM TEMPERATURE GRANULATED FLY ASH ON A FIXED BED AS SORBENT FOR ORGANIC CONTAMINANTS FROM WASTEWATER

G Rinaldi

F Medici

University of Roma La Sapienza

Italy

ABSTRACT. The removal of colour, turbidity and C.O.D. from the polluted waters containing organic contaminants, like humic acids and sodium ligninsulphonates was experimented: the results suggest an 'ecofriendly' technology for the polluting industries producing both: those producing fly ashes (i.e. coal thermal plants) and organic contaminated wastewaters. The process was developed in a fixed bed with use of by using a granulated fly ash as sorbent, a clear advantage in comparison to the use of fly ashes in the form of a slurry. Granulation was carried out by mixing a silico-aluminous coal fly ash with 10 % (w/w) of hydrated lime and water; curing of the granules was at room temperature. The process allowed the obtainment of granules with tailored porosity and good durability in water. Last, but not least, the exhausted bed can be regenerated by thermal treatment, and repeatedly used. The granulated fly ash has yet a sorption capacity at about the same extent in comparison with the original product.

Keywords: Granulation, Fly ash, Lime, Sorption, Organic pollutants, Fixed bed, Reuse.

Prof G Rinaldi, is Associate Professor of Aerospace Materials Science and Technology at the Faculty of Engineering, member of the Academic Team for the "Doctorate" (PhD) in Chemical and Environmental Engineering and Professor of Instrumental Chemical Analysis at University School on Industrial Safety and Protection of the University of Roma "La Sapienza" (Italy).

Prof F Medici, is Associate Professor of Materials Technology and Applied Chemistry at the Faculty of Engineering and member of the Academic Team for the "Doctorate" (PhD) in Chemical and Environmental Engineering of the University of Roma "La Sapienza" (Italy).

INTRODUCTION

In previous researches [1, 2, 3] room temperature granulation of fly ashes from coal thermal plants was carried out by mixing them with low quantities (up to 10 %) of hydrated lime. Good mechanical strength was obtained by curing in a carbon dioxide-enriched environment, or by seasoning in air the mixes containing some organic additives [4].

The resulting long-term durability of the granulated products was very interesting because their mechanical strength remained practically unchanged both in wet environments and in water; moreover there was not release of the heavy metals present in the original coal fly ash [3].

In the previous experimental studies we attempted to diminish the porosity level of the granules to enhance their compression strength [4], directly linked to the low porosity of the material. To do this, we synthesized some organic polyaminic additives, tailored to increase the extent of carbonation of the remaining free lime fraction into the mixes [5, 6]. The long-term durability in water of some porous granules, above all those from a silico-aluminous fly ash and hydrated lime without the presence of the additives, suggested the suitability of these "porous" granules as a sorbent medium for organic contaminants from wastewater; in fact, for the utilization as sorbent material, porosity would be very useful.

The use of coal fly ash alone, in the sorption of organic pollutants from wastewater, has been widely investigated: for the removal of colour, turbidity and COD from bleach plant effluents of a Kraft pulp mill [7], of glucose and starch [8], textile dyeing and printing plant effluents [9], carboxylic acids [10] and other organics [11, 12, 13, 14]. These experiments were executed by using the coal fly ash in the form of a slurry in the water to be purified, so that the process needed the final filtration or sedimentation of the slurry itself, for the disposal in landfill of the solid fraction containing the sorbed pollutants; another not negligible drawback is that the final wet organic product cannot be cheaply destroyed by incineration to recycle the sorbent fly ash.

Moreover for the purification of the effluents from a Kraft pulp mill [7] the fly ash slurry is very effective for non colour-imparting components, but for the removal of colour imparting lignin fractions, the slurry must be acidified. In fact, the last components react with the Fe^{+2}, Al^{+3}, Ca^{+2} cations of the fly ash dissolved in acidic medium to form sparingly soluble salts (chemical precipitation). The acidic environment nevertheless hinders the physical sorption process of the former components, so a pH of compromise must be used. Similar problems [11] arise for different bleach plant effluents and even for domestic sewage.

It would be very advantageous to use fly ash in form of a durable granulated product that could be thermally regenerated after saturation and then reused. In this paper a technique is presented for the obtainment of granules from coal fly ash and hydrated lime that, after room temperature curing, show a high degree of porosity. In the granules, the fly ash keeps unchanged its sorption capacity. A fixed bed of granules was employed for the sorption of organic pollutants in water (humic acids, sodium lignin sulphonate) with an output of a decontaminated effluent. The exhausted granules were then thermally regenerated (one hour at 400°C in air) and reused for many cycles with practically the same original sorptive capacity.

EXPERIMENTAL PART

Materials

Coal - fly ash: silico aluminous product from the Brindisi (Italy) coal (South Africa) thermoelectric plant. Chemical composition (w/w %): SiO_2 = 46.5, Al_2O_3 = 44.4, Fe_2O_3 = 10.1, CaO = 7, MgO = 1.1, $Na_2O + K_2O$ = 1.8, SO_3 = 1.5, CO_2 = 0.4, unburned = 5.2. The XRD spectrum is shown in Figure 1.

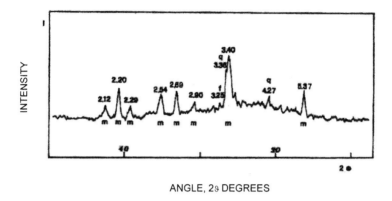

ANGLE, 2ᴤ DEGREES

Figure 1 XRD spectrum of the coal fly ash (m = mullite; q = quartz, f = feldspar)

The radioactivity of the product was also measured, in comparison with a natural Italian tuff (see Table 1), by using a gamma-spectrometer [Na (I,Tl) 3" × 3"].

Table 1 Activity of the coal fly ash in comparison with the activity of a natural tuff

ENERGY, KeV	ELEMENT	w.r.t. TUFF
214.5	Th (Th)	Half intensity
322.5	Ra – Th	Equivalent
577	Th	Equivalent
600	Ra	Equivalent
870	Th	Reduced
1105.5	Ra	Equivalent
1440	K – 40	Very reduced
2570	Th	Equivalent

Hydrated lime: commercial superventilated hydrated lime $Ca(OH)_2$ = 90 % (w/w).

Chemical pollutants: humic acid from BDH and sodium lignin sulphonate from C. Erba (Italy) were employed in distilled water in the form of a solution/suspension, to obtain two artificial samples of polluted water.

Granulation and Granules

The mixes comprising coal fly ash and hydrated lime, previously carefully homogenized, were poured into a rotating drum (Figure 2) with the gradual (two minutes) addition of tap water by means of a sprayer. At the beginning, the axis of the drum was kept horizontal (two minutes) then at 15° angle (five minutes); the rotating speed was 40 rpm.

Figure 2   Rotating drum (diameter = 30 cm, height = 20 cm)

The granules from different mixes were cured at room temperature and 50 % RH for 90 days. After curing the physical characteristics of the granules were measured: volumic mass 1100-1320 (kg/m^3), bulk density 750-800 (kg/cm^3). The granules were sieved to collect the granulometric fraction between 0.8 and 1.2 cm; mechanical strength was then measured. The composition of the different mixes and the mechanical strength of the granules are summarized in Table 2.

Table 2   Experimented mixes and mechanical strength of the granules
(rotation speed 40 rpm) after 90 days of seasoning

LIME, g	FLY ASH, g	WATER, g	Rc, MPa	IMPACT STRENGTH, m, falling
5	95	40	3.5	2
7	93	40	3.6	2
10	90	40	4.7	4
30	70	40	8.9	8

After that, a mix of 90 % coal fly ash (FA) and 10 % of hydrated lime (L) was granulated, utilizing a Water/ (L+FA) ratio equal to 0.5 by weight, with a 60 rpm rotation speed, to obtain a batch of granules with higher porosity. After curing (90 days at room temperature and 50 % RH) and sieving (0.8 – 1.2 cm), the granules were tested to evaluate mechanical and physical characteristics (see Table 3).

Table 3 Mechanical and physical characteristics of the porous granules
(rotation speed 60 rpm) after 90 days of curing

LIME, g	FLYASH, g	WATER, g	Rc, MPa	IMPACT STRENGTH, m, falling	WATER ABSORPTION, %	VOLUMIC MASS, kg/m³
10	90	50	2	3.2	30	1050

The XRD spectra of the 90:10 mix after 14 and 28 days of curing are reported in Figure 3 a) and b) respectively.

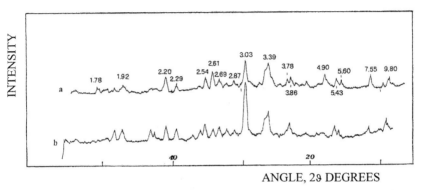

Figure 3 XRD spectra of the mix (fly ash: hydrated lime = 90:10) after 14 days (a) and 28 days (b) of curing in air (RH 50 %) at room temperature

Sorption tests

Polluted waters containing humic acid and, separately, sodium lignin sulphonate were prepared.

Humic acids: 10 g / 5000 g of water, stirred at room temperature for 24 hours. Colour and turbidity of the "polluted" water were measured by a photoelectric colorimeter at 480 nm (i.e. at the wavelength of maximum absorbance). The starting transmittance was 40 %.

1. Preliminary batch tests: 5 g of the original fly ash were poured into one litre of the polluted water. After 30 min of stirring the slurry was settled; after 6 hours the water reached 100 % transmittance. In a second test, 5.5 g of granules, with a diameter between 0.2 and 0.4 cm were placed in a large beaker containing one litre of the same suspension. After 30 min of stirring, the transmittance of the solution reached 100 %. In further experiments different granulometric fractions (diameter 0.3-0.5 cm, 0.5-0.6 cm, 0.5-1.0 cm) were employed with the same procedure; the results were practically the same, but the coarse fraction needed about one hour of stirring to obtain 100 % of transmittance in the solution.

2. Column operation: 120 g of dried granules (diameter 0.4-0.5 cm) were introduced into a glass column (diameter 3 cm); height of bed 40 cm. In further experiments different percolation speeds of the polluted water were experimented to identify the optimum sorption condition: 3.54 10^{-2} (cm³/cm³·min). When the transmittance of the eluate, from 100 % of the first fraction, diminished to 90 % of transmittance, the bed was deemed to have been exhausted. The exhausted granules, after washing with distilled water, were air-dried, then placed in an air oven at 400°C for 1 hour to thermally destroy the organic matter. The weight loss corresponding to the total sorption capacity was 1.0 g of humic acid for 100 g of granules. A specimen of the original granules showed no loss in weight under the same thermal treatment. Regenerated granules were reused in column operations several (five cycles); after the last cycle the sorption capacity had diminished to 90 % of the original value. After five cycles of treatment of one hour each at 400°C the compressive strength of the granules remained practically unchanged.

Sodium lignin sulphonate: 15 g / 10000 g of distilled water stirred at room temperature for 24 hours. The original "polluted" water had pH = 6.8 and a starting COD = 3312 ppm (dichromate reflux determination). The sorption tests were carried out in glass column (diameter 3 cm) containing 230 g of granules (diameter 0.3-0.5 cm); a bed height of 100 cm was used. The optimum speed of percolation was 0.014 (cm³/cm³·min); the bed was deemed exhausted when the COD of the collected samples (1 cm³ every 10 cm³ of eluate) raised to 96 ppm. The total sorption capacity was 14 mg/g granules on the basis of COD measurements. The pH of the eluate immediately increased to 7.2. Thermal regeneration was not executed (the combustion of the sodium lignin sulphonate leads to the formation of solid sodium sulphate).

DISCUSSION

From the XRD spectrum (Figure 1) it must be stated that the silico-aluminous fly ash is a mainly amorphous product (amorphous "halo" of the spectrum): the only appreciable crystalline phases are mullite and quartz; feldspar is also present. Moreover the chemical composition suggests that the product must have pozzolanic activity: in fact the mixes with hydrated lime and water, after curing in air at room temperature and 50 % RH, produce a solid material (see Table 2). The compressive and impact strength are attributable both to the carbonation of lime and to the pozzolanic reactions between lime and the silico-aluminous amorphous components of the ash.

From XRD spectra of the mixes after 14 and 28 days of seasoning (Figure 3 a) and b) respectively) it can be noted the general decrease in the amount of the mullite (2.12 Å - 2.20 Å - 2.29 Å - 2.54 Å - 2.69 Å - 2.90 Å - 3.40 Å - 5.37 Å), that reacts with hydrated lime (the reflection at 2.61 Å, present after 14 days is reduced after 28 days). Furthermore there is a gradual increase in calcite (2.09 Å - 2.49 Å - 3.04 Å and above all 3.86 Å); moreover the main reflection of hydrated calcium carboaluminate (7.55 Å) and those of the ettringite (3.88 Å - 5.61 Å - 9.73 Å) are detectable, the last originated from the sulphates present in the soluble fraction of the original coal fly ash.

Compressive strength gradually increases with hydrated lime content of the mixes: with 30 % of lime the strength reaches 8.9 MPa (Table 2), a value comparable to that of the lightweight commercial aggregates (Leca, Termolite etc.). These results suggest the possibility of employment in the concrete (light aggregate); moreover the radioactivity (Table 1) is similar to that of Italian natural tuffs, widely employed in building construction, above all as thermal insulating and sound absorbing material.

Nevertheless the aim of the research was to obtain fly ash-based granulated material with some chief requirements: it must absorb organic pollutants from wastewater; second the granules must have a high degree of porosity, i.e. high surface contact with polluted waters; the granules must evidently be durable in water. For employment in a reusable fixed bed as a water-purifier, the last requirement is the regenerability.

The increase of porosity (up to 30 %) was obtained by using a high water content and low quantity of lime (10 %) in the mixes and a higher speed of rotation of the drum; naturally the porous granules behave in a different manner (Table 3) relating to the compressive strength (2 MPa), volumic mass (1050 kg/m^3) and a bulk weight (650-700 kg/m^3). The best results in the sorption of humic acids were obtained with a medium percolation speed, with a bed of granules with diameter 0.4-0.8 cm: the sorption capacity was about 1 g of humic acids/ 100 g of bed. The regenerability of the granules by means of thermal treatment (one hour at 400°C in air) determines a net increase of the total output of the process; in fact the sorption capacity slowly decreases only to 90 % of the starting value after five sorption-regeneration cycles. It should be born in mind that the mechanical strength of the granules is not affected by thermal treatments.

Similar results were obtained by using different organic materials, previously investigated with coal fly ashes in the form of slurries: starch [8], phenols [9] and aromatic amines [11, 14]. Nevertheless the regeneration and the reuse of the granules can be carried out only with the organic pollutant that can be completely burned in air, without inorganic residues. In fact the tests executed with sodium lignin sulphonate showed a sorption capacity equivalent to 14 mg (COD) / g (granules), but the regeneration of the bed by means of the thermal treatment, is difficult because the treatment bears an inorganic residue of Na_2SO_4: the reuse of the granules needs careful and thorough washing with water.

The sorption of the humic acids is linked to the level of porosity of the granules: it is physical sorption; in fact the pH of the purified water stayed practically unchanged during sorption. On the contrary for the sodium lignin sulphonate, there is an abrupt increase of pH, from 6.8 to 7.2, suggesting a chemical interaction of the pollutant with the the sorbent, probably a reaction with the residual hydrated lime of the granules (even after the curing) to form the sparingly soluble calcium lignin sulphonate.

CONCLUSIONS

The results show that the sorption capacity of the silico-aluminous coal fly ash is not negatively affected by the granulation/stabilization with 10 % of hydrated lime. After granulation and curing the fly ash has yet a sorption capacity for some organic pollutants comparable to the non-granulated product. The granules can be employed in a fixed bed for the purification of wastewaters containing some organic pollutants, to remove colour, turbidity and COD. Moreover the regenerability of the granules is an advantage in comparison to the use of the fly ash in form of a slurry. Even if the exhausted granules of the bed were discarded, they could be easily conveyed and disposed in a landfill, owing to their acceptable mechanical strength and durability.

REFERENCES

1. RINALDI, G., Granulati artificiali a base di ceneri volanti e calce idrata, *Ingegneria Sanitaria-Ambientale*, July-August (1992), pp. 38-50.

2. MEDICI, F., PIGA L., RINALDI, G., Behaviour of polyaminophenolic additives in the granulation of lime and fly ash, *Waste Management*, Vol. 20, (2000), No. 7, pp 491–498.

3. MEDICI, F., RINALDI, G., Poly-ammino-phenolic additives accelerating the carbonatation of hydrated lime, *Environmental Engineering Science*, Vol. 19, (2002), No. 4, pp 271-276.

4. MEDICI, F., PANEI, L., PIGA, L., RINALDI, G., Room temperature granulation of fly ash with hydrated lime and a polyaminoalkylol additive, *Proceeding of 14th International Conference on Solid Waste Technology and Management*, Philadelphia, November 1 - 4, 1998, S6B.

5. MEDICI, F., RINALDI, G., Hydrated lime pastes containing coal fly ash and polyaminoaphenolic additives for the inertization of fly ashes from municipal solid wastes incinerators, In *Sustainable Concrete Construction*, Dhir R.K., Dyer T.D. and Halliday J.E. Editors, Thomas Telford Publishing, London (2002), pp 139-150.

6. MARRUZZO, G., MEDICI, F., PANEI, L., PIGA, L., RINALDI, G., Characteristics and properties of a mixture containing fly ash, hydrated lime and an organic additive, *Environmental Engineering Science*, Vol. 18, (2001), pp 159–166.

7. SHAH, N.C., BHATTACHARYA, P.K., Bleach plant effluent treatment by fly ash, *Chimicaoggi*, Vol. 59, (1988), pp 59–63.

8. BAVISKAR, J.R., MALIK, G.M., ANSARI, I.A., Judicious utilization of fly ash for treatment of starch and liquid glucose based effluent, *Oriental Journal of Chemistry*, Vol. 19, (2003), pp 143–148.

9. NALANKILLI, G., SIVAKUMAR, N., Utilisation of fly ash for the removal of colour from the dyeing and printing effluents, *Proceedings of the 2nd Fly ash Disposal and Utilization Conference*, New Delhi, Feb. 2 - 4, 2000.

10. RAMU, A., KANNAN, N., SRIVATHSAN, S.A., Adsorption of the carboxylic acids on fly ash and activated carbon, *Indian Journal of Environmental Health*, Vol. 34, (1992), pp 192-199.

11. BANNEJEE, K., HORNG, P.K., CHEREMISINOFF, P.N., SHEIH, M.S., CHENG, S.L., Sorption of selected pollutants by fly ash, *Proceedings of the 43rd Industrial Waste Conference*, Purdue University, Lafaiette, 1989, pp 397-406.

12. WADEKAR, A.P., PANDYA, S.R., COD reduction in textile mill waste by ecofriendly technique, *Indian Journal of Environmental Protection*, Vol. 20, (2000), pp 102–105.

13. BANNERJEE, K., CHEREMISINOFF, P.N., CHENG, S.L., Sorption of organic contaminants by fly ash in a single solute system, *Environmental Science and Technology*, Vol. 29, (1995), pp 2243–2251.

14. BANNERJEE, K., HORNG, P.Y., CHEREMISINOFF, P.N., SHEIH, M.S., CHENG, S.L., Sorbate characteristics of fly ash, *Int. Conf. Physiochemical Biol. Detox. Hazardous Wastes, 1988,* Wu Yeun Editor, Technomic Publisher, Lancaster (1989), pp 249–268.

ADMIXTURES' EFFECT ON MECHANICAL STRENGTH OF A CONCRETE MADE OF RECYCLED AGGREGATE

L Belagraa

M Beddar

M'sila University

Algeria

ABSTRACT. The needs of the construction sector are still increasing for concrete. However the shortage of natural resources of aggregate could be a problem for the concrete industry, in addition to the negative impact on the environment due to the demolition wastes. In the last decade a major interest has developed for the reuse of recycled aggregates that presents more than 70 % of the concrete volume. These should fulfil the requirements of lower cost and better quality, in order to establish its role in the concrete. The aim of this study is to assess the effect of the local admixtures on the mechanical behaviour of recycled aggregate concrete (RAC). Physical and mechanical properties of RAC were investigated including density and water reduction. In addition the non destructive test methods (pulse-velocity, rebound hammer) were used to determine compressive strength. The results obtained were compared to crushed aggregate concrete (CAC) using the normal compressive testing machine test method. Thus, the convenience of indirect tests in the case of a recycled aggregate concrete were demonstrated.

Keywords: Recycled concrete aggregate (RAC), Demolition waste, Non- destructive tests, Mechanical strength.

L Belagraa is a Senior Lecturer in the Civil Engineering Department at M'sila University. Former president of scientific committee, former research studies director. His research focuses on the blending cements and concrete repair materials.

M Beddar is a Senior Lecturer in the Civil Engineering Department. Former head of civil engineering institute. His research focuses on fibre reinforced concrete.

INTRODUCTION

Concrete is still the mostly used material by the construction industry and the highway construction sector. The industry need in this field for such a material has increased over the years. The conservation of natural resources and preserve against pollution with its negative impact on the environment has led the specialists in the civil engineering domain to focus their efforts on the management wastes resulting from demolition.

The reuse of recycled aggregate that comes from construction waste presents a major interest for users and researchers of concrete as it occupies 70 % of concrete. The study herein concerns an investigation of the properties of recycled aggregate concrete (RAC) incorporating admixtures to formulate a much more durable concrete. The experimental programme has the objectives of studying the effect of a super plasticizer (S120) on the mechanical strength (Rc) of RAC. The non-destructive test methods were carried out to assess this hardened property and to see if methods such as rebound hammer and ultrasonic techniques can be conveniently adapted in this case. Other physical properties like the density of RAC were studied. The third aim of this research study is comparison of the performance of recycled aggregate concrete and concrete based on ordinary crushed aggregate (CAC).

TEST PROCEDURES

The objective of the tests is to achieve a recycled aggregate concrete having the performance of a normal concrete with crushed stone aggregate. Recycled concrete specimens were cast using different percentages of admixtures 0, 1, 1.5 and 2 %. The ordinary concrete was prepared with similar aggregate size (8/16 and 16/25) and identical admixture dosages. Tests of the specimens at an age of 28 days using a compressive testing machine and non destructive methods were used to evaluate the mechanical response of concrete. In all mixes a constant workability of 50 mm was maintained using the slump test method.

MATERIALS AND EQUIPMENTS

Sand

The sand used in this study was a clean siliceous and fine sand of fraction 0/5 mm from Boussaada region. Its characteristics are reported in Table 1 and the grading curve is shown in Figure 1.

Gravel

Ordinary gravel was obtained from crushed limestone rock and delivered from the quarry of COSIDER El Euch region (Bordj Bou Arreridj).The gravel fraction used in this study was 8/16, 16/25 in proportions of 40 and 60 %, respectively. The characteristics are shown in Table 1 and the grading curve in Figure 1.

Recycled Aggregate

Pieces of old concrete specimens were crushed using a steel hammer provided by the civil engineering laboratory. The size of the particles maintained for this investigation was 8/16 an 16/25 at a percentage of 40 % and 60 %. The characteristics of the recycled aggregate are reported in Table 2 and Figure 2.

Table 1 Some characteristics of the sand and ordinary crushed aggregates used in the tests

	MATERIAL DENSITY, kg/l	BULK DENSITY SPECIFIC WEIGHT, kg/l	COMPACTNESS, %	POROSITY, %	Es, %
Sand	2.56	1.60/1.70	36.42/70.76	36.58/29.24	73.4
Gravel					
Gca 8/16	2.54	1.33	49.2	50.75	
Gca 16/25	2.57	1.31	48.87	51.12	

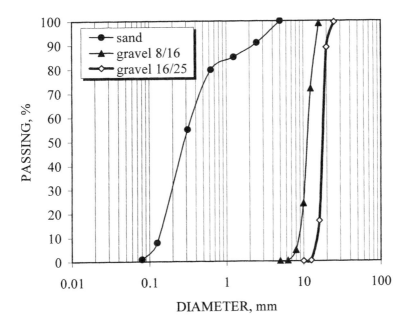

Figure 1 Grain size distribution of sand and crushed aggregate (ca)

Table 2 Some characteristics of the recycled aggregate

RECYCLED AGGREGATE	MATERIAL DENSITY, kg/l	BULK DENSITY SPECIFIC WEIGHT, kg/l	COMPACTNESS, %	POROSITY, %
Ra 8/16	2.40	1.24	48.14	51.86
Ra 16/25	2.34	1.12	53.14	46.85

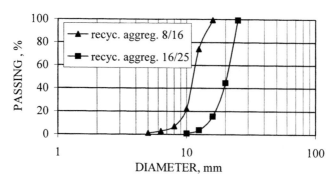

Figure 2 Grain size distribution of recycled aggregate (ra)

Cement

The cement used was type CPJ 45 delivered from Kebira and widely used in the construction sector in Algeria.

Admixtures

The admixtures is a local superplasticizer product Medaplast 120 manufactured by Granitex in Oued Smar near Algiers. It is a brownish emulsion having a relative density of 1.2 g/cm^3 and a pH ranging from 8 to 8.5.

Water

A tap water from the civil engineering laboratory was used for concrete batching.

Concrete mix

The concrete mix proportions used were for a grade 350 mix, determined by the absolute volume " Scramtaiev method", [4].

- Cement   324 kg/m^3
- Sand   565 kg/m^3
- Gravel   8/16 565 kg/m^3 (40 %)
-      16/25 760 kg/m^3 (60 %)
- Water (total)   180 litre (This quantity takes into account the degree of aggregate absorption)

TESTING

Workability

The method used to assess the workability for both ordinary and recycled concrete was the slump test method. A workability of about 50 mm was maintained for all mixes.

Compressive strength

Compression tests were carried out on cubic specimens ($150 \times 150 \times 150$) mm^3 Tests were done using the hydraulic press model "STRASSENTEST FHF". The specimens were centred on the tray of the press and a continuous load was applied. The ultimate compression load for each concrete cube was recorded at 28 days age.

Rebound hammer test

The specimens were placed in the centre of the hydraulic machine press, a continuous load was applied and maintained within a range of 10 to 20 kN. The rebound hammer test was carried out on five different points spaced at 2 cm intervals on both faces of the cubic specimens. The final result from the test was calculated using the following equation:

$$Rs = \sum_{i=n}^{n} S/n$$

with   n – Number of tests carried out on both faces of the cube
S- The recorded value of rebound hammer

Ultrasonic method

The pulse velocity test was carried out on the two opposite sides of the specimens ($150 \times 150 \times 150$) mm^3 using direct transmission. The transit time t in μs was recorded and the velocity V is measured as; $V = d/t$.

V - Velocity in km/s
d - The distance between the two transducers
t - The transit time in μs

ANALYSIS OF RESULTS

Presentation of results

In this study the concrete mixes have been prepared according to the method of absolute volume. The following percentages of admixtures 0, 1, 1.5, and 2 % were chosen. The cubic specimens ($150 \times 150 \times 150$) mm^3 were cast and cured in water to be tested at ages of 28 days. Prisms measuring ($100 \times 100 \times 400$) mm^3 were prepared in similar conditions to evaluate the flexural strength. Initially specimens measuring $150 \times 150 \times 150$ mm^3 with different admixture dosages were studied using the compressive testing machine. Additionally the non destructive tests (ultrasonic and rebound hammer) were carried out. The ultrasonic strength Ru has been assessed as a function of the compressive strength test (Rc) using the formula:

$$Ru = K \cdot \rho \cdot V^4/g \ (N/mm^2)$$

- Ru – the ultrasonic strength (N/mm^2)
- k -coefficient of calibration
- ρ - Concrete density (kg/m^3)
- g – the earth gravity (m/s^2)
- V – The velocity (km/s)

The value of compressive strength Rc is compared to the combined values of ultrasonic (V) method and the rebound hammer reading (S). The strength Rsu is then assessed according to the formula of Feret;

$$Rus = [S/n_0 + n_1 \cdot S - n_2 \cdot V]^2$$

$n_0 = 3.64$ S - Rebound hammer reading
$n_1 = 0.023$ Rus -Ultrasonic-rebound hammer resistance
$n_2 = 0.56$ V - Velocity

Table 3 Results of mechanical strength and density for crushed aggregate concrete (CAC)

ADMIXTURE %	BULK DENSITY ρ, g/cm^3	VELOCITY V, km/s	COEFF. K	REB-HAMMER S	STRENGTH Rs, N/mm^2	STRENGTH REB-ULTR Rsu, N/mm^2	COMP. STRENGTH Rc (cac), N/mm^2
0	2.311	4.13	3.72	22.36	11.80	17.0	25.62
1	2.306	4.2	4.10	27.20	20.00	24.40	29.98
1.5	2.345	4.5	3.96	28.30	21.80	25.14	39.10
2	2.355	4.70	3.02	28.43	21.50	37.20	34.50

Table 4 Results of mechanical strength and density for Recycled Aggregate Concrete (RAC)

ADMIXTURE %	BULK DENSITY ρ, g/cm^3	VELOCITY V, km/s	COEFF. K	REB-HAMMER S	STRENGTH Rs, N/mm^2	STRENGTH REB-ULTR Rsu, N/mm^2	COMP. STRENGTH Rc (cac), N/mm^2
0	2.395	4.43	3.41	23.40	14.50	19.00	32.00
1	2.434	4.83	4.76	28.63	21.80	32.00	37.20
1.5	2.455	4.74	3.11	27.03	20.00	28.20	38.96
2	2.455	4.89	2.66	28.26	21.60	33.20	37.77

RESULTS AND DISCUSSION

Compressive strength

According to Figure 3 the behaviour of RAC shows the same trend of strength development for all dosages of admixture at 28 days. However for a dosage of 1 % admixture the recycled concrete gives a lower compressive strength compared to normal concrete with crushed stone aggregate. For admixture contents over 1.5 % the recycled concrete showed similar comparative values to ordinary concrete. Thus, performance of RAC similar to ordinary concrete can be achieved with the incorporation of admixtures.

Ultrasonic tests

Figure 4 illustrates that what ever the type of aggregate used, strength development is similar for normal concrete and RAC. Furthermore the compressive strength of ordinary concrete is superior to the results given for RAC. The effect of admixtures on the mechanical strength is more advantageous for recycled aggregate concrete.

The contribution of the admixtures to increasing strength is marked. This is due to the improvement of workability of fresh concrete, as well as water reduction in concrete mix resulting in an improvement of the resistance of concrete matrix.

Rebound hammer results

Although the rebound hammer test results are affected by many factors, such as the mix characteristics (cement type, content and the aggregate type) or the member characteristics (mass, density, surface type, age, curing type and surface carbonation), only the main factor of aggregate type is considered in this study.

The rebound hammer test results in Figure 5 show that the 0 % admixtures dosage gives the lowest value of strength. It is noted that there is a slight increase of strength for higher percentages (1, 1.5 and 2 %). The recycled concrete RAC proved to give lower readings compared to normal concrete (CAC).

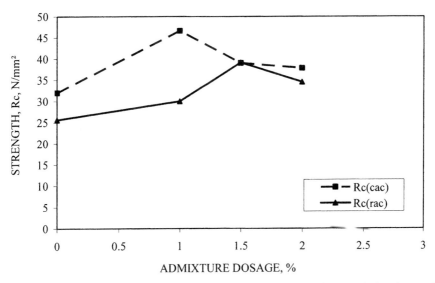

Figure 3 Compressive strength as a function of admixture content for recycled and normal concrete (compression test) at 28 days age

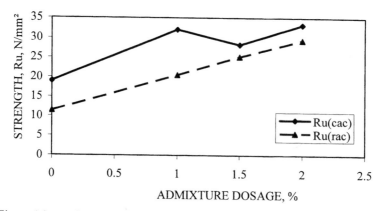

Figure 4 Strength Ru against admixture dosage of recycled and normal concrete (ultrasonic test) at 28 days of age

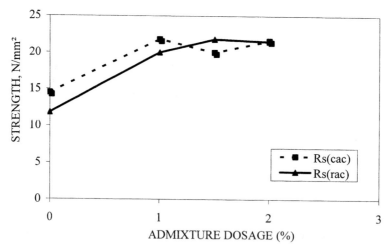

Figure 5 Strength against admixture dosage of recycled and crushed aggregate concrete (rebound hammer test) at 28 days age

Density

Figure 6 shows that the bulk density of CAC displays a slight increase in hardened specimens compared to recycled concrete aggregate (RAC). This may be related to the heavy density of the crushed aggregate type; which is more compact. This is more evident for dosages of 1, 1.5 and 2 % of admixture.

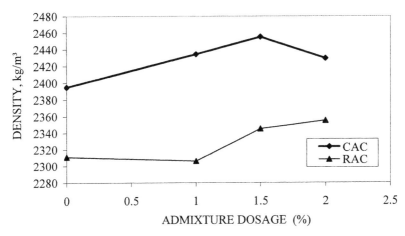

Figure 6   Density in function of admixture dosage of recycled and normal concrete

CONCLUSIONS

In light of this study with the objective of studying the effect of admixtures (Medaplast 120) on the mechanical response of RAC compared to crushed aggregate concrete (CAC), where the assessment of the strength used indirect tests (ultrasonic and rebound hammer) methods beside a compression machine test. The interpretation of these results leads to the following conclusions:

1- The compressive strength development is similar for both recycled aggregate concrete (RAC) and crushed aggregate concrete (CAC) with the same admixture dosages.

2- The recycled aggregate concrete shows the same performance compared to normal concrete for an admixture dosage of 1 %.

3- The density of crushed aggregate concrete presents a slight increase in comparison to recycled aggregate concrete. This is attributed to the type of recycled aggregate being less compact and so less dense.

4- Non-destructive tests can be used to assess the strength of RAC, but a correction coefficient is required to obtain a similar value to the compressive strength given by the compression machine test.

5- There is an increase of RAC strength when combined with admixtures compared to normal concrete without admixture incorporation.

REFERENCES

1   DELALJA, D., NEGRICHI, S., Étude et réponse mécanique de béton constitué de gravier recyclé issu de déchets de démolition, projet de fin d'étude,université de M'sila, juin 2002.

2   SELLAMI, N., SELLAMI, N., Étude de l'adaptation des essais non destructifs ultrason scléromètre dans le cas des bétons en fibres de polypropylènes, projet de fin d'études, département de génie civil, juin, 2000.

3   DREUX, G., FESTA, J., Nouveau guide du béton, édition Eyroles, 1995.

4   KOMAR, A., Matériaux et éléments de construction, édition Mir, Moscou, 1982.

5   MICHEL, V., Pratique des ciments et des bétons, édition du MONITEUR, 1976.

6   NEVILLE, A.M., Properties of concrete, 3rd edition, Longman Scientific & Technical, 1986.

THEME THREE:

MINIMISING ENVIRONMENTAL IMPACT

ENVIRONMENTAL ASSESSMENT OF CEMENT BASED PRODUCTS: LIFE CYCLE ASSESSMENT AND THE ECOCONCRETE SOFTWARE TOOL

A Josa

R Gettu

A Aguado

Technical University of Catalonia

Spain

ABSTRACT. Life Cycle Assessment (LCA) is an essential tool for the quantitative evaluation of the impact of a product on the environment. Since this impact can occur in any stage of its life cycle (from the mining of raw materials to ultimate disposal), its whole life has to be considered. The basic features of LCA are an inventory (LCI) of environmental interventions (inputs and outputs of energy and materials used, emissions produced, etc.) and the assessment of their effect (LCIA) in terms of different potential environmental impacts (greenhouse effect, ozone layer, acidification or depletion of raw materials and energy, among others). LCA can be used for a number of purposes, including the environmental improvement of processes and materials, the comparison of products, or for eco-labelling. For conducting LCA studies, specific software must be used; a recently-developed program is the EcoConcrete software, compiled by the European concrete-related industries. In this paper, the main aspects of the application of LCA to concrete products are presented, including a brief description of the methodology used, some examples of application and the EcoConcrete software.

Keywords: Life cycle assessment, Cement, Concrete, Environmental impact, LCI, LCIA

Dr Alejandro Josa is a Civil Engineer and Professor of Geotechnical Engineering in the School of Civil Engineering, Technical University of Catalonia (UPC). His research interests include cement utilisation, environmental impact of construction products and modelling of concrete behaviour.

Dr Ravindra Gettu is a Senior Researcher in the Department of Construction Engineering of the UPC. His research interests include concrete technology, self-compacting and fibre-reinforced concretes, and nonlinear behaviour of concrete.

Dr Antonio Aguado is a Civil Engineer and Professor of Concrete Technology in the School of Civil Engineering, UPC. His research interests include technology transfer, concrete technology and dams.

INTRODUCTION

The limitation of available resources, the importance of protecting nature and the need to reach sustainable development makes the evaluation and minimisation of the environmental impact of any activity increasingly important [1, 2]. This is also the case in the design, manufacturing and use of construction products. The undeniable economic importance of the construction sector [3] occurs along with significant environmental impacts. Examples of such impact are the consumption of renewable or non-renewable resources, air and water pollution; energy consumption during product manufacture and maintenance, and waste generation as a consequence of the demolition of different types of installations [4].

Besides these negative impacts, it is important to underline the positive effects of the construction sector on the environment, in terms of the contribution to the improvement of the habitability conditions, control of the spreading of diseases and protection against aggressive climatological agents or natural phenomena, among other aspects [5, 6]. Along with bearing in mind the impact caused, it is necessary to evaluate it quantitatively. The most suitable tool currently available for such evaluation is Life Cycle Assessment (LCA), which is briefly described in the following section.

LIFE CYCLE ASSESSMENT OF PROCESSES AND PRODUCTS

The life cycle assessment (LCA) methodology permits the evaluation of the environmental effects of a process or product, considering all the stages in its production from the extraction of resources to the processing of the wastes derived from it [7]. "This includes identifying and quantifying energy and materials used, and the wastes released to the environment, assessing their environmental impact and evaluating opportunities for improvement" [8, 9]. Obviously, there will always be certain limits in the phases considered, which must include the most relevant ones. This is not feasible with specific processes, such as the manufacturing of a specific element, or with intermediate products, such as cement or concrete, which have many subsequent applications. In these cases, the analysis ends in the considered process or in the manufacturing of the product ('gate to gate' or 'cradle to gate'). The ISO 207 Technical Committee is currently developing international standards for environmental management, and the committee LCA SC4 deals with LCA.

Figure 1 (adapted from [5] and enlarged) displays the generic life cycle of a concrete product. As seen in this figure, the life cycle begins in the acquiring of raw materials and ends when the product is dumped as waste or else when it is intended for other uses. In between, there are all the phases that the product goes through and where impacts on the environment can occur, including all the possible intermediate transportation phases.

The complete life cycle of a product includes phases that could be named "constructive" (i.e., manufacture, use, maintenance) and phases that could be named "deconstructive" (i.e., reuse, demolition, recycling). The concept of "deconstruction" is becoming increasingly important, since it facilitates the utilisation of previously-used materials and the reduction of the impact that their re-use, recycling or down-cycling may imply. In "pure" deconstruction, the process to follow would be the opposite of the process used in construction. Figure 2 shows the constructive and deconstructive phases of a product life, with a specific conception phase included both in construction and in deconstruction. Only by considering, from conception, that the products should minimise their global impact and be reintegrated in the environment, will it be possible to succeed in optimising these totally from this viewpoint.

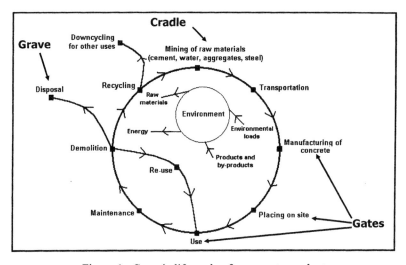

Figure 1   Generic life cycle of a concrete product

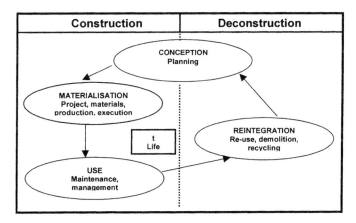

Figure 2   Phases in the generic life cycle of a product

The absence of a single environmental unit for different environmental loads (i.e., consumption of raw materials, emission of CO_2, water and air pollution, etc.) implies that the result of a LCA is made up by a series of values corresponding to each impact considered. Nevertheless, in some cases, using criteria in which social or political preferences must be assessed, such values may be reduced to just one. Generally speaking, this implies that, whenever alternatives are compared, the result cannot be definite if one or more alternatives have some, not all, higher impacts than the others.

LCA may be used, mainly, as an environmental analysis tool in order to improve industrial processes or to compare alternative processes or products, or else as a tool for environmental management to guide the development and research of new products and processes (i.e., granting of eco-labels, setting-up of environmental basis for comparison, etc.). On the other hand, LCA provides a common framework of reference to systematise the terminology and the methodology used, and to compare environmental surveys of different origins.

The fact that, at the moment, LCA is not totally standardised is a distinctive disadvantage since different methodologies [10-14] may lead to contradictory results if different hypotheses (i.e., system boundaries, time and location, etc.) are adopted [15]. In spite of these inconveniences, LCA has a great advantage: it is generally accepted by the different parties involved (i.e., administrations, NGOs and ecological organisations, the industry, etc.).

Usually, a LCA is composed of the following basic phases:

a) Definition of the objective and the scope of the survey [7] including system boundaries (i.e., phases included), and the origin and accuracy of data, among other aspects.

b) Inventory (LCI): In this phase, all the environmental interventions of the system (i.e., consumption of materials and energy, emissions, waste, etc.) are compiled [7], resulting, in general, in a very long list of figures (Figure 3 shows a relatively short example).

c) Impact assessment: The environmental interventions of the inventory are classified, characterised and normalised during this phase [17], and the results are duly added in different impact categories (i.e., consumption of resources, greenhouse effect, acidification, toxicity, etc.) after having applied the corresponding coefficients based on scientific evidence (i.e., comparison of the effects of different environmental interventions). In some cases, a single final value is obtained [18, 19] using assumptions based on social or political criteria.

d) Analysis of results and conclusions.

EXAMPLES OF APPLICATION OF LCA ON CONCRETE PRODUCTS

The Zaltbommel Road Bridge

This case, presented in [5] and [20], compares the environmental impact of two alternative bridges through the CML methodology [10]. The study was commissioned by the Dutch national highways authority and conducted by the University of Amsterdam. The proposed road bridge was planned across the River Waal, about 20 km north of Hertogenbosch, in The Netherlands. This bridge had to meet growing road traffic in the motorway and was intended to replace the existing steel bridge built in 1933.

Figure 4 shows a summary of the results obtained. In this figure the value of 100% has been given, for each impact category, to the worst alternative (concrete or steel), and the corresponding relative value (<100%) has been given to the other one. This way of representing LCA results does not allow the comparison of the relative importance of each impact category but provides the quickest way to see the results. It can be seen in this figure the good results of the concrete solution except in the depletion of natural resources. The latter result is due to the scarcity of some resources in The Netherlands. Taking into account these results, the Dutch authorities finally selected and constructed the concrete bridge.

Railway sleepers

This case is presented in [5] and [21], and compares the environmental impact of two alternative solutions for railway sleepers through the CML methodology [10]. The study was commissioned by the environmental office of the Dutch Railways and conducted by external consultants. Since 1950s, The Netherlands have used both prestressed concrete (about 35% of all sleepers) and treated timber sleepers. Since the Dutch Railways replaced sleepers on about 200 km of track every year, in addition to the construction of new tracks, the authorities were interested in using the best solution from the environmental viewpoint.

LCI	
Use of natural resources and materials	
Limestone	Slags
Clay	Explosives
Gypsum	Water
Iron oxides	Etc.
Natural gas	Diesel
Heavy fuel oil	Crude Oil
Coal	Biomass
Petcoke	Etc.
Emissions	
Particules	Hg
CO_2	Cd
CO	Tl
NO_x	Sb
SO_2	As
N_2O	Pb
HC	Cr
NH_3	Co
HCl	Cu
HF	Mn
CH_3	Ni
VOC	V
COD	Sn
Wastes	Etc.
Other	
Energy	Noise
Area	Etc.

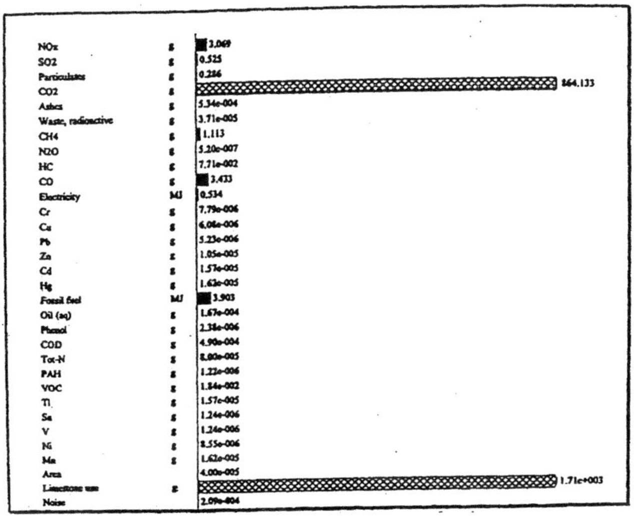

Figure 3 Environmental interventions for cement (left, [16]) and LCI for 1 t of cement (right)

Table 1. Example of impact categories [14]

IMPACT CATEGORIES

Global warming	Photochemical smog	Human toxicity air	Hazardous waste
Ozone depletion	Ecotoxicity water chronic	Human toxicity water	Radioactive waste
Acidification	Ecotoxicity water acute	Human toxicity soil	Slags/ashes
Eutrophication	Ecotoxicity soil chronic	Bulk waste	Resources (all)

Figure 5 shows a summary of the results obtained, with the same format as in Figure 4. It can be seen that the concrete solution also has good results except in the depletion of natural resources, for the same reason as in the previous case, and in non-chemical waste. The negative effects of the timber sleepers are largely due to the creosote treatment.

External Sewer

This case is presented in [22]. It analyses the environmental behaviour of an external concrete sewer and compares the results with those of several alternative solutions (i.e., three types of PVC and vitrified clay). The study was commissioned by the industrial sector and was conducted by Intron. In general, all results were favourable to the concrete solution. Figure 6 shows a summary of these results in the same format as in the previous cases.

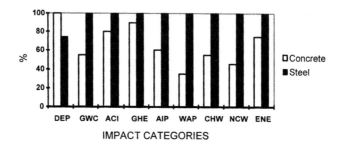

Figure 4   Results for the Zaltbommel road bridge [5, 20]
(DEP: depletion of non-renewable natural resources; GWC: groundwater consumption; ACI: acidification; GHE: greenhouse effect; AIP: emissions to air; WAP: emissions to water; CHW: chemical wastes; NCW: non-chemical wastes; ENE: energy consumption)

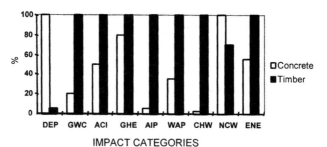

Figure 5   Results for the railway sleepers [5, 21]
(DEP: depletion of non-renewable natural resources; GWC: groundwater consumption; ACI: acidification; GHE: greenhouse effect; AIP: emissions to air; WAP: emissions to water; CHW: chemical wastes; NCW: non-chemical wastes; ENE: energy consumption)

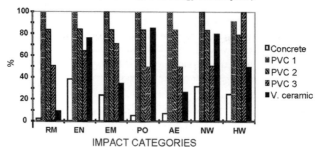

Figure 6   Results for the external sewer [22]
(RM: raw materials; EN: energy; EM: emissions; PO: photochemical oxidant formation; AE: aquatic eco-toxicity; NW: normal wastes; HW: hazardous wastes)

Pavements

This case is presented in [23], and compares the environmental impact of concrete and asphalt pavements for a specific application in a Nordic motorway. The study was commissioned by several industrial sectors, including those involved in concrete and asphalt, and was conducted by Technical Research Centre of Finland (VTT).

Table 2 shows a summary of the results. It can be seen that, in this case, each alternative scores better than the other in 5 out of 10 parameters, which means that no clear conclusions can be drawn. Other cases of pavement comparisons can be found in [24].

Table 2   Results for the pavements [23]

	CONCRETE PAVEMENT	ASPHALT PAVEMENT (Finnish maintenance)	ASPHALT PAVEMENT (Swedish maintenance)
CO_2, kg/km	940000	590000	670000
SO_2, kg/km	1700	2500	2800
NO_x, kg/km	4700	3000	3600
CO, kg/km	2000	610	670
VOC, kg/km	1000	1900	2100
Particulates, kg/km	650000	1200000	1200000
Hg, kg/km	0.0076	0.000042	0.000064
Non-renewable energy, GJ/km	11000	21000	25000
Noise (affected land) , ha/km	70	52	52

Comments on the Cases Presented

The cases presented in the previous paragraphs are a selection of those published in different references. Concrete solutions have frequently good results in relation to alternatives due to a number of reasons. On the one hand, the resources required (i.e., limestone, clay) are very abundant in nature and are easily mined. On the other hand, concrete products are chemically inert and very durable, require low maintenance and can be completely recycled. Finally, the emissions and wastes produced are non-toxic. On the negative side it should be said that cement manufacture requires important amounts of energy and produces high quantities of some gases, in particular CO_2. It is important to emphasize, again, that for analysing the results of an LCA it is always essential to take into account the assumptions adopted.

THE ECOCONCRETE SOFTWARE TOOL

The EcoConcrete software tool is a tailor-made software providing fully-fledged, peer reviewed results of LCA of 10 selected concrete applications, according to three different methodologies (CML [10, 14], EDIP [12], Eco-Indicator [13]) and ISO standards. The tool is based in a user-friendly development of Microsoft Excel and is co-owned by BIBM, the International Bureau for Precast Concrete, CEMBUREAU, the European Cement Association, EFCA, the European Federation of Concrete Admixtures Associations, ERMCO, the European Ready Mixed Concrete Organization, and EUROFER, the European Confederation of Iron and Steel Industries.

The ten functional units correspond to the following concrete applications: a flat slab, a continuous beam, a foundation pile, a motorway pavement, a bridge pylon, a separation floor, a load bearing wall, the elements of a solid wall, a column and pavement blocks. The first five correspond to ready-mixed concrete applications and the last five correspond to precast concrete applications.

These ten functional units are analysed from cradle to grave using inventory data provided by the software co-owners, and taking into account the use and maintenance phases and the end-of-life scenario of the different applications considered. An example of graphical results corresponding to the analysis of precast pavement blocks is shown in Figure 7. These results are also given numerically and can be processed further. All the results are split into the different stages of the life cycle allowing the analysis of the relative importance of each of them in the final impact.

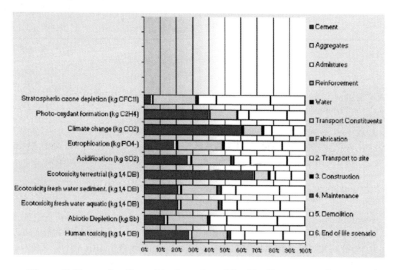

Figure 7. Example of graphical results of the EcoConcrete software tool

FINAL REMARKS

LCA methodology is widely accepted for assessing the effect that all kinds of products, and in particular those corresponding to the construction sector, cause on the environment. Since relevant effects can take place in any stage of the life cycle of the product, a cradle to grave approach must be considered. It is important to have this approach from the very beginning of the design in order to minimise the global effect on the environment including the end-of-life scenario.

Applying LCA to the field of concrete products enables the identification of the life cycle phases where it might be necessary to introduce environmental improvements, as well as comparing alternative solutions (i.e., processes, products) from the perspective of their impact on the environment.

Concrete based solutions frequently have a good environmental behaviour in comparison with alternatives due to a number of reasons. However, the results are highly dependent on the assumptions of the analysis conducted, which means that in many cases it is difficult to reach definitive conclusions.

EcoConcrete is a new tailor-made software tool for the use of the LCA methodology to 10 specific concrete applications. It provides fully-fledged and peer reviewed results applying three different methodologies following ISO standards.

ACKNOWLEDGEMENTS

The financial assistance for the work presented has been partially provided by the Spanish MCYT grants MAT02-04310 and MAT 2003-5530, and by several research projects funded by the Spanish Institute of Cement and its Applications (IECA).

REFERENCES

1. JOSA, A., GETTU, R. AND AGUADO, A. Evaluación medioambiental de productos de la construcción derivados del cemento (I and II). CIC Información (Barcelona), 1997, No. 299, pp. 30-35, No. 300, pp. 49-55.

2. JOSA, A., AGUADO, A. AND GETTU, R. Life cycle assessment of concrete products. 5 Ulusal Beton Kongresi – Beton Dayanikliligi (Dürabilite), Istanbul, 2003, pp. 279-293.

3. SEOPAN. Informe trimestral sobre el sector de la construcción. Asociación de Constructoras de Obras de ámbito Nacional, Spain, 2004, 12 pp.

4. STANNERS, D. AND BOURDEAU, P. (Ed.). Europe's Environment - The Dobris Assessment. European Environment Agency, 1995.

5. CEMBUREAU. Concrete: the benefit to the environment. 1995, 104 p.

6. JOSA, A. AND ALEIXANDRE, J.L.. El hormigón y el medio ambiente. IECA Levante, Spain, 1998, 77 p.

7. ISO 14040: Environmental management - Life cycle assessment - Principles and framework. International Organisation for Standardisation, Geneva, 1997.

8. CEMENT AND CONCRETE ASSOCIATION OF AUSTRALIA. The Sydney House - Life Cycle Assessment – Report, 2001.

9. SETAC. A Conceptual Framework for Life-Cycle Impact Assessment - Guidelines for Life-Cycle Assessment: a 'Code of Practice'. SETAC, Brussels, 1993.

10. CML. Environmental Life-Cycle Assessment of Products - Guide - Backgrounds. CML-TNO-B&G, NOH-9253/54, Leiden, 1992, 224 pp.

11. NORDIC. Nordic Guidelines on Life-Cycle Assessment. Nord, Stockholm, 1995.

12. WENZEL, H., HAUSCHILD, M., AND ALTING L. Environmental Assessment of Products. Chapman and Hall, 1996.

13. GOEDKOOP, M., AND SPRIENSMA R. The Eco-indicator 99, A damage oriented method for Life Cycle Impact Assessment. Pré Consultants, Amersfoort, Netherlands, 1999.

14. GUINEE J., GORREE M., HEIJUNGS R., HUPPES G. ET AL. Life Cycle Assessment, an operational Guide to ISO Standards, volume 1,2,3. Centre of Environmental Science, Leiden University (CML), Leiden, Netherlands, 2000.

15. JOSA, A., AGUADO, A., HEINO, A., BYARS, E. AND CARDIM, A. Comparative analysis of available life cycle inventories of cement in the EU. Cem. Concr. Res., Vol. 34, Issue 8, pp. 1313-1320, 2004.

16. CEMBUREAU. Cembureau LCI Format for Cement. Cembureau, Brussels, 1999.

17. ISO 14041: Environmental management - Life cycle assessment - Goal and scope definition and inventory analysis. International Organisation for Standardisation, Geneva, 1998.

18. AHBE, S., BRAUNSCHWEIG, A. AND MÜLLER-WENK, R. Methodik für Oekobilanzen auf der Basis ökologischer Optimierung. 133, BUWAL, Berne, 1990.

19. KORTMAN, J.G.M., KINDEIJER, E.W., SAS, H. AND SPRENGERS, M. Towards a single indicator for emissions - An exercise in aggregating environmental effects. IDES, CE, Infoplan, Amsterdam, 1995.

20. KORTMAN, J.G.M. AND LIM, R.G. Milieu-Vergelijking van twee Aanbruggen. University of Amsterdam, 1992.

21. HOEFNAGELS, F. AND DE LANGE, V. De millieubelasting van houten en betonnen dwarsliggers. CREM, Amsterdam, 1993.

22. INTRON. LCA study – Environmental profile and environmental measures of a concrete external sewer. Sewer Technology, 6th year, 1995.

23. HÄKKINEN, T. AND MÄKELÄ, K. Environmental adaption of concrete. Environmental impact of concrete and asphalt pavements. VTT, Tech. Res. Centre of Finland, Espoo, 1996.

24. JOSA, A., AGUADO, A., CARDIM, A. AND GETTU, R. Construcción y medio ambiente - Evaluación ambiental de productos derivados del cemento - Aplicación a pavimentos de hormigón. II Congreso Interamericano de Pavimentos de Concreto, Cartagena de Indias (Colombia), 2000, pp. 27-29.

FREEZE-THAW RESISTANCE OF POROUS CONCRETE MIXED WITH CRUSHED ROOFING TILE

M Sugiyama
Hokkai Gakuen University

Japan

ABSTRACT. This paper addresses the problem of recycling construction waste, namely, the use of crushed roofing tile as a component of concrete; the resulting compressive strength, static elastic modulus and freeze-thaw resistance were tested. The experiment was carried out in two series. Series 1 measured the compressive strength and the static elastic modulus of high volume fly ash concrete containing crushed roofing tile material replacing 20 to 100 % of the coarse aggregate volume. The results showed that compressive strength was increased when 20 to 60 % of the coarse aggregate volume was roofing tile material. Series 2 tested the freeze-thaw resistance of porous concrete containing crushed roofing tile material using two ASTM test methods: procedure-A, both freeze and thaw in water, and procedure-B, freeze in air and thaw in water. The result showed that the use of crushed roofing tile material caused no adverse effects to freeze-thaw resistance.

Keywords: Aggregate, Compressive strength, Crushed roofing tile, Freeze-thaw resistance, Porous Concrete.

Dr Masashi Sugiyama is a professor of the Faculty of Engineering, Hokkai Gakuen University, Sapporo, Japan. He received his doctorate in engineering from Hokkaido University in 1981. His doctoral dissertation was "The Effect of Drying on the Physical Properties of Concrete." His specialty is research and development to improve the durability of concrete. He received a prize for research into super-high-durability concrete from the Japan Concrete Institute in 1988. His interests include chemical admixtures, compressive strength, elastic modulus, freezing and thawing resistances and neutralization. He is a member of RILEM, ACI, JCI, and AIJ.

INTRODUCTION

Construction in Japan amounts to about 70 trillion yen annually, corresponding to about 14 % of the gross domestic product. One by-product is the production of about one hundred million tons of construction waste, about 25 % of the total amount of industrial waste in Japan [1, 2]. It is clearly important to promote the recycling of this construction by-product in order to suppress the generation of waste. The author has previously conducted research on various methods of recycling waste as a construction material [3, 4]. This paper presents the results of experiments to recycle roofing tiles. The tests were carried out in two series. Series 1 measured the compressive strength and the static elastic modulus of high volume fly ash concrete with 20 to 100 % of the coarse aggregate volume replaced by crushed roofing tile material. Series 2 used ASTM test method procedure-A, freeze and thaw in water, and procedure-B, freeze in air and thaw in water, to test the freeze-thaw resistance of porous concrete containing crushed roofing tile material.

SERIES 1: COMPRESSIVE STRENGTH AND STATIC MODULUS

Design of Experiments

Series 1 measured the compressive strength and the static elastic modulus of high volume fly ash concrete containing crushed roofing tile material.

Mix Proportions

The mix proportions are shown in Table 1. The water-cement ratio for all test specimens was 40 %. The crushed roofing tile material was used to replace 0 %, 20 %, 40 %, 60 %, 80 %, and 100 %, by volume, of the coarse aggregate. The materials used in the concrete are as follows: ordinary cement (density of 3.16), fly ash (density of 2.16 g/cm^3, loss on ignition of 1.7 %, specific surface area of 3780 cm^2/g, methylene blue absorption of 0.55 mg/g), fine aggregate (river sand, density of 2.69 g/cm^3), coarse aggregate (river gravel, density of 2.80 g/cm^3), and crushed roofing tile material (density of 1.30 g/cm^3). The concrete was cured in water at 20°C; the slump was in the range from 15 to 19 cm. The compressive strength tests were conducted after 7, 28, and 91 days of curing. The dimensions of the specimens were Ø10×20 cm.

Table 1   Mix proportions of high volume fly ash concrete containing crushed roofing tile

ROOFING TILE %	W/C+F %	WATER kg	CEMENT kg	FLY ASH kg	FINE AGG. kg	COARSE AGG. kg	ROOFING TILE kg
0	40	158	197	197	764	1112	0
20	40	158	197	197	764	890	103
40	40	158	197	197	764	666	207
60	40	158	197	197	764	445	309
80	40	158	197	197	764	221	413
100	40	158	197	197	764	0	516

Compressive Strength Results

The compressive strength test results, shown in Figure1, reveal that compressive strength increased when approximately 20 to 60 % of the coarse aggregate volume was composed of crushed roofing tile material.

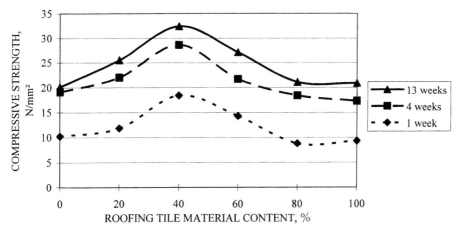

Figure 1   Compressive strength of concrete as a function of roofing tile material content

Static Elastic Modulus Results

The results of static elastic modulus tests are shown in Figure 2. Up to a roofing tile material content of 80 % by volume, the static elastic modulus remains relatively constant; above 80 %, the static elastic modulus somewhat decreases as the roofing tile material content is further increased.

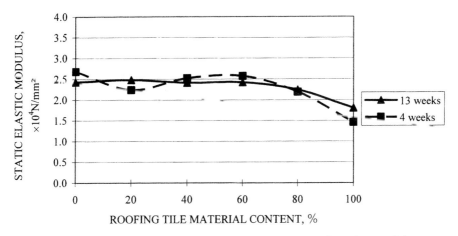

Figure 2   Static elastic modulus of concrete as a function of roofing tile material content

SERIES 2: FREEZE/THAW RESISTANCE OF POROUS CONCRETE

Design of Experiments

Series 2 evaluated the freeze-thaw resistance of porous concrete containing crushed roofing tile material using two ASTM test methods: procedure-A, freeze and thaw in water, and procedure-B, freeze in air and thaw in water.

Mix Proportions

The mix proportions are shown in Table 2. The water-cement ratio for all test specimens was 22 %. The crushed roofing tile material was added as 0 %, 20 %, 40 %, and 100 % of the coarse aggregate volume. The materials used in the concrete are as follows: ordinary cement (density of 3.16 g/cm^3), coarse aggregate (river gravel, density of 2.71 g/cm^3), and crushed roofing tile material (density of 1.30 g/cm^3). The coarse aggregate made from the crushed roofing tile material passed a 20 mm sieve and was retained on a 5 mm sieve. The dimensions of the specimens were 10×10×40 cm. The freeze/thaw tests were conducted after the concrete specimens were cured in water at 20°C for 28 days.

Table 2   Mix proportions of porous concrete containing crushed roofing tile material

ROOFING TILE %	W/C+F %	TARGET AIR VOID %	WATER kg	CEMENT kg	COARSE AGGREGATE kg	ROOFING TILE kg
0	22	30	53	240	1547	0
20	22	30	53	240	1234	148
40	22	30	53	240	930	296
100	22	30	53	240	0	742

Air Void Percentage and Compressive Strength Results

The results of compressive strength and air void percentage tests are shown in Table 3. The air void percentage was measured using the volumetric method of the Japan Concrete Institute [6].

Volumetric method formula:

$$A = (1-(W_2-W_1)/V) \times 100$$

where
A = Air void percentage of porous concrete (%),
W_2 = Weight in air,
W_1 = Weight in water, and
V = Volume.

The actual air void percentage determined by the volumetric method ranged from 19.1 to 27.6 %, reasonably close to the target of 30 %.

The compressive strength measured after the 28 days of curing ranged from 11.0 to 17.9 N/mm², sufficient for porous concrete. It should be noted that the compressive strength increased as the percentage of crushed roofing tile material was increased.

Table 3   Air void percentage and compressive strength of porous concrete

CRUSHED ROOFING TILE %	AIR VOID PERCENTAGE, %		COMPRESSIVE STRENGTH, AFTER 4 WEEKS OF CURING N/mm²
	TARGET AIR VOID	ACTUAL, VOLUMETRIC METHOD	
0	30	27.6	11.1
20	30	27.7	11.0
40	30	22.0	14.1
100	30	19.1	17.9

Freeze/Thaw Test Results: ASTM Procedure A

The test results using procedure A, shown in Figures 3 and 4, clearly show that the substitution of crushed roofing material for coarse aggregate had no adverse effect on the freeze/thaw resistance of porous concrete. Although all of the porous concrete specimens, regardless of the crushed roofing tile content, greatly deteriorated within 150 freeze/thaw cycles, RDEM (relative dynamic elastic modulus) decreasing below 70 %, the specimens containing no crushed roofing material performed the worst. The weight of specimens rapidly decreased after about 130 to 180 cycles, depending on the crushed roofing material content.

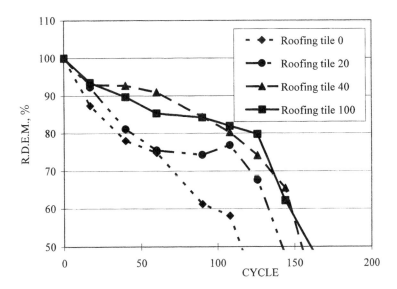

Figure 3   R.D.E.M. as a function of freeze/thaw cycles (Procedure A)

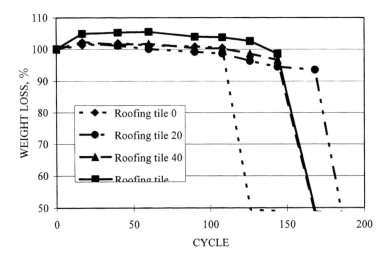

Figure 4   Weight loss as a function of freeze/thaw cycles (Procedure A)

Freeze/Thaw Test Results: ASTM Procedure B

As for the procedure A test results above, the test results using procedure B, shown in Figures 5 and 6, show that the substitution of crushed roofing tile material for coarse aggregate always improved the durability of porous concrete against freeze/thaw cycles. It should be noted that none of the porous concrete specimens, regardless of crushed roofing tile content, showed much deterioration after 200 cycles of freezing and thawing; RDEM (relative dynamic elastic modulus) only decreased to between 83 and 90 %, and the associated weight loss was minimal.

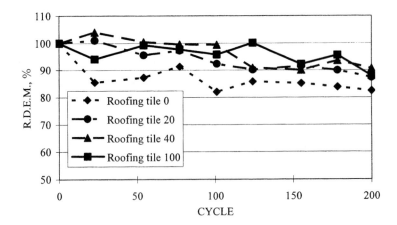

Figure 5   R.D.E.M. as a function of freeze/thaw cycles (Procedure B)

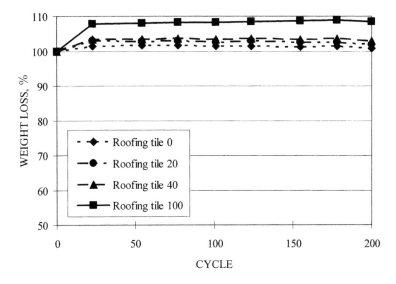

Figure 6   Weight loss as a function of freeze/thaw cycles (Procedure B)

CONCLUSIONS

This study examined the effects of substituting crushed roofing tile material for various amounts of coarse aggregate in concrete by measuring the compressive strength and the elastic modulus of high volume fly ash concrete and the freezing and thawing durability of porous concrete.

The conclusions are shown in the following.

1. The compressive strength of high volume fly ash concrete can be increased by substituting crushed roofing tile material for approximately 20 % to 60 % of the volume of the coarse aggregate. In addition, substituting crushed roofing tile material up to approximately 80 % has little effect on the static elastic modulus; however, at 100 % substitution, the static elastic modulus is somewhat decreased.

2. The presence of crushed roofing tile material has no adverse effect on the freezing and thawing durability of porous concrete.

REFERENCES

1. Present state and future trend of reuse technology for construction by-products, Architectural Inst. of Japan, Kanto branch, 1996, 3.

2. Present state and future view of zero emission in construction projects, Architectural Inst. of Japan, Kinki branch, 1999, 8.

3. SUGIYAMA, M., The Experiment on Compression Strength and Freeze-Thaw Resistance of Concrete which Mixed Tile Chip, Proceedings of the International Symposium of Recycling and Reuse of Glass Cullet, Concrete Technology Unit and University of Dundee, Scotland UK, 19-20 March, 2001, pp 189-194.

4. SUGIYAMA, M., The Freeze-Thaw Test Results of Porous Concrete with Crushed Scallop Shell Material Added, Proceedings of the International Symposium of Recycling and Reuse of Waste Materials, University of Dundee, Scotland, UK, 9-11 September, 2003, pp 459-464.

DURABILITY OF REINFORCED CONCRETE APPLYING SOME EXPERT SYSTEMS FROM THE WORLD WIDE WEB

J Gomez Dominguez

Monterrey TEC

C E P Garay Mendoza

Mexican Federal Government

Mexico

ABSTRACT. A concrete structure was designed to be built near the sea in Mexico. Two concrete strengths were considered, a conventional concrete 25 N/mm^2 and a high strength concrete 41 N/mm^2; additionally mixes were redesigned to allow the use of silica fume with 5% and 10% cement replacement by weight, generating four more mixes. The basis for predicting durability in the concrete structure was related to the initiation of corrosion of the reinforced concrete, prediction of such phenomenon was done with the help of a series of programs available via the WWW. The chloride ion penetration prediction taken as the time of initiation of corrosion was estimated with Fick's 2nd law of diffusion for a semi-infinite media. From an economical analysis performed over a 75 year period, it was found that all the possible projects are financially feasible, however the project that yielded the higher rate of return was the building made with high strength concrete and 10% silica fume with a 10.71% rate of return. Therefore this building is by far the best investment.

Keywords: Durability, Reinforced concrete, Corrosion, Silica fume, Diffusion, Rate of return.

Dr Jorge Gomez Dominguez, Professor of Civil Engineering at Monterrey TEC, Monterrey Campus, Nuevo Leon, Mexico. He graduated with honors with a B.Eng. from Mexico State University at Toluca, Mexico. He holds a M.S.C.E. and a Ph.D. from Purdue University at West Lafayette, Indiana. His areas of interest are concrete technology, construction materials, durability and repair of concrete structures.

MC Carlos E P Garay Mendoza, he graduated with a BEng from Nuevo Leon State University at Monterrey, Mexico. He graduated with a M.C. from Monterrey TEC in 2002. Currently he is Director of Building Construction at the Secretary for Public Education in the Mexican Federal Government.

INTRODUCTION

Advances in concrete technology normally lead construction practice. At least in Latin America, it seems that little is done to incorporate such advances in the everyday concrete job, and the technological experience developed with years and years of research done all over the world remains hidden in the literature. Recently the World Wide Web (WWW) has become a powerful source of high quality research accessible to everybody for free! This paper is a theoretical exercise for showing our construction industry the monetary justification of incorporating concrete technology advances to produce a durable concrete structure exposed to a warm coastal environment. This theoretical analysis would have been practically impossible to realize without the virtual support given by the WWW sites consulted. The information consulted is so valuable that it sets an example of how money invested in such research can feed into so many other studies, this is a real example of a sustainable synergy for research, why to spend more money to do the same research?

The economic impact caused by corrosion consumes between 2 and 5 percent of the gross domestic product in some countries [1], and it has been found that such costs could have been avoided by 15 to 25 percent if existing technology had been applied. There is a lack of conclusive statistics in Mexico, but with more than 10,000 km of coasts, one can assume that corrosion should also be of concern. A recent study of economic impacts relating to this problem shows that more than 90 percent of industries in Mexico face some sort of damage due to corrosion [2]. It is well known that many of the structures that failed when one of the most destructive hurricanes, Gilberto (1988), hit the Yucatan coast exhibited corrosion damage.

WWW EXPERT SYSTEMS

Expert systems that can be found in the Wide World Web are specialized programs for solving a wide variety of problems. A classical expert system includes data storage and algorithms for processing data and retrieving answers. The expert system utilizes a base of knowledge developed by a group of scientists that previously had carried out research involving validation of the model or models used in the system.

OBJECTIVE

The Objective of this work is to generate convincing economic information to illustrate the construction industry in Mexico, about the necessity to consider better performance concrete when building new structures near the sea in warm climates, since in these conditions durability is a major drawback for conventional concrete, which has been a common practice in Mexico. There exists the need to show that at the end, better performance concrete warranties an economic return when considering all the construction expenses that have to be made without leaving out maintenance costs.

APPROACH

In this theoretical work, the lack of a fully equipped laboratory to do research on durability, especially on long term exposure for initiating corrosion on reinforced concrete, led us to use virtual laboratories available in the Wide World Web. After the age for initiation of corrosion was found on a given type of concrete, an economical analysis was performed to find out the rate of return for the money invested. The time-frame considered for the economic analysis was set arbitrarily to 75 years and maintenance to counteract corrosion was included in the analysis after the age for initiation of corrosion was reached.

CONCRETE CONSTRUCTION

The concrete structure considered for cost analysis consisted of a small one floor commercial building, 100 squared meters and 6 meters in height located about 100 meters away from the sea. The building was fully designed according to ACI 318-95 code, considering also some local provisions for coastal wind. The structure frame consisted of isolated footings, squared columns, beams and a grid slab for the roof, partition walls and other finishing concepts were also considered.

Materials

Concrete considerations were critical, since these influence durability. Therefore two concrete strengths were considered for structural design, since a unique strength was used in the whole frame, two different buildings should be kept in mind. One building considered a 25 N/mm^2 concrete strength, which is common in most of the constructions in Mexico; this concrete will be referred to as conventional. The other option considered a 41 N/mm^2 concrete strength. The later option is by no means a High Performance Concrete (HPC) based just on strength, but is a good start to develop this type of concrete as it will be described later. The reinforcing steel considered in all cases had a 412 N/mm^2 yielding stress.

Concrete mix design

Concrete mix design was accomplished with the help of a pair of sections of a WWW site [3] or computer integrated knowledge system (CIKS). Two basic mixes were considered as controls, a conventional concrete (25 N/mm^2) designed as per ACI 211.1-91 and a high strength concrete (41 N/mm^2) designed as per ACI 211.4R-93. Those mixes were further redesigned including 5 and 10 percent silica fume replacement by cement mass. Mix design proportions are shown in Tables 1 and 2; results were used to calculate the production of concrete. Other considerations that should be kept in mind are: cement type I was considered, severe concrete exposure to salts or sulphates was also assumed, and for obtaining the desired slump (90 mm) especially for mixes that included silica fume, an admixture was considered to give the desired fresh consistency.

Table 1 Mix proportions for conventional concrete (25 N/mm^2), in kg/m^3

	CONVENTIONAL	MIX	PROPORTIONS
Silica Fume Rep.	0%	5%	10%
Water	211	211	211
Cement	463	440	416
Silica Fume	0	23	46
Coarse Aggregate	1012	1012	1012
Fine Aggregate	718	710	701

Table 2 Mix proportions for high strength concrete (41 N/mm^2), in kg/m^3

	HIGH STRENGTH	MIX	PROPORTIONS
Silica Fume Rep.	0%	5%	10%
Water	204	204	204
Cement	466	445	419
Silica Fume	0	23	47
Coarse Aggregate	1128	1128	1128
Fine Aggregate	574	565	537

ANALYSIS

Durability of the concrete structure was related in this work to the initiation of corrosion on the reinforcing steel, it is well known that such phenomenon depends on the presence and level of chloride concentration, as well as moisture and oxygen. Without disregarding the influence of carbonation especially in contaminated environments, it is believed that chloride ion diffusion in concrete becomes a viable way of predicting the possible age at which corrosion will be initiated once steel is reached. Answers to not only this problem but many other problems related can be found at CIKS [3]. CIKS provides access to run a computer program for predicting the service life of reinforced concrete structures subjected to certain environmental conditions. The service life depends on the level of chloride concentration surrounding the reinforcing steel and the corrosion threshold value.

Initially CIKS allows the determination of the chloride ion diffusivity of the concrete considered based on a few concrete mix parameters. Subsequently the life prediction can be calculated with the use of another program based on Fick's second law solving for the time involved (t) in the following equation:

$$\frac{C_{corr}}{C_{ext}} = erfc \frac{x}{2\sqrt{D\,t}}$$

Where:
   C_{corr} = concentration of chloride ions needed to initiate steel corrosion
   C_{ext} = external chloride ions concentration
   erfc = Gauss error function of depth (x)
   x = depth of the reinforcement
   D = chloride ion diffusivity
   t = predicted service life

Life Prediction

Tables 3 and 4 show data and prediction results after running CIKS. The program estimates first the diffusivity and then a subroutine yields the predicted life after solving the equation above. All this was done for each concrete included in the study. It should be indicated that in solving the equation, CIKS considers a normal distribution of the depth of reinforcement, so the program requires an average cover depth (x) and a cover depth standard deviation (x_σ) which are included in the following tables.

Table 3 Data and life prediction for conventional concrete (25 N/mm^2)

	CONVENTIONAL	MIX	LIFE	PREDICTION
Silica Fume Rep.	0%	5%		10%
C_{corr}, kg/m^3	3.024	2.207		1.237
C_{ext}, kg/m^3	5.34	5.34		5.34
x, m	0.03	0.03		0.03
x_σ, m	0.008	0.008		0.008
D, m^2/s	1.311×10^{-12}	0.254×10^{-12}		0.067×10^{-12}
t (service life), yrs	7	19		34

Table 4 Data and life prediction for high strength concrete (41 N/mm^2)

	HIGH STRENGTH	MIX	LIFE	PREDICTION
Silica Fume Rep.	0%	5%		10%
C_{corr}, kg/m^3	3.178	2.354		1.327
C_{ext}, kg/m^3	5.34	5.34		5.34
x, m	0.03	0.03		0.03
x_σ, m	0.008	0.008		0.008
D, m^2/s	0.909×10^{-12}	0.154×10^{-12}		0.033×10^{-12}
Service life, yrs	13	36		74

Economical Analysis

First, the total construction costs were determined for each of the concrete options considered, involving materials, labor and land value. Then, an economical analysis was performed to determine which option would generate the best investment. Analysis was done with the aid of an electronic spreadsheet program written for the purpose; this analysis contemplated a 75 year time-frame and considered maintenance costs after initiation of corrosion, and inflation effects through the life of the structure. The spreadsheet computes a flow of money, the money invested versus the money recovered through rent of the building. After each run, the rate of return of the money invested becomes the output of a series of iterative computations the program makes. Since in Mexico it is not uncommon to use concrete made at the job site, additional runs were carried out to contrast the cost of this type of concrete against the cost of using premixed concrete.

Results of the economical analysis are presented graphically in Figures 1 to 4. Figures 1 to 3 show how money behaves with time, in all cases the present worth increases until corrosion gets into the picture, after that, maintenance is required and present worth decreases. It then increases again once the structure has been fixed. It can be seen from either Figure 1 or Figure 3 that the structure made with high strength concrete (10% silica fume cement replacement) delays the longest time to initiate corrosion (74 years!), so it would become by far the best structure in the assumed environmental conditions.

It should be pointed out that this last structure reported the best rate of return for the money invested, namely 10.71%, this happens to be the structure where concrete produced at the jobsite is used. The next best solution would be the structure made with high strength premixed concrete (10% silica fume replacement) with a 10.44% rate of return; rates for other options were lower.

Figure 4 illustrates the differences in maintenance expenses among the mixes used to build the structure; this figure illustrates particularly the differences between the best mixes for conventional and high strength concretes as it can be seen from the contrast in areas under the correspondent lines, namely A_1 and A_2.

DISCUSSION

The construction industry must be aware of the real cost of building non-durable concrete structures. In our case a simple 25 N/mm^2 conventional concrete proved not to be a wise decision for a structure near the sea, since corrosion would start in 7 years! On the other hand, a better performance concrete can be produced if silica fume is used as a cement replacement. This solution has proven to reduce concrete porosity, which consequently reduces the diffusivity of chloride ions and retards the initiation of corrosion [4]. In our case study, Figure 3 clearly indicates that the reinforced concrete structure made with high strength concrete enriched with silica fume (10% cement replacement) would last sufficiently long without corrosion problems, to allow for making profit from this durable structure.

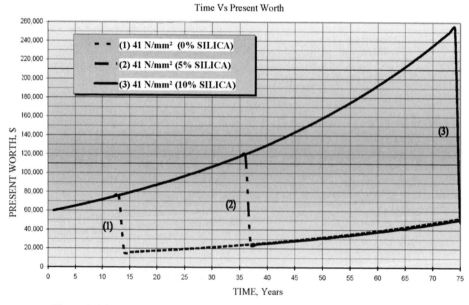

Figure 1 Money present worth with time in years for high strength concrete

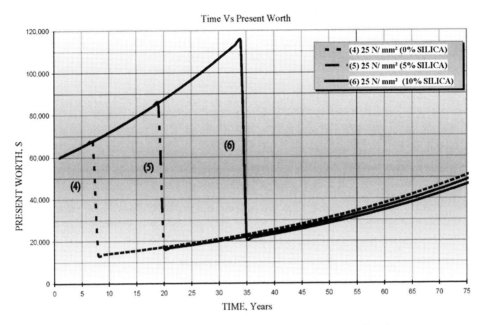

Figure 2 Money present worth with time in years for conventional concrete

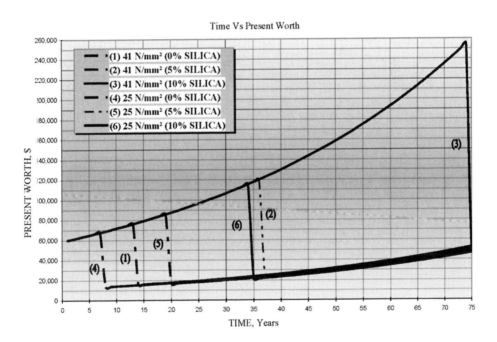

Figure 3 Money present worth with time in years for contrasting all concrete options.

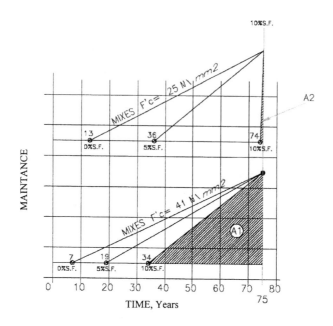

Figure 4 Maintenance with time for contrasting all concrete options

CONCLUSIONS

It can be concluded that money well-spent at the beginning to make a concrete structure durable will pay dividends while it fights corrosion. Once corrosion starts the money will be at hand to remedy the problems this will bring. A light at the end of a tunnel appears for those researchers without economical resources to do experimental work on durability of concrete, and many other areas, thanks to all the institutions that disseminate knowledge through the WWW.

REFERENCES

1.   BABOIAN, R., Corrosion- a problem of international importance. Corrosion Testing and Evaluation: Silver anniversary volume ASTM STP 1000, Philadelphia, 1990, pp 7-13

2.   AVILA, G.J., Mas alla de la herrumbre. La ciencia desde Mexico, No. 9, Fondo de Cultura Economica, 1986, pp 39-41

3.   http://ciks.cbt.nist.gov/~bentz/welcome.html

4.   BENTZ, D.P., Influence of silica fume on diffusivity in cement-based materials II. Multi-scale modeling of concrete diffusivity. Cement and Concrete Research, Vol. 30, 2000, pp. 1121-1129

ACCELERATED CARBONATION OF PORTLAND CEMENT MORTARS PARTIALLY SUBSTITUTED WITH A SPENT FLUID CATALYTIC CRACKING CATALYST (FCC)

E Zornoza   **P Garcés**

J Monzó   **M V Borrachero**

J Payá

Universidad Politécnica de Valencia

Spain

ABSTRACT. Accelerated carbonation tests have been carried out on Portland cement mortars and pastes partially substituted with a spent fluid catalytic cracking catalyst (FCC). This study has revealed that the incorporation of FCC on mortars produces a significant reduction of the alkaline reserve. Furthermore, it has been observed that this pozzolan also produces a reduction in mortar porosity. When the FCC/cement system is carbonated, both portlandite and CAH/CASH are transformed, yielding calcium carbonate. When the water/binder ratio (w/b) is low, the carbonation rate of mortars is negligible and is not modified by the presence of FCC. When the w/b ratio is medium or high, the carbonation rate rises sharply. In these cases, the presence of FCC accelerates the carbonation process suggesting that the reduction in porosity does not compensate for the reduction in portlandite which acts as a chemical barrier.

Keywords: Cement, FCC, Carbonation, Pozzolan, Mortar.

E Zornoza, PhD Student. Effect of industrial waste admixtures on the durability of Portland cement mortars and concrete.

Dr P Garcés, Professor of Building Materials. Director of Civil Engineering Area. Effect of carbonaceous materials and industrial waste admixtures on the properties of Portland cement mortars and concrete.

Dr J Monzó, Professor at Department of Construction Engineering, Universidad Politécnica de Valencia (Spain). His research activities and publications are in the fieds of analytical chemistry, building materials and recycling.

Dr M V Borrachero, Assistant Professor at Department of Construction Engineering, Universidad Politécnica de Valencia (Spain). Her research interest includes industrial by-products used as replacement of Portland cement.

Dr J Payá, Professor at Department of Construction Engineering, Universidad Politécnica de Valencia (Spain) and director of GIQUIMA. His main research interest is the use of solid industrial by products and wastes in cements and concrete production.

INTRODUCTION

The fluid catalytic cracking catalyst (FCC) is an aluminosilica-based powdered material used in petrol refineries. When this catalyst loses its catalytic properties, it must be replaced by a new catalyst. The wasted catalyst is then thrown out, and managed as an inert residue.

In recent years some authors have investigated the properties of this waste material as a pozzolan replacing cement in mortars or concrete [1-5]. Some previous works have shown that FCC is able to improve mechanical properties in mortars or concretes due to a densification of the cementitious matrix caused by pozzolanic reaction.

FCC is characterized by a high initial reactivity and consumes a large quantity of calcium hydroxide which has been released by cement hydration [6]. Both phenomena (densification and calcium hydroxide consumption) produce opposite effects on the carbonation resistance of the mortar. On the one hand, the precipitation of new hydration products as a consequence of the chemical reaction of portlandite and the aluminosilica framework of FCC, diminishes the size of pores and the connectivity and tortuosity of pores and capillaries in the cemented matrix. On the other, the decrease of portlandite content due to its consumption in the pozzolanic reaction and the dilution effect by replacing cement, diminishes the chemical capacity of the cement matrix for reacting with carbon dioxide according to the following process:

$$Ca(OH)_2 + CO_2 \Rightarrow CaCO_3 + H_2O$$

The aim of this research is to assess the role of this pozzolanic material in order to improve or reduce mortar's resistance to carbonation.

EXPERIMENTAL

In the experiments, samples consisted of 4x4x16 cm mortar specimens. The samples were made changing two parameters: the level of replacement of cement by FCC, 0 and 15%; and the water/binder (w/b, b being the sum of cement and FCC) ratio, ranging from 0.3 to 0.7. The aggregate used was normalized sand (EN 196-1) and the aggregate/cement ratio was 3:1. In mortars prepared containing w/b ratio lower than 0.5, superplasticizer was added (Sika Viscocrete).

After demoulding, the specimens were cured under water for a 28-day period. After curing, the specimens were maintained in the laboratory environment for a 15-day period in order to favour the evaporation of water present in the pores of the specimens. After this conditioning step, the samples were placed in a carbonation chamber containing an atmosphere consisting of 100% CO_2 and 65±5% relative humidity, for a three week period. During this period, two parameters were monitored: the mass of the samples by weighing and the depth of the carbonation front by spraying phenolphthalein solution and measuring the carbonation front, according to UNE/112011:1994. When the carbonation period finished, porosimetry analysis was performed in order to analyze the changes in the porous structure of the material. The mercury intrusion porosimeter used was an Autopore IV 9500 V1.05 supplied by Micromeritics Instrument Corporation. The analyses were performed from a pressure of 0.5 to 33000 psia.

Additionally, thermogravimetric analyses of cement/FCC pastes were performed in order to evaluate the consumption of calcium hydroxide and the production of calcium carbonate in the carbonation process. The substitution levels, in this case, were 0 and 20% and the w/b ratio was 0.8. These pastes were cured under water for a 28-day period, and then a sample was taken to analyse before carbonation process. After that, the pastes were subjected to attack in the carbonation chamber for a 28-day period and additional samples were taken after the carbonation process. Thermogravimetric analyses were performed in a TGA 850 Mettler Toledo thermobalance. The program used consisted of a heating rate of 20 °C/min from 35 to 1000 °C. Analyses were carried out in an alumina crucible and a dry air atmosphere. Thermogravimetric curves will be represented as differential thermogravimetric curves (DTG) in order to make more evident the differences between samples with and without FCC, and carbonated or not carbonated.

RESULTS AND DISCUSION

Thermogravimetric Analysis of Pastes

Figure 1 shows the DTG curves for cement pastes after the 28 day curing period, before the carbonation process. The development of peaks corresponding to the dehydration of CAH and CASH phases, which appear between 200 and 300 °C, are higher for the FCC substituted paste due to these being the typical compounds formed in the pozzolanic reaction of the FCC (mainly hydrated gehlenite). At about 500 °C, the peak corresponding to the decomposition of portlandite released in the cement hydration process can be observed. Evidently, the paste with 20% FCC has consumed a large amount of portlandite because of the pozzolanic reaction. Finally, at about 800 °C, weight loss corresponding to the decomposition of calcium carbonate can be observed. This part of the curves shows that there was a small quantity of calcium carbonate included with the raw materials.

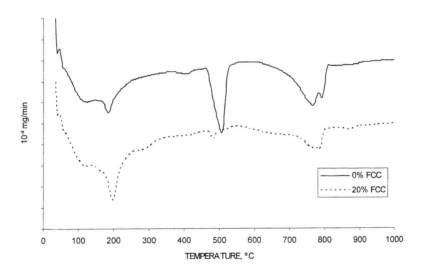

Figure 1   DTG curves of cement/FCC pastes after 28 days of curing. Atmosphere: dry air. Heating rate: 20 °C/min.

Figure 2 shows the DTG curves for cement/FCC pastes after the carbonation process. First of all, it is observed that those peaks corresponding to the dehydration of CAH and CASH phases have disappeared, because they evolved due to the reaction with carbon dioxide to form calcium carbonate. It is also observed that the calcium hydroxide has reacted with the carbon dioxide to form calcium carbonate, thus portlandite was not found after carbonation.

Finally, there is a big peak between 800-900 °C that shows a large formation of calcium carbonate. As it has been pointed out, there are two sources of calcium that can react with the carbon dioxide: CAH/CASH phases, and portlandite. The quantity of calcium carbonate formed in the paste without FCC is higher, due to the ease with which carbon dioxide can react with portlandite and large quantities of this phase in the paste.

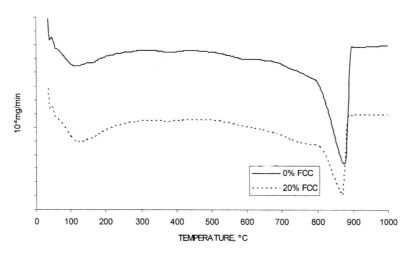

Figure 2   DTG curves of cement/FCC pastes after carbonation. Atmosphere: dry air. Heating rate: 20 °C/min.

Mercury Intrusion Porosimetry Studies on Mortars

It has been described by other authors that porosity with diameters between 0.05 and 3 μm is the most affected by changes in the w/b ratio and has the bigger effect on the permeability of mortars [7].

Figure 3 shows the porosity within this range for the cement/FCC mortars. In this figure, it can be observed that the higher the w/b ratio, the bigger the volume of porosity in this range. It can also be observed that the presence in the mortar of FCC produces a significant decrease in this type of porosity due to the additional formation of binder products from the pozzolanic reaction. This reduction was more evident for pastes with lower w/b ratio (0.3 and 0.4).

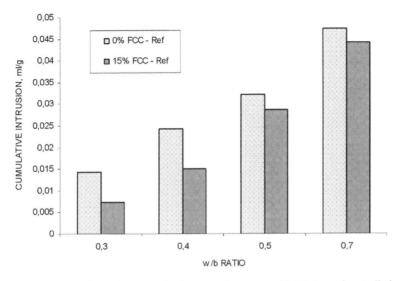

Figure 3   Cumulative intrusion in the pore size range of 0.05-3 μm for studied mortars before carbonation.

Figure 4 shows the pore size distributions for mortars with a 0.5 w/b ratio, before and after the carbonation process. It is observed that the carbonation process produces a reduction in the porosity for both FCC and non-FCC mortars. This behaviour is due to the formation of calcium carbonate from the calcium hydroxide, making a denser final structure.

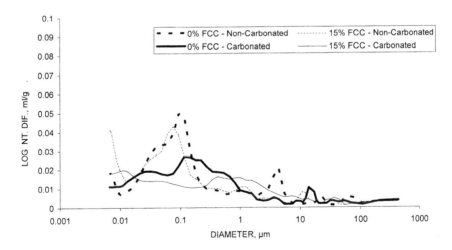

Figure 4   Pore size distributions for cement/FCC mortars with 0.5 w/b ratio, before and after the carbonation process.

Carbonation Rate Studies on Mortars

Figure 5 shows the evolution of the carbonation depth in 4x4x16cm mortar specimens during the carbonation process. Those mortars with low w/b ratio (0.3 and 0.4) showed a very low carbonation depth after the 21-day carbonation period. These mortars do not show differences between those that are not substituted and those that have a 15% cement replacement by FCC. As the w/b ratio is raised, the carbonated layer becomes larger. Mortars with 15% of FCC, and w/b ratio of 0.5 and 0.7 showed faster carbonation than plain cement mortars (without FCC).

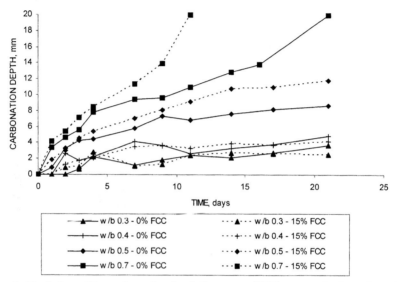

Figure 5   Evolution of the carbonation depth for cement/FCC mortars in the carbonation chamber. Atmosphere: 100% CO_2, 65% RH.

Experimental data from Figure 5 can be fitted to the equation that correlates the carbonation depth, X, with the exposure time, t:

$$X = V_{CO_2} t^{1/2}$$

where V_{CO_2} is the constant that relates the carbonation depth with the exposure time and is usually named as the carbonation rate (mm/year$^{0.5}$). In Figure 6, the carbonation rate obtained in 100% CO_2 atmosphere tests for cement/FCC mortars are represented. The lowest w/b ratio mortars showed very low carbonation rates. The replacement of cement by FCC, diminishing the portlandite content of the cementing matrix by pozzolanic reaction and by dilution effect, did not negatively affect behaviour in carbonation tests. Apparently, the refinement in pore size due to the pozzolanic reaction in FCC/cement mortars compensates the partial loss of the chemical portlandite barrier. However, for the highest w/b ratio, mortars containing FCC displayed the fastest rate of carbonation, suggesting that the pozzolanic role of FCC in these mixtures was not enough to compensate for the decrease in portlandite.

CONCLUSIONS

The partial replacement of cement by FCC, a very reactive pozzolanic waste, produces a decrease of portlandite content in Portland cement mixtures due to dilution and pozzolanic effects. The carbonation of FCC/cement pastes affects the remaining portlandite and also the CAH/CASH hydration products.

Two main effects produced by the incorporation of FCC as a pozzolan in cement mortars: reduction of the alkaline reserve and of porosity, have been studied when specimens were subjected to CO_2 attack. On one hand, when the w/b ratio is low (0.3 and 0.4) the reduction in porosity produces a very slow diffusion of carbon dioxide regardless of the low alkaline reserve of the mortar when FCC is present. On the other hand, when the w/b ratio is 0.5 or higher, the porosity of the mortar rises, and the carbonation rate is increased. In this case the reduction of the alkaline reserve produced by the presence of FCC has an important effect on the carbonation rate, accelerating the process.

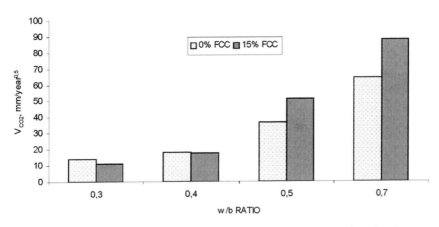

Figure 6   Carbonation rate in cement/FCC mortars from the carbonation chamber test.
Atmosphere: 100% CO_2, 65% RH.

ACKNOWLEDGEMENTS

This research was supported by Ministerio de Ciencia y Tecnología, Spain (Project MAT-2001-2694). Emilio Zornoza would like to thank Ministerio de Educación, Cultura y Deporte (Spain) for his doctorate grant (Programa FPU, AP2002-3421).

REFERENCES

1.   SU, N, FANG, H-Y, CHEN, Z-H AND LU, F-S. Reuse of waste catalysts from petrochemical industries for cement substitution. Cement and Concrete Research. Vol. 30, 2000. pp 1773-1783.

2.   PAYÁ, J, MONZÓ, J AND BORRACHERO, M V. Physical, chemical and mechanical properties of fluid catalytic cracking catalyst residue (FCC) blended cements. Cement and Concrete Research. Vol. 31, 2001. pp 57-61.

3.   PAYÁ, J, MONZÓ, J AND BORRACHERO, M V. Fluid catalytic cracking catalyst residue (FCC). An excellent mineral by-product for improving early-strength development of cement mixtures. Cement and Concrete Research. Vol. 29, 1999. pp 1773-1779.

4.   PAYÁ, J, MONZÓ, J, BORRACHERO, M V, AMAHJOUR, F, GIRBÉS, I AND ORDÓÑEZ, L M. Advantages in the use of fly ashes in cements containing pozzolanic combustion residues: silica fume, sewage sludge ash, spent fluidized bed catalyst and rice husk ash. Journal of Chemical Technology and Biotechnology. Vol 77, 2002. pp 331-335.

5.   PACEWSKA, B, BUKOWSKA, M, WILINSKA, I AND SWAT, M. Modification of the properties of concrete by a new pozzolan. A waste catalyst form the catalytic process in a fluidized bed. Cement and Concrete Research. Vol. 32, 2002. pp 145-152.

6.   PAYÁ, J, MONZÓN, J, BORRACHERO, M V AND VELÁZQUEZ, S. Evaluation of the pozzolanic activity of fluid catalytic cracking catalyst residue. Thermogravimetric analysis studies on FC3R-lime pastes. Cement and Concrete Research. Vol. 33, 2003. pp 1085-1091.

7.   PARROT, L J. A review of carbonation in reinforced concrete, Cement and Concrete Association, Ed. Slough, 1987.

FLY ASH BLENDED CEMENT:
AN EFFECTIVE MATERIAL FOR ADDRESSING
DURABILITY RELATED ISSUES

N Shafiq

F Nuruddin

University Technology Petronas

Malaysia

ABSTRACT: This paper presents the experimental results of the investigation of various properties related to the durability and long-term performance of mortars made of Fly Ash blended cement, FA and Ordinary Portland cement, OPC. The properties that were investigated in an experimental program include; equilibration of specimen in different relative humidity, determination of total porosity, compressive strength, chloride permeability index, and electrical resistivity. Fly Ash blended cement mortar specimens exhibited 10% to 15% lower porosity when measured at equilibrium conditions in different relative humidities as compared to the specimens made of OPC mortar, which resulted in 6% to 8% higher compressive strength of FA blended cement mortar specimens. The effects of ambient relative humidity during sample equilibration on porosity and strength development were also studied. For specimens equilibrated in higher relative humidity conditions, such as 75%, the total porosity of different mortar specimens was between 35% and 50% less than the porosity of samples equilibrated in 12% relative humidity, consequently leading to higher compressive strengths of these specimens. A valid statistical correlation between values of compressive strength, porosity and the degree of saturation was obtained. Measured values of chloride permeability index of fly ash blended cement mortar were obtained as one fourth to one sixth of those measured for OPC mortar specimens, which indicates high resistance against chloride ion penetration in FA blended cement specimens, hence resulting in a highly durable mortar.

Keywords: Fly ash, Chloride permeability index, Equilibrium condition, Electrical resistivity.

Dr Nasir Shafiq is an associate professor in the Department of Civil Engineering, University Technology Petronas, Tronoh, Perak, Malaysia. He received his Ph.D. from University of Leeds, UK in the area of Concrete Durability and Service Life design. His research interest includes, concrete durability, microstructure, cement hydration, corrosion in concrete and prediction modeling.

Dr Fadhil Nuruddin is an associate professor in the Department of Civil Engineering, University Technology Petronas, Tronoh, Perak, Malaysia. He received his Ph.D. from University Science, Malaysia in the area of Carbonation of Concrete. His research interest includes multiple blended cement concrete, concrete carbonation and durability.

INTRODUCTION

Over the last few years the use of agricultural and industrial wastes with pozzolanic reactivity such as fly ash, FA, and silica fume, SF, as a partial replacement of cement are becoming very popular in producing high strength and high performance cement mortar and concrete. Many studies have shown that the addition of pozzolanic materials in concrete tighten the pore structure and hence reduced the total porosity. This tight pore structure of concrete and mortar increases its resistance against the penetration of aggressive fluid and ions, which results in a high performance concrete. Porosity and pore structure perhaps more than any other characteristics affect the behaviour of concrete [1]. Total porosity, pore size and their distribution directly control both the engineering and transport properties of concrete and mortar and therefore set the performance criteria of its durability. Porosity and pore structure, in turn, are influenced by the original packing of cement, mineral admixtures, and the aggregate particles; the water-to-solid ratio; the rheology; and the curing conditions [2].

The compressive strength of concrete is not the sole parameter to define the quality of concrete as stated in some codes and specifications. However, there are other parameters, such as the transport properties, which are very important to assess the durability and long-term performance of concrete structures.

Corrosion of embedded steel reinforcement in concrete is one of the major threats to its durability. Cabrera and Ghoddoussi [3] stated that the rate of corrosion is controlled by the ease with which ions can pass through the concrete cover from a cathodic region to an anodic one. Hence a larger potential gradient associated with a low concrete resistivity will normally result in a high rate of corrosion. The humidity of concrete and the presence of ions in the pore solution of concrete affect its electrical resistivity.

In this study total porosity, compressive strength, electrical resistivity and the chloride permeability index of mortar samples made of different mixes composed of 100% OPC, 50% OPC + 50% FA, and 60% OPC + 40% FA were investigated. All samples were cured for 28 days in fog room then dried in 75%, 65%, 40% and 12% relative humidity until equilibrium conditions were attained, which took nearly 12 weeks. All samples used for measuring of total porosity, compressive strength and electrical conductivity were tested at equilibrium condition, whereas the chloride permeability index test was conducted on fully saturated samples. The experimental tests were carried out partly in the department of civil engineering, University Technology PETRONAS and partly at the University of Leeds, UK.

EXPERIMENTAL INVESTIGATIONS

Materials and Mix Proportions

Mortar mixes were obtained after wet sieving the coarse aggregate for concrete mixes. These were prepared with OPC and/or OPC + FA, sand and gravel using a weight ratio of 1:2.33:3.5. The OPC used complied with the requirements of BS 12 [4]. A Quartzitic sand and gravel conforming to BS 882 [5] were used as fine and coarse aggregates respectively. All mortar mixes were designed to have the same workability, i.e. a targeted slump of 55±5 mm. Details of all mixes used in this investigation are given in Table 1.

Table 1   Concrete/mortar mix proportions

MIX TYPE	OPC, kg/m^3	FA, kg/m^3	SAND, kg/m^3	GRAVEL, kg/m^3	W/C
OPC	325	0	757	1137	0.55
PFA40	195	130	757	1137	0.49
PFA50	162.5	162.5	757	1137	0.48

Specimen Casting, Curing and Equilibration

Cubes of dimensions 50mm × 50mm × 50mm were cast for compressive strength and electrical resistivity tests. All cubes were cured for 28 days in a fog room, followed by equilibration in 75%, 65%, 40% and 12% relative humidity (RH) for nearly 12 weeks. Similarly 600mm × 250mm × 40mm thick mortar planks were cast and cured for 28 days in a fog room, then 100mm diameter and 25mm thick circular discs were cored out for equilibration in different relative humidity conditions.

Determination of Compressive Strength

Equilibrated mortar samples of 50mm × 50mm × 50mm dimensions were tested using a universal testing machine in accordance with the British Standards BS 1881: Part 116, 1983 [6].

Measurement of Total Porosity

In this investigation, a vacuum saturation apparatus was used, which is similar to that developed by RILEM [7] for measuring the total porosity. The mortar samples equilibrated in different RH were weighed in the air, W_i. Then the specimens were vacuum saturated in water, in a fully saturated condition; the specimens were weighed in the air, W_{SA} and in the water, W_{SW}. Finally, they were dried in an oven at 105°C to constant weight, W_d. The total porosity, P was calculated as:

$$P = \frac{W_{SA} - W_d}{W_{SA} - W_{SW}} 100$$

Measurement of Chloride Permeability Index

After equilibration of mortar specimens in different relative humidity conditions, they were saturated using vacuum saturation before the test was conducted. The Chloride Permeability Cell developed at the University of Leeds [8] was used in this investigation. Two samples from each of the mortar mixes were tested at an applied voltage of 30 volts for 6 hours to determine the chloride permeability index. The samples were fixed between the two compartments, the one bearing sodium hydroxide (NaOH) solution and the other bearing sodium chloride (NaCl) solution, the cell was tightened using long screws and nuts within the holes provided at the corners of the compartments, and the edges and gaps were completely sealed using silicon rubber in order to prevent any leakage during the test. The test was conducted according to the procedure as illustrated by Cabrera and Lynsdale [8].

Measurement of Electrical Resistivity

The moisture within concrete and the presence of ions in the pore solution, primarily affect the resistivity of concrete. Tutti [9] mentions that the electrical resistance of concrete is more sensitive to the equilibrium relative humidity than any other parameters. Cabrera and Ghoddoussi [3] reported that all those factors controlling the permeability of concrete also affect its resistivity. Moist concrete behaves essentially as an electrolyte with a resistivity in the order of 100 ohm-m, a value in the range of semi conductors. Oven-dried concrete has a resistivity in the order of 10^9 ohm-cm, like a good insulator. For very high resistivity of concrete and mortar, the corrosion-induced current is very small; therefore, the chances of significant corrosion are unlikely to occur [10]. In general, Vassie [11] suggested that if the resistivity is greater than 120 ohm-m, significant corrosion is unlikely to occur.

Berke et al [12], reported that from the measurements made on 36 trial concrete mixes containing micro silica, severe corrosion occurred in two mixes which exhibited resistivity values of 480 ohm-m and 730 ohm-m. Besides moisture content, the resistivity of concrete is a function of other factors, including temperature and water-cement ratio, which are directly proportional to resistivity. Bamforth and Pocock [13] investigated the effect of fly ashes on the resistivity of concrete, they found that at 2 years of age, fly ash concrete exhibited resistivity values in the order of 100 ohm-m or greater and the OPC concrete showed resistivity in the range of 100 to 300 Ohm·m.

In this investigation, the relative electrical resistance of the mortar specimens was measured using an instrument based on Wheatstone Bridge, at a frequency of 1 kHz and 10 Volts.

RESULTS AND DISCUSSIONS

Total Porosity of Mortar Samples

Table 2   Total porosity, P (%) of mortar mixes

RH, %	TOTAL POROSITY, P, % OF MORTAR MIXES		
	OPC	40% FA	50% FA
75	18.51	16.73	17.73
65	19.83	17.36	17.84
40	20.56	17.60	18.06
12	21.10	18.08	19.05

Table 2 shows the experimental results of total porosity, P of the mortar samples equilibrated in different relative humidity conditions. As shown in Table 2, the total porosity of fly ash blended cement mortar was 10% to 15% lower than the corresponding OPC mortar specimens.

It is also noted that the ambient relative humidity also significantly affected the total porosity of the samples. Porosity was 7% to 9% lower in mortar samples equilibrated at 75% RH as compared to samples equilibrated at 12% RH.

Compressive Strength of Equilibrated Mortar Samples

Table 3 shows the values of the compressive strength of mortar cubes equilibrated at different relative humidity. Since the compressive strength is a function of the total porosity, therefore a similar trend for fly ash blended cement mortar and OPC mortar was observed. The mortar cubes equilibrated at 75% RH showed the lowest porosity. Therefore, such mortar cubes displayed the highest compressive strength. Since both compressive strength and total porosity were significantly affected by the ambient relative humidity, compressive strength may be statistically correlated with a parameter termed as open porosity, V_e, which is defined as the fraction of unsaturated pores, and mathematically expressed as:

$$V_e = p\left(1 - \frac{S}{100}\right)$$

Where, P is the total porosity and S, is the degree of saturation of the equilibrated samples, and V_e is the open porosity.

The compressive strength, f_{cu} of different mortar mixes was plotted against the corresponding values of open porosity, V_e and a valid statistical correlation was obtained as:

$$f_{cu} = 62.27 - 0.805V_e$$
$$R^2 = 87.9\%$$

Table 3   Compressive strength, f_{cu} (MPa) of equilibrated mortar samples

RH, %	COMPRESSIVE STRENGTH, f_{cu}, MPa		
	OPC	40% PFA	50% PFA
75	58.1	61.6	60.1
65	53.2	57.1	54.3
40	50.7	54.9	52.5
12	46.7	51.4	49.7

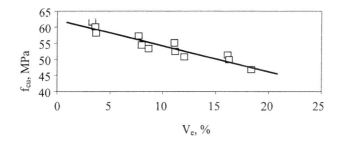

Figure 1   Compressive strength, f_{cu} versus open porosity, V_e of equilibrated specimens

Chloride Permeability Index (CPI)

It is noted from Figure 2 that that the fly ash blended cement lowered the chloride permeability index to one fourth to one sixth as compared with that the values obtained for specimens made of 100% OPC mortar. Therefore fly ash blended cement mortar offers high resistance against the penetration of chloride ions. Although the specimens were fully saturated after reaching the equilibrium condition in different relative humidity, the effects of ambient relative humidity are prominent for all mixes as shown in Figure 2. Chloride permeability index values nearly two times higher were obtained for samples equilibrated in 12% relative humidity compared to values obtained for samples equilibrated in 75% relative humidity. One of the obvious reasons is the higher porosity of samples equilibrated in dry ambient air.

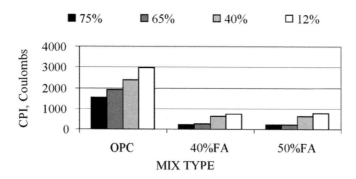

Figure 2   Chloride permeability index (CPI) of equilibrated mortar samples

Electrical Resistivity of Mortar

As resistivity is inversely proportional to the chloride permeability index a similar trend was observed for electrical resistivity of the different mortar samples, as shown in Table 4. The specimens made of fly ash blended cement mortar displayed 4 to 7 times higher resistivity as compared to the resistivity values measured for corresponding OPC mortar samples. Similarly, the effects of equilibrium conditions were also significant as observed in Table 3, for OPC mixes, samples equilibrated in 75% RH showing twice the resistivity of that samples equilibrated in 12% relative humidity, whereas, for fly ash blended cement mixes, electrical resistivity of samples equilibrated in 75% increased by more than 3 times the resistivity of those samples equilibrated in 12% relative humidity. Electrical resistivity, ρ was plotted against the chloride permeability index, CPI as shown in Figure 3 and a valid statistical correlation was obtained as:

$$\rho = 2.0x10^7 \left(CPI\right)^{-0.97}$$
$$R^2 = 0.98$$

Table 3   Electrical resistivity, ρ (Ohm·cm) of equilibrated mortar samples

RH, %	ELECTRICAL RESISTIVITY, ρ, Ohm·cm		
	OPC	40% PFA	50% PFA
75	12106	91860	84452
65	9879	75884	79333
40	7896	29582	30800
12	6309	25053	24467

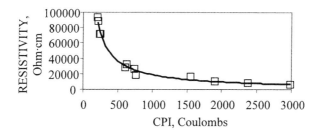

Figure 2   Electrical resistivity, ρ versus chloride permeability index, CPI

CONCLUSIONS

From the above results and discussions, the following major conclusions were made:

1. Partial replacement (40% and 50%) of cement with fly ash refines pore structure, enhances the compressive strength, and increases resistance to chloride ion penetration, resulting in a durable and high performance cement mortar.

2. A partial replacement of 40% fly ash content produces better result than 50% partial replacement. Therefore, 40% partial replacement of cement with fly ash is most suitable.

3. The effects of ambient relative humidity during equilibration of mortar samples were significant on all the characteristics of mortar investigated. Equilibration in wet conditions enhanced the properties of concrete, whereas dry ambient air caused an increase in the porosity of mortar, resulting in lower performance mortar.

4. Valid statistical correlations were obtained between compressive strength and open porosity and between chloride permeability index and electrical resistivity of mortar samples.

ACKNOWLEDGEMENTS

The authors would like to extend their acknowledgements to the department of Civil Engineering, University Technology PETRONAS, for offering an excellent laboratory facilities and analytical tools, which enabled authors to accomplish a part of experimental investigation this research and dissemination of the results output. The authors would like to acknowledge the Civil Engineering Materials Unit at the University of Leeds, UK for furnishing a part of experimental data of this study.

REFERENCES

1. COOK, R.A., HOVER, K.C., (1993), "Mercury porosimetry of cement based materials and associated corrosion factors", ACI Materials Journal, Vol. 90, No. 2, pp 152-158.

2. BROWN, P.W., SHI, D., (1993), "Porosity-Permeability Relationships", Materials Science of Concrete, Vol. 2, Skalny, J and Mindess, S, (eds.), pp 83-109.

3. CABRERA, J.G., GHODDOUSSI, P., (1994), "The Influence of Fly Ash on the Resistivity and Rate of Corrosion of Reinforced Concrete", ACI SP 145, Durability of Concrete, Nice (France), pp 229-244.

4. BRITISH STANDARDS INSTITUTION, BS 12, (1996), "Specifications for Portland Cement", BSI, London.

5. BRITISH STANDARDS INSTITUTION, BS 882, (1992), "Specifications for Aggregates from Natural Sources for Concrete", BSI, London.

6. BRITISH STANDARDS INSTITUTION, BS 1881 PART 116, (1983), "Method of Determination of Compressive Strength of Cubes", BSI, London.

7. RILEM, CP 113, (1984), "Absorption of Water by Immersion under Vacuum", Materials and Structures, Research and Testing, No. 101, pp 393-394.

8. CABRERA, J.G., LYNSDALE, C.J., (1988), "Measurement of Chloride Permeability in Superplasticised Ordinary Portland Cement and Pozzolanic Cement Mortars", Int. Conference, "Measurements and Testing in Civil Engineering", Lyon-Villeurbranne, pp 279-290.

9. TUTTI, K., (1982), "Corrosion of Steel in Concrete", Swedish Cement and Concrete Research Institute (CBI), Stockholm, No. 4.82.

10. MALHOTRA, V.M., CARINO, N.J., (1991), "Non-destructive Testing of Concrete", CRC Press, pp 217-225.

11. VASSIE, P.R.W., (1978), "Evaluation of Techniques for Investigating the Corrosion of Steel in Concrete", TRL: Supplementary Report 397.

12. BERKE, N.S., HICKS, M.C., (1993), "Predicting Chloride Profiles in Concrete", Corrosion Engineering, Vol. 50, No. 3, pp 234-239.

13. BAMFORTH, P.B., POCOCK, D.C., (1990), "Minimising the Risk of Chloride Induced Corrosion by Selection of Concreting Materials", Proceedings of the Conference on, "Corrosion of Reinforcement in Concrete", Society of Chemical Industry, Elsevier Applied Science, pp 119-131.

THE USE OF COAL COMBUSTION BOTTOM ASH IN LIGHTWEIGHT MASONRY UNITS

J Groppo   B Phillips

R Rathbone   R Perrone

University of Kentucky

C Price

Charah Incorporation

United States of America

ABSTRACT. The production of coal combustion bottom ash exceeds 20 million tons in the United States and at one time was a major component in the manufacture of masonry units (a.k.a. "cinder block"). Quality issues such as staining along with substantial variations in size distribution have resulting in its being supplanted by other aggregates. This research has focused on the use of bottom ash as a major component in high value lightweight aggregate. Bottom ash from utilities consuming sub-bituminous as well as, high sulfur and low sulfur bituminous coals have been examined using a laboratory-scale pneumatic cylinder press and batch block machine to evaluate a variety of mix designs. It was found that by controlling the gradation and using fly ash as a filler at significant concentration that these materials were capable of producing excellent products. The results to date indicate bottom ash can again be an important source of aggregate in the US if processed to control size consist and remove impurities.

Keywords: Coal combustion ash, Bottom Ash, Concrete masonry units, Lightweight aggregate.

John Groppo is a Senior Mining Engineer at the University of Kentucky Center for Applied Energy Research in Lexington, Kentucky, USA.

Benjamin Phillips is a Graduate Student in Civil Engineering at the University of Kentucky, Lexington, Kentucky, USA.

Robert Rathbone is a Geologist at the University of Kentucky Center for Applied Energy Research in Lexington, Kentucky, USA.

Roger Perrone is a Research Scientist at the University of Kentucky Center for Applied Energy Research in Lexington, Kentucky, USA.

Charles Price is President of Charah, Inc., Louisville, Kentucky, USA.

INTRODUCTION

Concrete masonry units (CMU's) or blocks are extensively used throughout the world for a variety of construction applications. The heavy weight of such units can be problematic; a standard 20 cm × 20 cm × 40 cm (8 inch × 8 inch × 16 inch) hollow concrete blocks made with conventional standard weight aggregate weighs approximately 17 kg (37.5 lbs.). Since the aggregate is the single largest weight component of CMU's, using lightweight aggregate can provide substantial weight reduction, provided unit strength is not compromised.

Natural lightweight aggregates, such as pumice or scoria, have been successfully used, provided they are available. In recent decades, artificial lightweight aggregates such as expanded shales and clays have become available and are extensively used in producing lightweight blocks [1, 2]. While these types of lightweight aggregates can indeed reduce unit weight, their use is frequently limited by their cost. Another potential lightweight aggregate is coal combustion bottom ash. Bottom ash is the coarse residue remaining after coal combustion. In the US it is typically discarded as waste, but with proper gradation and processing, represents a significant potential for use as lightweight aggregate [3].

EXPERIMENTAL

Bottom ash samples were collected from three different utilities burning different types of coals in pulverized coal combustion boilers; high and low sulfur bituminous and low sulfur sub-bituminous. The size distributions and bulk densities of the as-received bottom ash samples are summarized in Table 1.

The bottom ash generated from burning low sulfur bituminous coal had the lowest bulk density while sub-bituminous coal generated the highest bulk density ash. All of the bottom ash samples were below the maximum loose bulk density requirements of lightweight aggregate for use in concrete masonry units (CMU's) as outlined in ASTM Designation C331. As well, each of the bottom ash substrates evaluated were within the grading requirements for use as lightweight aggregate, which is also outlined in ASTM C331. However, initial attempts to produce lightweight CMU's required unacceptably high levels of cement to meet strength requirements and appeared unusually porous compared with conventional blocks. It was determined that improving strength characteristics without increasing cement usage would require further evaluation of gradation criteria.

An approach was devised to evaluate several size distributions in order to maximize compaction, thereby reducing voids in block to minimize the amount of cement required to fill the voids. Each of the substrates was screened into various size fractions and the fractions were recombined in differing proportions to achieve the maximum packing arrangement. A 0.45 power chart was used for plotting gradation [4]. In this approach, the sieve sizes are raised to the 0.45 power and plotted vs. the cumulative percent passing and the densest mix results in a straight line. The various mixes we compared with the 0.45 power curve and compaction was verified experimentally using a Proctor hammer. Thus, a single size distribution for each substrate was selected to provide maximum compaction.

Rather than evaluating mix designs by producing commercial-size CMU's, a laboratory-scale apparatus was fabricated to produce test cylinder specimens. The device (Figure 1) operates by compressing a test mix into a standard plastic 7.62 cm (3 inch) diameter cement mold while vibrating. The compression and vibration are controlled separately. Compression is by a double action 11.43 cm (4.5 inch) diameter pneumatic cylinder mounted above a steel plate. The cylinder drives a steel piston into the mold located below it. A pneumatic vibrator is bolted to the same plate. This mechanism is mounted to a steel base plate by four isolation mounts which allow it to vibrate. Regulated air to the pneumatic cylinder and vibrator is controlled by the two spring lever control valves mounted to the base plate.

Test cylinders were stored at room temperature in a curing cabinet (Figure 2) at 100% humidity for 24 hours before stripping the molds from the cylinders. Each cylinder was weighed and measured so that unit weights could be calculated prior to capping with sulfur capping compound and returned to the curing cabinet. Strength determinations were made on 3 cylinders at 3, 7, 14, and 28 days in accordance with ASTM procedures and reported as averages.

Table 1 Size Distribution and Bulk Density of As-Received Bottom Ash Samples

		PLANT NAME			
		Belews Creek	Henderson	Rockport	ASTM C331 Specs
Type of Coal Burned		Low Sulfur Bituminous	High Sulfur Bituminous	Low Sulfur Sub-bituminous	
As-Received Bulk Density	kg/m^3	689	977	927	1120
	lb/ft^3	43.0	61.0	57.9	70
Size Distribution (Cumulative Weight Percent Passing)					
mm	mesh				
4.75	4	94.2	83.0	84.2	85-100
2.36	8	79.6	57.6	64.2	---
1.18	16	54.1	25.9	48.6	40-80
0.595	30	35.0	13.0	38.3	---
0.297	50	23.1	6.6	26.8	10-35
0.150	100	12.0	2.8	11.9	5-25

RESULTS AND DISCUSSION

Initial test cylinders were produced with each substrate and what was considered a minimum amount of cement to produce cylinder specimens that would hold together. Results showed that the compressive strengths achieved for all of the substrates were well below the target 6.8 MPa after 3 days (Figure 3). In an effort to increase strength without increasing cement addition, another series of cylinders were produced whereby fly ash was substituted into the mix design as a fine aggregate. The rationale for this approach was to further reduce the void volume of the cylinders with fine fly ash. For each substrate, fly ash from the same utility was used. The results obtained for the Belews Creek ash are shown in Figure 4. Compressive strength increased for fly ash substitutions up to 30% while further increasing the fly ash to 45% did not provide additional strength gain. At 30% substitution, strengths were above the target 6.8 MPa even after only 3 days of curing. Since pozzolanic reactions primarily occur after longer curing times (i.e. 28 days), the early strength gains that occurred with fly ash are attributed to decreasing the void volume, thus reducing the amount of cement required to fill the voids. Results obtained with the other substrates are shown in Figures 5 and 6 and similar trends were observed. Of particular note is the very high strengths that occurred when using Rockport ash, a class C fly ash which provided both early strength development and additional cementitious material.

Figure 1 Apparatus for Producing Test Cylinders to Evaluate Mix Designs

Figure 2 Capped Cylinder Specimens in Curing Cabinet

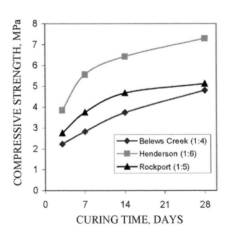

Figure 3 Effect of Curing Time on Compressive Strength For Cylinders Made with Bottom Ash from Different Sources.

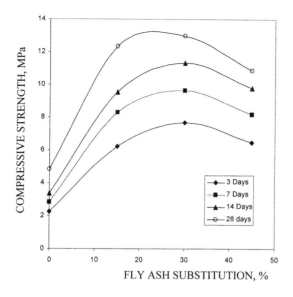

Figure 4 Effect of Fly Ash Substitution as Fine Aggregate on Compressive Strength for Belews Creek Bottom Ash (Cement:Aggregate = 1:4)

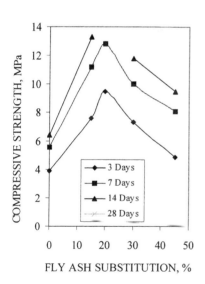

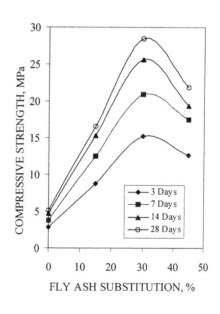

Figure 5 Effect of Fly Ash Substitution as Fine Aggregate on Compressive Strength for Henderson Bottom Ash (Cement:Aggregate = 1:6)

Figure 6 Effect of Fly Ash Substitution as Fine Aggregate on Compressive Strength for Rockport Bottom Ash (Cement:Aggregate = 1:5)

Although results showed that the addition of 20% to 30% fly ash as fine aggregate significantly improved compressive strength, it is also recognized that any reduction of cement content would be of economic benefit to potential commercial development of this utilization approach. A additional series of mix designs were evaluated using 20% to 30% fly ash as fine aggregate and reducing the cement:aggregate ratio. The results obtained for Belews Creek ash are shown in Figure 7. As expected, reducing the cement:aggregate ratio from 1:4 to 1:6 resulted in diminished strength. A ratio of 1:4 provided strength of 7.7 MPa after 3 days of curing while a ratio of 1:5 required 7 days of curing to achieve similar strength development and 1 a ratio of 1:6 required as along as 14 days.

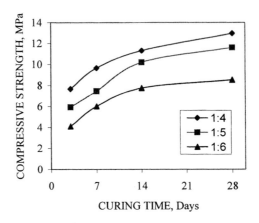

Figure 7 Effect of Curing Time and Cement to Aggregate
Ratio on Compressive Strength for Belews Creek Bottom
Ash with 30% Fly Ash Substitution as Fine Aggregate

For the Henderson ash, (Figure 8), acceptable strength was achieved in as little as 3 days with a lower cement:aggregate ratio of 1:6 while using only 20% fly ash as fine aggregate. The lower cement and fly ash requirements for this substrate compared with Belews Creek are attributed to the physical differences in two ashes. The Belews Creek bottom ash had a low bulk density of 689 kg/m^3 compared to 977 kg/m^3 for the Henderson ash. The differences in bulk density are readily attributed to the differences in porosity. The Belews Creek ash was very porous and "frothy" while the Henderson ash was smooth, vitreous and "slag-like". Both fly ash and cement will migrate into the porous ash, thus reducing the amount of these materials available to form a cementitious matrix along grain boundaries. This is less likely to occur with a more dense ash and does not occur at all with conventional aggregate. Cementitious material is available of reaction along grain boundaries, thus facilitating strength development.

The results achieved with Rockport ash are shown in Figure 9. Acceptable strengths occurred in 3 days even at cement:aggregate ratios as low as 1:8. The early strength development is partially attribute to the cementitious properties of the Class C ash used with this aggregate. As previously shown, strength improved with as much as 30% fly ash substituted as fine aggregate. Even though the bulk density of the Rockport ash is similar to that of the Henderson ash, the Rockport ash is relatively porous and it is likely that some of the fly ash migrated into the pores of the bottom ash during mixing.

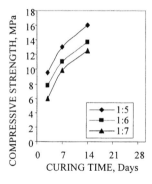

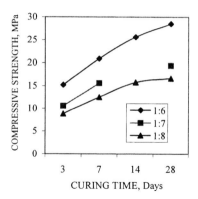

Figure 8 Effect of Curing Time and Cement to Aggregate Ratio on Compressive Strength for Henderson Bottom Ash with 20% Fly Ash Substitution as Fine Aggregate

Figure 9 Effect of Curing Time and Cement to Aggregate Ratio on Compressive Strength for Rockport Bottom Ash with 30% Fly Ash Substitution for Fine Aggregate

SUMMARY AND CONCLUSIONS

A laboratory technique has been developed for the purpose of evaluating mix designs for the production of CMU's. The technique utilizes a pneumatic piston and vibration to produce test cylinders with mechanisms that mimic those used in commercial CMU production. Gradation has been shown to be an essential component to achieving compaction, a critical element to achieving acceptable strength.

Fly ash addition was effective for reducing cement requirements to attain acceptable compressive strength. In general, the strength increased with increasing fly ash addition to a maximum of 20% to 30% by weight. The amount of fly ash that could be used was substrate dependent and attributed, in part, to the porosity of the bottom ash used as aggregate.

A comparative summary of cylinder test results along with target results is shown in Table 2. Acceptable strength and unit weight was achieved with the Rockport ash in a little as 3 days using 30% fly ash as fine aggregate and a cement:aggregate ratio of only 1:8. The early strength development at low cement requirement was attributed in part to the cementitious properties of the Rockport Class C fly ash. The Belews Creek bottom ash, required a higher proportion of cement (1:5) to achieve target strength in 7 days.

The higher cement requirements are attributed to the porous nature of this ash which provided surface roughness and porosity for fly ash particles to fill, thus reducing the amount of fly ash available to fill voids in the compacted mix. The Henderson ash, having smoother surface texture, achieved maximum strength with the addition of only 20% fly ash and a cement:aggregate ratio of 1:7.

Table 2 Summary of Cylinder Test Results

PARAMETER	PLANT NAME		
	Belews Creek	Henderson	Rockport
Fly Ash Substitution, weight %	30	20	30
Cement:Aggregate ratio	1:5	1:7	1:8
Target Strength, MPa	6.90	6.90	6.90
3 Day Strength, MPa	5.94	6.10	8.65
7 Day Strength, MPa	7.47	9.93	12.52
Target Unit Weight, kg	0.42	0.42	0.42
Unit Weight, kg	0.36	0.42	0.42

REFERENCES

1. BUILDEX Inc., Unique Combination of Quality and Economy, Buildex Publication, No. 141, P.O. Box No. 15, Ottawa, Kansas.

2. Maschinenfabrik GmbH., Mix Composition for the Production pf Hollow Concrete Blocks, KNAUER Information No. 2/74, Germany.

3. GROPPO, J.G., ROBL, T.L., Commercial-Scale Recovery pf Lightweight Block Sand from Stored Coal Combustion Ash, Proceedings: Sustainable Concrete Construction Dundee, Scotland, UK, 2002, pp 485-493.

4. AMIRI, B., KRAUSE, G.L., TADROS, M.K., Lightweight High-Performance Concrete Masonry-Block Mix Design, ACI Materials Journal, Sept-Oct. 1994, pp 495-501.

DOES RECYCLING FIT WITH SUSTAINABLE USE?

C F Hendriks
G M T Janssen
Delft University of Technology
The Netherlands

ABSTRACT. This contribution deals with the importance of sustainability in the building and construction sector. The basic idea is that sustainability and durability should always be considered together. Different models are presented to determine the degree of sustainability: the degradation factor, life cycle assessment, the Delft ladder of priorities, high grade applications, design for recycling, design for disassembly and the eco-cost-value ratio.

Keywords: Recycling, Sustainability, Waste management, LCA, EVR, Design for recycling, Design for disassembly, Pollution

Professor Dr Ir C F Hendriks was, until he passed away on November 13th 2004, Professor in Materials Science and Sustainable Construction at Delft University of Technology, Faculty of Civil Engineering and Geosciences, researching durability and sustainability of construction materials.

Ir G M T Janssen is a commercial Engineer and Master of Total Quality Management. She specialises in waste management and recycling. She is Assistant Professor of the Delft University of Technology, Director of the Consultancy Enviro Challenge and Director of The Dutch and Flemish Associations of Mobile Recycling Companies.

INTRODUCTION

The main reasons to recycle materials are reduction of waste and reduction of the demand for natural raw materials taking into account the disturbance of landscape as a result of this.

There are however also objections against recycling or at least aspects that have be clarified. A general question concerns the influence on the quality of the building or structure in which recycled materials are applied: performance, maintenance, durability and lifetime. For the answer to this question some decision models are developed, which will be presented in this contribution. The question concerning quality is also related to environment, health and working conditions. The best method in this case to quantify these effects are Life Cycle Assessment (LCA). In this case, energy consumption can also be taken into account.

Finally the image of recycled materials is of importance. This has mainly to do with the fear of risks in relation to the status of waste. In many cases recycled materials are considered as waste because of legislation. Recently the authors published a new approach which makes clear that when recycled materials meet the same specification for application as natural raw materials without increase of risk, they should be considered as raw materials. Because decisions of whether to recycle or not recycle should be based on the specific conditions which are relevant for a project, the authors developed some tools which are dealt with in this contribution.

Life Cycle Thinking (LCA method)

Life cycle thinking is based on the fact that decisions taken in one phase (design, implementation, management, maintenance, demolition and reuse) should always be set against the background of the consequences for the following phases. The intention is to minimise the environmental interventions over the entire life cycle. The most suitable method for this is life cycle analysis (LCA), which has now received both national and international (ISO: International Organisation for Standardisation) recognition.

Table 1 Overview of environmental problems in the LCA method

DEPLETION	POLLUTION	DETERIORATION
Depletion of raw materials	Greenhouse effect	Deterioration of landscape
Abiotic and biotic depletion	Deterioration of ozone layer	Deterioration of ecosystems
	Human toxicity	Deterioration of habitats
	Ecotoxicity	Victims
	Smog	Dehydration
	Acidification	Fragmentation
	Eutrophication	
	Radiation	
	Rejected heat	
	Odour	
	Noise	
	Working conditions	

Waste Management

Until recently the Dutch approach was based on the so-called 'Lansink's ladder', i.e. prevention – reuse of building elements – reuse of materials, incineration with energy recovery – incineration – disposal.

Lansink's ladder dates back to the end of the 1970s, when the legislation concerning waste products was first being written. However, looking back with hindsight, we notice several things. Firstly, the number of opportunities has increased, i.e. the scale of these opportunities (construction, building elements and materials) and there are new insights regarding immobilisation (using chemical or physical methods to prevent the spread of contaminants). Secondly, Lansink's ladder includes a clear order of preference. Methods are now available concerning life cycle analysis, eco-costs (costs of ensuring that a system or process meets sustainability criteria, including the so-called 'hidden costs') and value assessment. In particular, the results of the LCA methods and the EVR (eco-costs/value ratio) are applicable for making the best choice as to which rung of the ladder should be used in a specific situation. The result is thus an 'eco-ladder' or 'sustainability ladder' for the building sector, in which the order up the steps is 'tailor made'. This new dynamic ladder is depicted below, under the name 'Delft ladder' [1].

Table 2 The 'Delft ladder'

THE 10 STEPS	CONSEQUENCES
Prevention	'Design for recycling' (DFR), recovery, based on remaining lifespan (technical and economic)
Reuse of constructions	DFR, oversizing, selective dismantling, remaining lifespan
Reuse of building elements	DFR, selective dismantling, reprocessing, return system
Reuse of materials	DFR, selective dismantling, reprocessing, return system, leaching and content of contaminants
Useful application as residue	Quality equal to reference (with regard to leaching)
Immobilisation with useful application	Leaching and content of contaminants
Immobilisation without useful application	Dumping conditions
Incineration with energy generation	Emission limitation
Incineration	Emission limitation
Dumping	Dumping conditions

The return system refers to building elements or building materials that are returned to the original supplier and are processed (cleaned, repaired etc.), thus allowing them to be reused for the same purpose.

The Dutch Building Materials Decree specifies acceptable compositions and leaching, while the Dumping Decree defines the regime with which dumping sites must comply. Immobilisation means that the dosage rate for leaching (leaching rate) must be drastically reduced once the material has been treated (thermally, chemically or by adding a binder).

High-Grade Reuse

The term 'high-grade reuse' is becoming clearer. It is often said that, although a lot of construction and demolition waste is reused, this is mostly low-grade waste that is only suitable for constructing roads. High-grade therefore means reuse in concrete or other building products, as the economic return is much higher for roads than for concrete buildings. Based on the Delft ladder, and with the exception of the most desired option, how can this high-grade level be defined? The key can be found by answering the following 10 questions as accurately as possible. Performing this exercise for all recycling options means that the results can easily be compared. However, just as when using the LCA to select a material, we see that none of the options receives a perfect score, but weighing the criteria further usually leads to the desired choice.

Table 3 High-grade levels for reuse

HIGH-GRADE APPLICATIONS: DEFINITION OF THE FUNCTIONAL UNIT AS PER THE LCA	CHOICES OR ACTIONS
Applications meet all existing requirements, according to both private and public law	Yes or no
Technical lifespan is greater than economic lifespan	Yes or no
Applications require little maintenance	Express in mass quantities of maintenance materials per functional unit
Maximum potential for future reuse	State the various opportunities for reuse
Minimum degradation during the usage phase	No longer usable due to degradation
Applications are material-extensive	Materials required per functional unit
Applications are energy efficient	Energy required per functional unit
Applications have minimum harmful emissions	Emissions per functional unit
Applications have minimum non-reusable residual waste	Amount of unusable waste in kg per functional unit
Applications conform to market standards	Yes or no

Design for Recycling (DFR)

Most primary construction materials are natural substances and are homogeneous, e.g. sand, gravel, clay, oil (including the production of plastics), wool, cotton and wood (for various applications in the building sector, but also for paper), and ores (for metal production). This homogeneity often makes it possible, with relatively simple preparation, to use these raw materials in the production process. However, this is not true for secondary materials, i.e. mainly discarded products or material conglomerates, the composition of which is often so complex that reuse is only possible after extensive processing. Recycling secondary materials generally requires a series of processes, of which recognition, sorting, size reduction, and separating are the most important. Housing or construction demolition rubble must first have the iron, wood, mastic, bitumen, cardboard, chipboard, plastics, aluminium, zinc, copper etc. removed before the aggregate from the rubble crusher is suitable for use as secondary material. Asphalt and concrete from broken-up road pavement is also contaminated with road dirt, salt and oil residues etc. that must be removed before the recycling process can begin.

Over the past decades, mechanical 'reprocessing' systems have largely replaced manual processes, primarily in the metal sector but also for construction and demolition waste (CDW), so that CDW can now be used on a large scale in the construction sector.

Not only the recycling sector, but also the government, has come under increasing pressure to implement 'design for recycling' (DFR) methods, i.e. designing items in such a way that they can be recycled. DFR means:

- using fewer materials;
- using parts that can be easily dismantled;
- marking all the plastic parts used so that they can be readily identified during dismantling and can be easily separated for further processing.

Governments, both in the Netherlands and elsewhere, are promoting DFR through certification, subsidies, 'return to supplier' obligations and levies on non-recyclable materials or scarce primary raw materials. A complete DFR design means that the following categories of materials must be capable of being separated and recycled.

Concrete
- according to VBT concrete specifications (meets national standards);
- non-standard compositions (for use in concrete).

Masonry
- bricks, sand-lime brick and concrete brick with mortar (for use in roads or concrete);
- aerated concrete;
- crushed brick used as such or in brick (without mortar);
- crushed sand-lime brick for use in sand-lime brick;
- crushed concrete brick for use in concrete brick.

Bituminous materials
- asphalt without modified binders;
- asphalt with modified binders;
- roofing materials.

Wood
- hardwood;
- unpreserved wood;
- preserved wood.

Plastics
- thermosetting and thermoplastic agents should be used separately.

Design for Disassembly (DFD)

The criterion used is that, regardless of whether or not materials can be disassembled in an economic fashion, they can be separated for later reuse, if necessary. The following provides a basis for this strategy:

- inspection and sampling;
- specifications;
- maintenance inventory;
- conditions of use.

Some elements and materials should have been previously removed, e.g. flue pipes, gypsum blocks, gypsum walls, aerated concrete, bituminous roofing containing tar, asbestos and materials mixed with dangerous waste. Mixing with (contaminated) soil should be avoided where possible. With regard to the various categories of materials that should be recycled and separated, the comments in the section 'Design for recycling (DFR)' concerning concrete, masonry, metals and wood, also apply here. Glass should be separated independently or in combination with concrete or masonry. Plastics should be split into thermosetting and thermoplastic agents. When demolishing roads and hydraulic engineering constructions the DFD method is fairly simple as the materials are usually applied in layers. Pollution of materials is an important aspect for the proper demolition and recycling procedure [2]. Table 5 shows the 'threatening components' (both for the environment and technique) in buildings and Table 6 shows a similar approach for industrial complexes. Pollution can diffuse into building materials [2].

Table 5 Threatening components in buildings

BUILDING COMPONENT	THREATENED COMPONENTS FOR RECYCLING	
	TECHNICAL	ENVIRONMENTAL
foundation inner wall	bituminous (tar) layer gypsum breeze blocks cellular concrete sandwich panels (composite)	bituminous (tar) layer paint
external wall	insulation sandwich panels (composite) bituminous (tar) layer hydrophobe layers/coatings	bituminous (tar) layer paint hydrophobe layers/coatings impregnating agents
chimney	soot and tar	soot and tar
roofs	sandwich panels bituminous finishing plastic finishing insulation	bituminous finishing
floors	anhydrite isolation resin coatings bituminous (tar) coatings	resin coatings paint bituminous coatings floor coverings with asbestos
roofs	sandwich panels bituminous (tar) layer plastic roofing insulation	bituminous (tar) layer corrugated asbestos
window frames/doors	adhesives	preservatives paint/stain adhesives

Table 6 Possible pollution of non-residential buildings per industry sector

INDUSTRY SECTOR	Volatile aromatic hydrocarbon	Volatile clorinated hydrocarbon	Nitrite	Nitro-samine	Oil /benzene alphatic	Asbestos	Dioxin	Bromine / bromine hydrocarbons
paint	++	++						
metal	+	++		+	++	+		+
garages	+	+			+++			
graphical	+++	+++						
cleaning / decontamination	+	+++			+			
mineral oil/ coal	++	+		+	+++		+	+
food (luxury food)		++	++	+	+	+		+
textiles	++	++						+
leather and rubber	++	++	+	+				
wood and paper		+	+		+		+	
plastics		++	+				+	
electrotechnical		+			+			+
medicine / laboratories		+			+			
energy, general	++				++			
construction / materials	+							+
agriculture				+	+			+
market gardening gardeners				+	+			++
cattle breeding				+				
ceramics			+					
glass		+	+					
cosmetics		+		+				
photography								
incineration/destruction					+		++	
chemical, general			++	+	+			

possibility of pollution:
+ realistic, ++ major, +++ extremely major (arbitrary)

Table 6 Possible pollution of non-residential buildings per industry sector (continued)

INDUSTRY SECTOR	Heavy metals	Cyanide	Fluoride	PAH's	PCB's	Pesticides (general)	Strong bases	Strong acids	Aldehydes / ketones
paint	+++				+		+		+
metal	+++	+	+	+	+				
garages	+				+				
graphical	++				+				
cleaning/ decontamination	+			++		++	+	+	
mineral oil and coal	++	++	++	+++	+		+	+	
food (+ luxury food)	+			+				+	+
textiles	++			+	++	+			
leather and rubber	++	+		+	+				
wood and paper	+		+	+	++	+	+	+	
plastics	+	+		+	+				
electrotechnical	++			+	++			+	
medicine/laboratories	++	+		+				+	
energy, general	++		+	++	+				
construction materials	+				+				+
agriculture	++					++			
market gardening + gardeners	+					++			
cattle breeding	+								
ceramics	+++		+						
glass	++								
cosmetics			+						
photography	++	+							
incineration/ destruction	+			++					
chemical, general	++	++	++		+	++	++	+	+

possibility of pollution:
+ realistic, ++ major, +++ extremely major (arbitrary)

Eco-costs/Value Ratio (the EVR model)

This model describes 'sustainability' (in the ecological sense) from a management systems approach, in which sustainability means the strive for eco efficiency, as defined by the World Council for Sustainable Development: 'the delivery of competitively priced goods and services that satisfy human needs and bring quality of life while progressively reducing ecological impacts and resource intensity through the life cycle, to a level at least in line with earth's estimated carrying capacity' [3].

This business-oriented definition connects the thinking of the modern manager ('the delivery of competitively priced goods and services with quality of life') to the need for a sustainable society ('while progressively reducing pollution to the earth's carrying capacity'). But what does this definition mean in practice for managers and designers, with regard to the many decisions taken daily? A numeric model has now been designed based on this definition, whereby well-founded choices can be made during the design phase with respect to the sustainability of the design, which is then used as a basis for setting out an entrepreneurial strategy.

The first section of the definition describes the 'value' of a product, and the second section describes the impact on the environment, or 'eco-costs'. The basic idea behind the model is to couple the 'value chain' to the 'product chain' as defined in the life cycle assessment. The added value and the costs are defined in the value chain for every step (from raw material to client, the usage phase and the End of Life phase). The environmental impact is also defined in monetary terms for each step in the chain, i.e. the eco-costs. If the usage phase and End of Life phase are also added to the chain, this gives the 'Total Costs of Ownership' (this approach is common in the car industry and for structural and civil engineering and civil projects). The 'Total Eco-costs of Ownership' are therefore defined in a similar way in the model.

The eco-costs are 'virtual' costs, i.e. the hidden costs required to manufacture and use a product in a sustainable manner, or 'in line with the earth's carrying capacity'. These costs are estimated based on the costs of technical measures to prevent undesired emissions and based on depletion of fossil energy and raw materials. As our society is nowhere near sustainable, the eco-costs are 'virtual', or not yet integrated into existing costs of the product chain (the existing life cycle costs). The eco-costs therefore consist of costs that have yet to be made to reduce the environmental impact (emissions and depletion) to a sustainable level. The eco-costs/value ratio (EVR) is defined as: EVR= eco-costs/value. The composition of eco-costs, costs and value is shown in Figure 1.

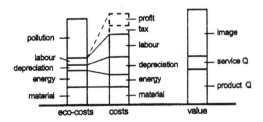

Figure 1   The value, costs and eco-costs of a product

REFERENCES

1.   HENDRIKS, Ch.F., Sustainable Construction, Æneas Technical Publishers, Boxtel, 2001, 158 pp.

2.   HENDRIKS, Ch.F. and JANSSEN, G.M.T. A New Vision on the Building Cycle, Æneas Technical Publishers, Boxtel, 2004, 251 pp.

3.   VOGTLÄNDER, J.G. and HENDRIKS, Ch.F., The Eco-costs/Value Ratio, Æneas Technical Publishers, Boxtel, 2004, 378 pp.

DEVELOPMENT OF TECHNOLOGY FOR PROCESSING OPEN-HEARTH SLAG INTO COMPOSITE STRUCTURAL MATERIALS

S I Pavlenko   **Ye V Filippov**   **U N Anashkina**

Siberian State University Industry

N S Anashkin

Steel KSC Ltd

Russia

ABSTRACT. An investigation into slag and waste sand from open-hearth furnaces of the "KSC Steel" company has been conducted with the purpose of processing for use in concrete as binder and aggregate. The materials studied were of two types: slag taken from the running production process (fresh slag) and sand stored in landfill (the amount exceeds 25 million tons). Physical characteristics, structure, chemical and mineralogical compositions, radionuclides and toxic element content have been examined. Recommendations for further research into the creation of the technology for their treatment have been made. The feasibility of utilizing the slag, after grinding and chemical separation, as fine (0 to 5 mm size fraction) and coarse (5 to 10; 10 to 20 and 20 to 40 mm fraction) aggregates for concrete and also as a binder component, after the decomposition of calcite ($CaCO_3$) and mechanochemical treatment, have been established. After the proper treatment and the addition of foundry sand and free calcium oxide, the fresh slag can be synthesized into a binder in planetary mills.

Keywords: Open-hearth slag, Structure, Chemical and mineralogical compositions, Silicate and ferruginous dissociation, Radionuclides, Composite binder.

S I Pavlenko is the Head of the Department of Civil Engineering of SSUI,   Doctor of Technical Sciences, professor. Novokuznetsk, Russia.

Ye V Filippov is a postgraduate student SSUI.

U N Anashkina is a student at SSUI.

N.S. Anashkin is Director of Steel KSC Ltd. Novokuznetsk, Russia.

INTRODUCTION

In 2003, under a contract to the "KSC Steel" company, a group of scientists of the SSUI in conjunction with the institutes of the Siberian branch of the Russian academy of Sciences (AGG&M and ICS&M) conducted investigations on open-hearth slag from the Kuznetsk Steel Combine (both from the running production process and dumped slag) and a "burnt sand" (waste mounding sand from foundry) with the purpose of processing into structural materials to be used for various purposes.

METHODS AND EQUIPMENT

The laboratory stage of the study included preparation of specimens, making analyses, calculations and interpretation of the results obtained. The instrumental analyses were performed using the equipment of the AIGG&M SB RAS Analytic centre and the laboratory of the mechanochemical reactions of the ICS&M SB RAS.

Micromorphology and the quality composition of several phases were examined using a JEOL JSM-35 electron microscope with the KEVEX energy-dispersion attachment.

Rock-forming components were analysed by means of the X-ray fluorescent silicate analysis using a SPM-2 multichannel spectrometer monitored by a computer. The preparation of samples was based on the method of fusing analysed material with an active flux. The phases were identified by a diffractometric analysis using a DRON-3 device.

To determine the content of toxic components in the solid substances, the following apparatus were used: Pue-Unikam SP-9 atomic-absorption spectrometer (AAS) and a Perkin-Eilmer 3030 B device for the air-acetylene and nitrogen-acetylene flame variant, as well as a 3030 Z spectrophotometer with an electrothermal atomizing and Zeiman background correction for the electrothermal variant.

The analysis of the materials for radionuclides was made at the State Sanitary Epidemiologic Inspection, Novokuznetsk, in accordance with the State Standard 30108 [1] by a spectrometric method using a "Gamma-01C" spectrometer. Chemical analysis was made in accordance with the State Standard 8269.1-97 [2]. Differential-thermal (DTA) and thermal-gravimetric (TGA) analyses were performed at a Paulik-Paulik-Erdey Q-1500 D derivatograph.

To determine the reaction ability of secondary mineral resources (SMR), planetary mills designed by the JCS&M SB RAS were used including an AGO-3 laboratory cyclic mill and a semi-industrial continuous mill. Their data and description are given elsewhere [3, 4]

EXPERIMENTAL

Physical properties, structure, chemical and mineralogical composition, radionuclides and toxic element contents of the SMR have been studied using the above mentioned methods, and recommendations for further for creating the technology of their processing have been defined.

Open-hearth slags were studied after passing through crushing-sorting plant and magnetic separation. The magnetic portion (scrap) is usually used for steelmaking and a non-magnetic one is separated into two size fractions (up to 80 & above 80 mm) and is disposed of. The non-magnetic part was studied with the purpose of developing new composite materials and innovative technologies.

Waste foundry sand ("burnt" sand) was also subjected to magnetic separation (magnetic fraction content up to 10 %) and then examined along with open-hearth slag.

In the process of working with a customer, a need has arisen to examine the so-called slag sand (in fact, the product of decay, containing silt and argillaceous materials whose physical and chemical characteristics over a long period of storage) sorted out during a primary processing of the waste slag and discharged to a new landfill. The possibility of its utilization has been considered.

Physicomechanical Properties of the Waste Slag Sand

1. Its bulk density was determined according to the State Standard (8736-93-3) [5] and was 1.64 g/cm^3.

2. Modulus (M_c) 2.92.

3. Silt and dusting materials were determined in accordance with the State Standard by elutriation compiled 34%.

According to the requirements of the State Standard 8736-93-3, this slag sand cannot be used as structural sand due to the last two indices which exceed the standard regulations (up to 2.7 M_c and up to 5% silt and argillaceous material).

Chemical Analysis of Secondary Mineral Resources

Chemical analysis of the slag used up sand was made in the Structural Materials and structures experimental centre, "Kuzbasscertification", in accordance the State Standard 8269.1-97 (2). The results of the analysis are given in Table 1.

The chemical analysis showed a high content of oxides in the slag: 23 to 26% ferric oxide (Fe_2O_3 + FeO), even after magnetic separation, 14 to 15% magnesium oxide, 22 to 27% calcium oxide, 11% aluminum oxide and small amounts of silicon oxide (especially in the waste slag). Creation of reactive minerals capable of acting as cement the additional incorporation of microsilica (preferably amorphous) and free calcium oxide as well. Waste sands from foundry activities, microsilica from ferroalloy production and ashes from thermal power plants may serve this purpose.

Due to their chemical composition, steelmaking slag may be basic or acid [6]. From the CaO-to-SiO_2 ratio, basic slags are classified into three types: they are low-basic if this value is below 1, the value of 1.6 to 2.5 characterizes medium-basic slag; the higher ratio corresponds to high-basic slag.

Table 1 Chemical Analysis of Open-Hearth Slag and Waste sand from Foundry, %

OXIDES	MATERIALS		
	Fresh slag	Waste slag	Burnt sand
SiO_2	15.16	8.05	89.6
Al_2O_3	10.65	11.04	6.1
Fe_2O_3	17.82	16.19	1.5
FeO	8.20	7.29	-
MnO	5.68	4.47	0.05
CaO	27.20	21.7	0.03
free CaO_{incl}	0.41	0.21	-
MgO	14.99	14.60	0.25
TiO_2	1.06	1.06	0.29
$Na_2O + K_2O$	-	-	1.02
SO_3	0.67	0.15	-
Loss-on-Ignition	0.16	1.18	0.5

Acid slag are mainly composed of SiO_2 (50 to 60%) and small amounts of FeO (10 to 20 %) and MnO (10 to 30 %). The acidity of slag is determined by the SiO_2: (FeO + MnO) ratio. With CaO contents of 3 to 15%, the acidity is evaluated using the SiO_2: (FeO + MnO + CaO) ratio.

The slags investigated are referred to as medium-basic (1.6 to 2.5).

$$waste \frac{21.7}{8.05} = 2.06 \qquad\qquad fresh \frac{27.2}{15.16} = 1.8.$$

Structure and Mineralogy of Secondary Mineral Resources

Unprocessed slag is non-uniform in its mineral composition with a grain size ranging from 0 to 80 cm. To create new compositions and technologies, it has to be ground into a disperse powder which is able to interact mechanochemically with various minerals.

Burnt sand from the Kuznetsk Steel Combine was subjected to a magnetic separation prior to analysis and grinding in planetary activator mills.

The material possessed a uniform grain structure with a grain size of 0.5 to 1 mm (Figure 1a). The amorphous state of the structure can be seen in Figure 1b. The X-ray diffractogram of the burnt sand (Figure 2) shows the presence of 50% radioamorphous phase, 45% quartz and 5% iron. No traces of feldspar ($2\theta \approx 26$–28^0) or argillaceous minerals ($2\theta \approx 3$–13^0) typical of natural sands were observed.

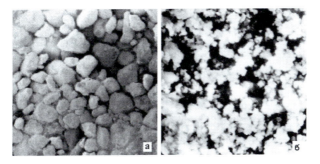

Figure 1   Burnt sand from Kuznetsk Steel Combine: a – initial, b – crushed

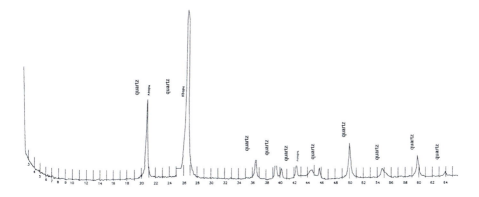

Figure   2 X-ray diffractogram of the burnt sand from the KSC

The results of X-ray diffraction carried out on the two samples of the open-hearth slag are shown in (Figures 3 and 4) made in the AIGG&M SB RAS Analytical centre.

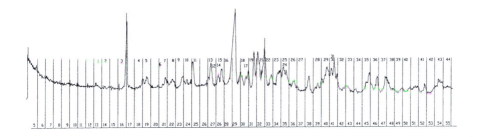

Figure 3   X-ray diffractogram of the dumped open-hearth slag

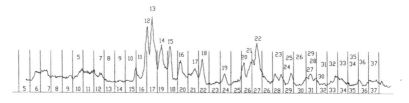

Figure 4   X-ray diffractogram of the fresh open-hearth slag

The fresh open-hearth slag contained dicalcium silicate (Ca_2SiO_4) and alumosilicate (Al_2SiO_5) as its main phases, a small amount of the MnO and Mn–Fe–Si–O combination, a portion of magnetite (Fe_3O_4) and pure iron (αFe). The sample contained parties of iron which were not picked out by magnetic separation and an unknown phase. A small amount of free calcium oxide was present which has to be bound with silica by mechanochemistry in order to improve binding properties.

The waste slag had calcite ($CaCO_3$) formed as a result of carbonization, dicalcium silicate (Ca_2SiO_4), aluminium silicate (Al_2SiO_5), Mn–Fe–Si–O combinations, small amounts of $MgMnSi_2O_6$, MnO, Fe_3O_4 and quartz (SiO_2). A small amount of unknown phase was present. Previous investigation shows a high content of magnesium (up to 15%) in the slag, a appropriate material for creating a new composite binder. Manganese oxide (4.5 to 5.7%) coddle extracted in a pure state which could be used for melting manganese steels.

As a result, technology for processing of secondary mineral resources should be created in accordance with the suggested trends [7, 8].

Testing for Deterioration

The determination of the resistance of coarse aggregate to (in the case of its utilization for concrete or pavements) for the fresh and waste open-hearts slag was performed in accordance with the State Standard 9758-86, position 25 [9]. The mass loss of a rubble portion before and after steam curing and cooling in a vessel with a lid (non-autoclave method) was taken into consideration.

Containers with the portion of the aggregates were placed into a vessel which was filled with water so that it did not reach the container bottom 10 to 20 mm. A lid was put on the vessel. The water was heated to boiling. The aggregate was steam-cured for 3 h, and then the container was taken out of the vessel and submerged in a water bath at room temperature for 3 h. The alternative steam curing and cooling to a room temperature was repeated three times. After this procedure, the aggregate portion was dried to a constant weight and screened.

The mass loss on steam curing (M_c) in percent was calculated by means of the following formula:

$$M_c = \frac{m_1 - m_2}{m_1} \cdot 100,$$

where m_1 is an aggregate portion before the test, g;
m_2 is the mass of the residue on the sieve after the test, g.

The mass loss was calculated as an arithmetical mean of the results of the two parallel definitions for each type of the slag (aggregate). The tests results were as follows: $m_1 = 3100$ g, $m_2 = 2841.75$ g.

Hence, $M_c = 8.33\%$.
The dumped slag rubble: $m_1 = 3050$ g, $m_2 = 2935$ g.
Hence, $M_c = 3.75\%$.
The State Standard regulation for M_c is 5%.

The results of testing show that the rubble in the form of a coarse aggregate from the fresh open-hearth slag cannot be used for concretes and pavements without previous treatment. The waste slag, due to its silicate decomposition result, meets the requirements of the State Standard for these purposes.

The resistance of the slag to a ferruginous decomposition was assessed in terms of the rubble weight loss before and after a 30-day storage in water in accordance with the above State Standard, position 26.

Containers with the aggregate portion were placed in a vessel with distilled water for 30 days. Then the specimens were dried to a constant weight and screened. The residua on the sieve were weighed. The mass loss in percent was calculated by the following formula:

$$m_f = \frac{m_1 - m_2}{m_1} \cdot 100,$$

where m_1 is the aggregate portion mass before testing, g;
m_2 is the mass of the residue on the sieve after testing, g.

The tests results were as follows:
The slag rubble produced (fresh): $m_1 = 1550$ g, $m_2 = 1508.15$ g.
Hence, $M_f = 0.27\%$.
The dumped slag rubble: $m_1 = 1525$ g, $m_2 = 1516.16$ g.
Hence, $M_f = 0.58\%$.
The State Standard regulation for M_f does not exceed 8%.

These data indicate that both "fresh" and dumped slag can be utilized without any restrictions.

Contents of Radionuclides, Carcinogenic and Toxic Elements

The radionuclides content was determined in the radiation control laboratory of the State Sanitary Epidemiological Inspection, Novokuznetsk, Kemerovo region, using a "Gamma-01C" spectrometer. The results of the tests carried out in accordance with the State Standard 30108-94 are given in Table 2.

Table 2   Radionuclides Content in SMR

RADIATION VALUE	RADIONUCLIDES CONTENT, Bq/kg			ERROR OF MEASUREMENT
	Fresh open-hearth slag	Waste open-hearth slag	Burnt sand	
Caesium-137	1.07	1.31	1.3	
Potassium-40	26.45	33.46	26.2	
Radium-226	55.89	39.66	22.2	±9
Thorium-232	12.4	14.67	17.1	
A_{eff}	96.57	83/39	66.8	

Evaluation criterion: from the radionuclide content, all secondary mineral resources meet the requirements for the first class ($A_{eff} < 370$ Bq/kg) and the radiation safety regulations RSR-99 BC 2.6.1.758-99 and may be used without any restrictions. The conclusions of the "State Sanitary Epidemiological Inspection" Centre are enclosed.

The contents of heavy metals, carcinogenic and toxic elements were determined in the accredited Analytical centre of the SB RAS by the atom-absorption method (see Section "Methods and equipment").

In all three types of materials studied, carcinogenic and toxic combinations or elements were not discovered. The organic combinations burn out at high temperature of treatment and do not get into landfills.

The contents of some heavy metals (mercury, arsenic, cadmium, tin, antimony, copper, zinc) were discovered but their quantity does not exceed the standard regulations for the influence on a human organism.

The concentrations of metals are given in Table 3.

Table 3   Content of Heavy Metals in Slag and Burnt Sand, mg/kg

HEAVY METALS	MATERIALS		
	Fresh open-hearth slag	Waste open-hearth slag	Foundry sand burnt sand
Hg	0.058±0.003	0.058±0.003	0.042±0.003
As	<1	<1	<1
Cd	0.16±0.01	1.9±0.1	0.046±0.002
Pb	22±2	69±2	7.8±0.2
Sb	1.2±0.5	6.1±0.5	0.76±0.5
Cu	100±5	118±5	71±2
Zn	102±5	1090±10	14±1

CONCLUSIONS

Previous study of the two types of the open-hearth slag (fresh & waste) and spent moulding sand from foundries comprised the following work:

1. Urgency and objectives for the investigation have been determined.

2. The primary measures for the use of the magnetic fraction of slag have been taken by the KSC Steel Company.

3. Methods and equipment for conducting the investigation have been developed.

4. The first stage of the research included investigations on:
   4.1. Physical characteristics and chemical composition of the secondary mineral resources.
   4.2. Structure and mineralogy of the SMR.
   4.3. Silicate and ferruginous decomposition.
   4.4. Contents of radionuclides, carcinogenic and toxic elements.
   4.5. Contents of heavy metals.

The conclusions arrived at from the first stage of the work may be stated as follows:

1. Chemical analysis showed a high content of oxides in the slag (23 to 26% Fe_2O_3 + FeO) even after the primary magnetic separation. It indicates the necessity of double or triple grinding-sorting and magnetic treatment of the slag. The content of magnesium oxide is rather high (14 to 15%) which opens up a perspective for developing composite and fireproof structural materials to be used in hydrotechnical concretes. The content of aluminium oxide (11%) could be used to advantage in the design of a composite binder with a higher fire resistance. The presence in the slag of 4.5 to 6% manganese oxide suggests the prospect of the development of a technology for its separation and use in manganese steels. The amounts of free calcium oxide and silicon oxide are not enough for producing cementless binder without admixtures. To create reactive minerals capable of binding, radioamorphous silica and free calcium oxide have to be introduced. Moulding sands from foundry activities, microsilica from ferroalloy plants and ash and slag from thermal power plants may serve as silica components. Proportions and properties will be determined at the second stage of the work.

2. Mineralogical study of the slag and foundry sand has shown the presence of dicalcium silicate which has binding properties and is a component of Portland cement clinker. Technology for producing a cementless (without clinker) binder is necessary. Its proportioning will be performed at the second stage. It should be noted that the amount of dicalcium silicate is greater in a fresh slag and much smaller in a waste slag due to the silicate decomposition into CaO and SiO_2 and the transformation during carbonization ($CaO + CO_2$) into calcite ($CaCO_3$). The task is to disintegrate calcite into reaction-capable ingredients.

3. Tests for silicate and ferruginous decomposition showed that:
   3.1. Rubble in the form of a coarse aggregate from the fresh open-hearth slag cannot be used in concrete and pavements without previous treatment (or storing in dumps).
   3.2. The dumped slag rubble meets the State Standard requirements for use in concrete and pavements.
   3.3. Fresh and dumped slag can be utilized without any restrictions.

4. Due to the radionuclides content, all three materials (two types of slag and the used up moulding sand) meet the requirements of the State Standard 30108-94 and the radiation safety regulations RSR-99 BC 2.6.1.758-99. They are referred to as the first class materials and may be used in construction or other fields without any restrictions.

The contents of heavy metals in slag and burnt sand are considerably lower than the Standard safety regulations and may be used in the production of structural materials.

REFERENCES

1. STATE STANDARD 30108-94, Structural materials and articles, Determination of the effective activity for natural radionuclides, Gosstroy RF. Izdatelstvo State Standard, 1994.

2. STATE STANDARD 8769.1-97, Rubble and gravel from dense rocks and industrial by-products for construction, Methods of chemical analysis, Gosstroy RF. Izdatelstvo State Standard, 1994.

3. AKSYONOV, A.V., PAVLENKO, S.I., AVVAKUMOV. YE.G., Mechanochemical synthesis of a new composite binder from secondary mineral resources, Novosibirsk, SB RAS publishing house, 2002. 64 pp.

4. COLLECTED PAPERS OF THE SSUI POSTGRADUATE STUDENTS ON "STRUCTURAL MATERIALS AND PRODUCTS". Advisor d.t.s., prof. S.I. Pavlenko, Novokuznetsk, Izdatelstvo SSUI, 2003, 144 pp.

5. STATE STANDARD 8736-93-3, Sand for construction, Technical specification, Gosstroy RF. Izdatelstvo State Standard, 1994.

6. ALEKHIN, YU.A., LYUSOV, A.N., The economic efficiency of utilization of secondary resources in the production of structural materials, V. Stroyizdat, 1988, pp 342.

7. TULEYEV, A.G., KULAGIN, N.M., PAVLENKO, S.I., Theoretical prognoses for the complex hydrochemical extraction of metals from secondary mineral resources, "Izvestia Vusov" journal, Chemistry and chemical technology, Ivanovo: Izdatelstvo ISCTU, Ministry of Education of the RF, 2002, № 7. pp 26–32.

8. TULEYEV, A,G, PAVLENKO, S,I, TKACHENKO, V,V., Deep complex processing of the mining by-products as one of the trends for solving ecological problems, Conference proceedings on "Perspectives of the development of mining industry in the III millennium", Russia Coal and Mining-2000, Messe Düsseldorf International, 6–9 June, 2000; Novokuznetsk, International Fair-Exhibition // Novokuznetsk, Izdatelstvo SSUI, 2000, pp 222–226.

9. STATE STANDARD 8735-88 (ST SEA 5446-85), Sand for construction, Methods of testing, Gosstroy USSR, Izdatelstvo State Standard, 1999.

INDUSTRIAL WASTE UTILIZATION IN BUILDING ELEMENTS

M M Prasad
S Kumar
National Institute of Technology
India

ABSTRACT. An investigation into blended concrete mixes containing 15%, 30% and 45% fly ash and 30% blastfurnace slag as partial replacements of cement and coarse aggregate has been reported. An M25 grade concrete with proportions 1:1.57:3.37 (C:F.A.:C.A.: w/c=0.49) has been considered as the reference mix. All together 144 specimens have been tested at 7, 28, 91 and 365 days of maturity for compressive strength, split-tensile strength and flexural strength. A theoretical investigation on the axial load carrying capacities of bamboo reinforced blended concrete composite (BRBCC) columns of size 250 mm, 300 mm and 350 mm in diameter has also been analysed. Test results show that the blended concrete containing 15% fly ash and 30% blastfurnace slag as partial replacement of cement and coarse aggregate respectively gives about the same strength as conventional concrete at 365 days of maturity. The axial load carrying capacities of BRBCC columns have been found to be about 0.87 times those of conventional columns of the same size. These columns can be used to carry axial loads from beams in prefabricated building systems particularly for low cost housing.

Keywords: Industrial wastes, Blended concrete, Fly ash, Blastfurnace slag, Bamboo strips, Compressive strength, Split-tensile strength, Flexural strength, Composite column.

Mr Sanjay Kumar is presently Lecturer in the Department of Civil Engineering at the National Institute of Technology, Jamshedpur, India. He obtained M.Sc. Eng. (Structural Eng.) from N.I.T., Jamshedpur in 2000. His fields of interest are research and development of new building materials, low cost building and repair/rehabilitation of old concrete structures.

Dr M M Prasad is Professor and Head of Civil Engineering Department at the National Institute of Technology, Jamshedpur, India. He obtained Ph.D. degree from Indian Institute of Technology, Roorkee, India in 1990. His research field includes low cost building technology, analysis and design of reinforced concrete structures and performance evaluation of new building materials in different environments.

INTRODUCTION

The building industry is one of the largest consumers of material resources. It is high time to give more emphasis on research and development for waste utilization as a resource material in the production of blended concrete. In India, the annual production of fly ash is about 100 million tonnes, but about 5 percent of total fly ash production is being utilized, which is very low. Owing to its pozzolanic contribution and other properties, the use of fly ash makes an excellent additive for Ordinary Portland Cement concrete mixes. In order to bring down the cost of disposal and to reduce environmental pollution, it is an imperative to increase the quantity of fly ash utilization [1]. Helmuth [2] has reported that in addition to economic and ecological benefits, the use of fly ash in concrete improves its workability, reduces segregation, bleeding, heat evolution and permeability, inhibits alkali aggregate reaction and enhances sulphate resistance. Ganesh Babu [3] reported that fly ash concretes show a lower early strength due to the fact that the compressive strength and other mechanical properties of these concretes will depend on the pozzolanic reactivity of fly ash, the richness of the mix, the character and grading of the aggregates, the water content of the mix and the curing conditions. Similarly, the steel industry in India, is producing about 12 million tonnes of blastfurnace slag. The slags are composed of calcium and magnesium silicates. The steel industry slags having desirable qualities, and can be used as partial replacement of coarse aggregate in concrete construction. Graf and Grube [4] have reported that the Ground Granulated Blastfurnace Slag concrete cured properly has lower permeability. Dehuai and Zheoyam [5] have indicated that the incorporation of fly ash and blastfurnace slag in concrete leads to many technical advantages. When the two mineral admixtures are used together, better results can always be achieved. The use of such industrial by-product or waste material having desirable qualities can result in saving of energy and conventional materials. With increase in population, the demand for construction of residential and public buildings is also increasing.

Pre-cast, pre-fabricated building system both for urban as well as rural areas using low cost materials is the only option to fulfil the demand for houses. Studies have been carried out on natural reinforcing material like bamboo, which has yielded good results. Nevertheless, non-ferrous natural reinforcing material still remains a dynamic field of further study and research. The bamboo available in abundance in rural areas is one of the cheapest materials utilized in concrete construction as a replacement of steel for different building elements, such as beams, slabs, columns etc. As reported by Cook et al. [6] on a cost basis, a comparison of bamboo and steel reinforcement in concrete columns of the same load capacity indicates that bamboo is some 49% cheaper than steel. This paper highlights present investigations on blended concrete by using three different types of mixes, with varying quantities (15%, 30% and 45%) of fly ash and a constant amount (30%) of blastfurnace slag as a partial replacement of cement and coarse aggregate respectively in a design mix of M25 grade concrete having mix proportions 1:1.57:3.37 (C:F.A.:C.A.: w/c = 0.49) as a reference mix. The theoretical investigation of axial load carrying capacities of bamboo reinforced blended concrete composite columns of size 250 mm, 300 mm and 350 mm in diameter have been also considered. Comparison of load carrying capacities of these columns with conventional steel reinforced columns has been also done. Testing of prototype models of such columns in finding the effect of various parameters involved in the determination of ultimate capacities of these columns has been also suggested. These columns can be used to carry axial loads from the beams in prefabricated building system particularly for low cost housing.

EXPERIMENTAL PROGRAMME AND TEST RESULTS

Materials

Ordinary Portland Cement 43-grade satisfying the requirements of IS:8112-1989 [7] having specific gravity 3.1 and fineness 3.25% on 90µ IS sieve was used. Locally available sand from river Kharkai, Jamshedpur, India passing through 4.75 mm IS sieve was used. It was clean and free from impurities having specific gravity 2.59, fineness modulus 2.61 and conforming to zone II as per IS: 383-1970 [8]. Locally available angular crushed dolorite chips of normal size 20 mm down conforming to IS: 2386-1963 [9] having specific gravity 2.75 g/cm^3, water absorption 0.5% and fineness modulus 7.73 was used. The fly ash obtained from Jojobera power plant, Jamshedpur conforming to IS: 3812-1966 [10] and having the following physical properties: fineness 1.5% (retained) on IS 90µm sieve and specific gravity 2.04 was used. Air cooled blastfurnace slag obtained from TISCO, Jamshedpur of size 20 mm down and 4.75 mm retaining having a fineness modulus of 6.45, specific gravity 2.88 g/cm^3 and crushing value of 20% has been used. Ordinary tap water was used for both mixing and curing of concrete.

Concrete Mix and Specimen Details

There is no standard method of designing blended concrete mix incorporating flyash and blastfurnace slag, so the method of mix design proposed in IS: 10262-1982 [11] has been adopted. The mix proportion 1:1.53:3.37 (w/c = 0.49), by weight, has been used in a conventional M25 grade of concrete after several trials. The replacement of 30% coarse aggregate by blastfurnace slag has been considered to yield good workability and strength [12]. A complete set of mix proportions with flyash and blastfurnace slag as partial replacements of cement and coarse aggregate respectively is shown in Table 1. The specimens used for the tests include cubes (150 mm × 150 mm × 150 mm in size), cylinders (∅150 mm × 300 mm) and prisms (100 mm × 100 mm × 500 mm) [13]. A total of 144 specimens were cast, cured and then tested at 7, 28, 91 and 365 days of maturity. An average of three specimens was obtained.

Table 1 Designation of Concrete Mix

COMPOSITION, %	MIX			
	M1	M2	M3	M4
W/(C+F)	0.49	0.49	0.49	0.49
Cement	1	0.85	0.7	0.55
Fly ash	--	0.15	0.3	0.45
Fine aggregate	1.53	1.53	1.53	1.53
Coarse aggregate	3.37	2.6	2.6	2.6
Blast furnace slag	--	0.77	0.77	0.77
Fly ash / blast furnace slag	0/0	15/30	30/30	45/30

Test Results

Test results for compressive strength, split-tensile strength and flexural strength conducted as per IS codes [12, 13] at 7, 28, 91 and 365 days of maturity have been tabulated in Table 2.

Table 2 Strengths of Concrete at test ages 7, 28, 91, and 365 days

MIX	COMPRESSIVE STRENGTH, MPa				SPLIT-TENSILE STRENGTH, MPa				FLEXURAL STRENGTH, MPa			
	7	28	91	365	7	28	91	365	7	28	91	365
M1	30.77	37.92	42.81	44.40	2.30	2.60	2.75	3.37	2.13	2.67	3.73	4.53
M2	25.63	32.77	40.15	42.21	1.64	2.49	2.75	3.32	1.87	2.53	3.00	4.46
M3	18.59	29.63	31.76	38.96	1.57	2.43	2.73	3.25	1.87	2.40	2.67	4.33
M4	11.70	19.40	31.0	37.92	1.03	2.37	2.68	3.07	0.80	2.33	2.53	3.33

THEORETICAL INVESTIGATION ON COLUMNS

Details of Composite Columns

The bamboo reinforced concrete composite column consists of one precast concrete pipe. The inside of the pipe is reinforced with 6 to 10 good quality bamboo strips of size 12mm x12mm. Steel wires and strips made from mild steel rod of 6 mm diameter have been used to keep the bamboo in position. The remaining inside space of the pipe and around the bamboo strips is filled with M25 grade concrete. Figures 1 and 2 show the typical cross- section of the composite column and other details. The precast pipe is of thickness 25 mm with hoop tension reinforcement in the form of thin wires to provide strength during concreting and handling. For connection with footing or with another column, dowel bars of sufficient length can be provided. All bamboo strips have been coated with anti-termite coating and then protective coating (TOP COAT) to preserve them against the action of insects and fungus and to act as a waterproof coating to improve the durability of the bamboo. The axial load carrying capacities of bamboo reinforced concrete composite columns having outer diameters 250 mm to 350 with increments of 50 mm have been calculated. Table 3 indicates dimensions of the composite columns studied. In conventional columns, the bamboo strips have been replaced by 12 mm dia Fe415 steel bars with same nos. Tables 3 and 4 indicate dimensions of the composite column studied.

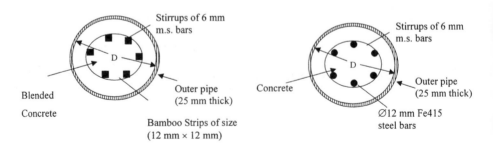

Figure 1 Typical cross-section of BRBCC column

Figure 2 Typical cross-section of Conventional column

Table 3 Dimension of Composite Columns

COLUMN TYPE	OUTER DIAMETER, mm	SIZE OF BAMBOO STRIPS, mm × mm	NO. OF BAMBOO STRIPS	TOTAL C/S AREA OF COLUMN, A_g mm^2	TOTAL AREA OF BAMBOO STRIPS, A_b mm^2	AREA OF CONCRETE, A_c mm^2
I	250	12 × 12	6	49087	864	48223
II	300	12 × 12	8	70686	1152	69534
III	350	12 × 12	10	96211	1440	94771

Table 4 Dimension of Conventional Columns

COLUMN TYPE	OUTER DIAMETER, mm	DIAMETER OF MAIN STEEL BAR, mm	NO. OF MAIN STEEL BAR	TOTAL C/S AREA OF COLUMN, A_g mm^2	TOTAL AREA OF MAIN STEEL BAR, A_b mm^2	AREA OF CONCRETE, A_c mm^2
I	250	12	6	49087	678.58	48223
II	300	12	8	70686	904.77	69534
III	350	12	10	96211	1130.97	94771

Theoretical Investigation

Normally, in the case of prefabricated residential building systems, the span of beams supporting floors varies from 2.0 m to 4.0 m. To find worse case loading on beams, grids of beams with spans equal to 2 m, 3 m and 4 m have been considered. A live load intensity equal to 3.0 km/m, a floor finish of 1 kN/m^2 and a slab thickness equal to 100 mm have been assumed. The slab panel having size 0.5 m × 1.0 m is used for roof as well as for floor finish. As per IS 456-2000 [14] for 100 mm thick R.C.C. slab, assuming floor-to-floor height equal to 3 m and wall thickness 100 mm, the maximum axial load for three storied building is less than 600 kN for a live load intensity of 3.0 kN/m^2. Assuming a maximum length of composite column equal to 3.0 m and considering the column effectively held in position but not restrained against rotation at both ends, its effective length will be 3.0 m. The slenderness ratio corresponding to a column type I to III is 12 which is treated as a short column. Considering the ultimate compressive strength of bamboo is 96 N/mm^2 and a factor of safety equal to 4, the permissible compressive stress in bamboo is 24 N/mm^2. Considering basic assumption that the strain in bamboo strips in compression is equal to strain in concrete in compression, the axial load capacity of a bamboo reinforced blended concrete composite column can be worked out by,

$$P_b = \sigma_{cc} A_c + \sigma_{cb} A_b \tag{1}$$

where, σ_{cc} = Permissible stress in blended concrete in axial compression,
  = 6 N/mm^2 for M25 grade concrete
  A_c = Blended concrete area in compression
  σ_{cb} = Permissible compressive stress in bamboo = 24 N/mm^2
  A_b = Area of bamboo strips in compression

Considering conventional columns reinforced with Fe415 steel and having the same diameters as those of BRBCC columns of type I to III, the axial load carrying capacities using the working stress and limit state methods [14], can be calculated by the following formulae.

$$P_c = \sigma_{cc} [(A_g + 1.5m - 1) A_{sc}] \tag{2}$$

$$P_{uc} = 0.4 \, f_{ck} \, A_c + 0.67 \, f_y \, A_{sc} \tag{3}$$

Where, m = modular ratio = 11 for M25 grade of concrete
   f_{ck} = characteristics strength of concrete = 25 N/mm² for M25 grade concrete
   A_c = concrete area in compression
   f_y = Yield strength of steel = 415 N/mm² for Fe415 steel bar
   A_{sc} = steel area in compression

Considering linear analysis, the load carrying capacity of conventional steel reinforced concrete column is given by

$$P = \sigma_{cc} \, A_c + \sigma_{sc} \, A_{sc} \tag{4}$$

where, σ_{sc} = permissible stress in steel = 190 N/mm² for Fe415 steel bar

Comparing equations (2) and (3), it is found that the limit state method gives an increase in permissible stress in concrete by 1.67 and in steel by 1.46. If this concept is applied for finding the ultimate axial load capacity of BRBCC column, P_{ub} will be given by

$$P_{ub} = 1.67 \times \sigma_{cc} \, A_c + 1.46 \times \sigma_{cb} \, A_b \tag{5}$$

In such a case the ultimate capacity of bamboo in compression becomes 1.46 × 24.0 = 35 N/mm². Thus using equation (5), the ultimate capacities of BRBCC columns of type I, II and III were found to be 513.43 kN, 737.05 kN and 1000.00 kN respectively. The axial load carrying capacities of BRBCC and Conventional columns with comparison were presented in the Table 5.

Table 5 Axial Load Carrying Capacities of BRBCC Columns and Conventional Columns

COLUMN TYPE	LOAD CAPACITY OF BRBCC COLUMS, kN		LOAD CAPACITY OF CONVENTIONAL COLUMNS, kN, BY WSM BY LSM		P_b/P_c	P_{ub}/P_{uc}
	P_b	P_{ub}	P_c	P_{uc}		
I	310.00	513.43	357.63	672.76	0.86	0.76
II	444.84	737.05	508.25	722.66	0.87	1.02
III	603.18	1000.00	682.44	1265.26	0.88	0.79

RESULTS AND DISCUSSIONS

Strength of Blended Concrete

From the test results, (Table 2), it has been observed that the compressive strength of all blended concrete are lower than the conventional M25 grade concrete at all ages. At 7 days of maturity the compressive strengths of blended concrete for 15%, 30% and 45% flyash with 30% blast furnace slag as partial replacements of cement and coarse aggregate, show compressive strengths 17%, 40% and 62% lower than the conventional concrete. Similarly at 28, 91 and 365 days the corresponding strengths are respectively 14%, 22%, 49%; 7%, 26%, 28%; and 5%, 13%, and 15% less than the conventional concrete. This is in agreement with the earlier findings [3], since the pozzolanic reactions do not start at early ages and the main strength of concrete is due to the contribution of the hydrated fraction of cement only. Since the cement content of the mix goes down due to replacement with fly ash, the strength development at the early age is low but at latter ages, that is, at 28, 91 and 365 days, there is a considerable increase in compressive strength of blended concrete compared to early age (7 days) strength. This may also be due to the fact that there is a pozzolanic effect of fly ash and gel formation fills in the pores generated by the liberation of $Ca(OH)_2$ during the hydration of OPC concrete. All these activities enhance binding inside the concrete matrix and strength increases. Blended concrete contains 15% fly ash and 30% blastfurnace slag exhibits behaviour comparable to the reference mix to a great extent. It is clear that at 7 days of maturity the split-tensile strengths of blended concrete having 15%, 30% and 45% fly ash and 30% blastfurnace slag as coarse aggregate are 29%, 32% and 56% less than the conventional concrete but at 28 days the corresponding split-tensile strengths have been found to be 5%, 7% and 9% less. At 91 days and 365 days there is very little variation in split-tensile strengths. Thus split tensile strengths of blended concrete at 91 and 365 days are comparable to the conventional M25 grade concrete. The increase in strengths may be due to the pozzolanic effect of the fly ash at latter ages i.e. 91 and 365 days of maturity. It has also been observed that the flexural strength of blended concrete is less than conventional concrete at all ages. However, at 28 days there is remarkable increase in the strength of blended concrete. Thus, the flexural strength of blended concrete is comparable with the conventional concrete as their strength variations are in the range of 6 to 13% only. At latter ages, i.e. at 91-day and 365-day the flexural strength decreases in the range of 11 to 25% and 2 to 27% respectively when there is replacement of cement by fly ash from 15% to 45%. It is also clear that the blended concrete with a replacement of 15% cement by fly ash and 30% coarse aggregate by blastfurnace slag gives comparable results.

Composite Column and Its Application

In the case of residential buildings where the span of beams supporting slab varies from 2 m to 4m and for a live load intensity of 3 kN/m^2, the BRBCC columns can be used to carry axial loads from beams in prefabricated building systems. The axial load capacities obtained from linear analysis are on the conservative side because the permissible compressive stress level selected for bamboo strips is based on an arbitrary assumed factor of safety equal to 4. The load shared by bamboo strips in BRBCC columns are 6.68%, 6.21% and 5.73% for column types I, II and III respectively. The load carrying capacities of conventional columns reinforced with a possible minimum area of Fe415 steel, have been worked out using the working stress and limit state methods. Since it is difficult to predict approximate values of modular ratios for BRBCC columns, a linear analysis approach has been used for finding load carrying capacities of these columns. Research on prototype BRBCC columns is needed to find modular ratios and other factors affecting the strength. However, ultimate capacities of BRBCC columns have been found out using equation (5) in which ultimate compressive stress in bamboo strips was considered as 350 N/mm^2 giving a factor of safety equal to 2.74. From Table 5, the axial capacities of BRBCC columns are about 0.87 times those of conventional columns of types I to III respectively found using the by working stress method. The water proofing coating (TOP COAT) controls shrinkage and swelling of the bamboo strips, and the bond between concrete and treated bamboo surface will be improved. This will help more redistribution of forces and hence the load carrying capacity will improve. These prefabricated pre-cast bamboo reinforced composite columns can be utilized along with precast structural component such as footings, beams, slab panels etc. for the construction of small residential and public buildings, row houses, mass housing projects in earthquake affected areas, temporary small bridges, and for low cost housing, in particular. Also these columns once the outer surfaces are treated can be suitably used as friction and bearing piles for the foundations of structures in soft and medium solids.

CONCLUSION

An investigation conducted using a mix proportion 1:1.53:3.37, w/c = 0.49 of M25 grade concrete and blended concrete mixes containing 15%, 30% and 45% fly ash with 30% blast furnace slag as partial replacement of cement and coarse aggregate respectively has been concluded. The blended concrete mix obtained after replacing, 15% cement by fly ash and 30% coarse aggregate by blastfurnace slag gives the best results. This optimum blended concrete may be used for low cost housing as it has considerable strength. The axial load carrying capacities of BRBCC columns having size 250 mm, 300 mm and 350 mm dia are 310 kN, 444 kN and 603 kN, respectively, whereas these values for conventional columns are 358 kN, 508 kN and 682 kN, respectively. The corresponding axial load carrying capacities for BRBCC columns are 0.87 times those of conventional columns of the same size found using the working stress method. The ultimate capacities of BRBCC columns are 0.76, 1.02 and 0.79 times than those of conventional columns. These short columns can be used to carry axial loads from beams in prefabricated building systems for low cost housing.

REFERENCES

1.   KUMAR, S., PRASAD, M.M., Performance of Bamboo Reinforced Flexural Members, Proc. Of the International Symposia, "Celebrating Concrete People and Practice", 3-4 Sept., 2003 at Concrete Technology Unit, University of Dundee, U.K.

2.   HELMUTH, R., Fly ash cement and concrete, Skokie, III: Portland Cement Association, 1987.

3.   GANESH BABU, K., SIVA NAGESHWARA RAO, G., Early strength behaviour of fly ash concretes, Cement and Concrete Research, Vol. 24, No. 2, pp 277-284.

4.   GRAF, H., GRUBE, H., The influence of curing on the permeability of concrete with different composites, Proc. RILEM Seminar on Durability of Concrete Structures Under Normal Out Door Exposures, Hanover University, 1984, pp 84-87.

5.   WANG, D., CHEN, Z.Y., On predicting compressive strengths of mortars with ternary blends of cement, GGBFS and fly ash, Cement and Concrete Research, Vol. 27, No. 4, 1997, pp 487-495.

6.   COOK, D.J., PAMA, R.P., SINGH, R.V., The Behaviour of Bamboo – Reinforced Concrete Columns Subjected to Eccentric Loads, Magazine of Concrete Research, Vol. 30, No. 104, Sept. 1978, pp 145-151.

7.   IS: 8112-1989, Specifications for 43 grade ordinary Portland cement B.I.S., New Delhi.

8.   IS: 383-1970, Method of test for coarse aggregate B.I.S., New Delhi.

9.   IS: 2386-1963, Method of testing aggregate in preparation of concrete, B.I.S., New Delhi.

10. IS: 3812-1966, Specification of fly ash for use as admixture of concrete, B.I.S., New Delhi.

11. IS: 10262-1982, Indian standard on recommended guidelines for concrete mix designs, B.I.S., New Delhi.

12. SANJAY, K., PRASAD, M.M., Use of blended concrete for low cost housing, Proc. of International Conference on "Recent Trends in Concrete Technology and Structures (INCONTEST 2003)", K.C.T., Coimbatore, India, 2003, pp 224-230.

13. IS: 516-1959, Method of test for strength of concrete, B.I.S., New Delhi.

14. IS: 5816-1970, Methods of test for split-tensile strength for concrete, B.I.S., New Delhi.

15. IS: 456-2000 "Code of practice for plain and reinforced concrete" B.I.S., New Delhi.

STAINLESS STEEL IN CONCRETE FOR EFFICIENCY AND DURABILITY

D J Cochrane

Nickel Institute

United Kingdom

ABSTRACT. Achieving long term durability and sustainability in the built environment is one of the principal construction targets of governments around the world. A limiting factor to durability in reinforced concrete structures is corrosion of the steel reinforcement, and structures exposed to coastal environments and/or subject to the application of de-icing salts, are particularly susceptible to damaging chloride ions from these sources. The consequence of corrosion is more frequent repair or a reduction in the design life. Corrosion damage world-wide has been estimated in billions of US dollars. Stainless steel reinforcement can provide the durability commensurate with the design life of the structure. It has been approved for use in highway bridges by the UK Highways Agency, the Federal Highways Administration in the USA, Ministry of Transportation in Canada, and Scandinavian Road Authorities. New British and USA Standards for stainless steel rebar have been issued, and guidance on the application of stainless steel rebar was issued by the Highways Agency in their Design Manual for Roads and Bridges in 2002. This paper will discuss the new documentation outlined and illustrate that increased durability can be obtained using stainless steel reinforcement for only a small increase in capital cost. Whole life costs will be shown to offer significant cost savings.

Keywords: Bridges, Chloride ion, Corrosion, Cost-effectiveness, Durability, Initial costs, Reinforcing bar, Stainless steel, Standards, Whole life costs.

David J Cochrane, served an engineering apprenticeship in the aeronautical industry before joining the steel industry. A former technical marketing structural engineer with British Steel and the Steel Construction Institute, he has been associated with stainless steel since 1970. He is the principal of Technical Publication Services and has been a consultant to the Nickel Institute since 1987, specialising in the application of stainless steel in architecture, building, and construction. He has lectured worldwide on the subject on behalf of the Institute, served on a number of British Standards Institution committees, and written numerous articles and publications on stainless steel.

INTRODUCTION

Austenitic stainless steels have been shown to have a high resistance to chloride ion [1,2]; the principle cause of corrosion to carbon steel reinforcement in reinforced concrete structures. Marine environments and structures subject to road de-icing salts are at highest risk to the ingress of chloride ion through the concrete cover.

Studies have shown that stainless steel can provide the resistance required for the design life of 120 years in highway bridges [3]. While stainless steel is more expensive than carbon steel reinforcement it can offer significant benefits through reductions in maintenance and repair, while the higher strength of stainless steel rebar can reduce the weight of steel reinforcement.

The new British Standard for stainless steel reinforcement and the guidelines for using stainless steel reinforcing bar, issued by the UK Highways Agency, will be discussed in this paper and the effects on initial and whole life costs considered using typical bridge designs.

Stainless steel is now being widely used as reinforcement in structures around the world for new build and repair [4] and standards activity and regulations are following the UK lead. For example, it is now mandatory to use stainless steel reinforcement in bridge parapets in Sweden and Denmark, and new guidelines on its application have been issued in France.

BS6744:2001 STAINLESS STEEL BARS FOR THE REINFORCEMENT OF AND USE IN CONCRETE [5]

The revised standard contains a wider range of austenitic stainless steels than its predecessor and includes guidance on where each would be appropriate depending upon the expected service conditions. Stainless steel rebar is produced to two strength levels, 500N/mm^2 and 650N/mm^2 in the diameters shown in Table 1. Stainless steel reinforcing bar is produced in the material grades shown in Table 2 with the recommended service conditions.

HIGHWAYS AGENCY GUIDELINES BA84/02 [6]

The new guideline document BA84/02 was prepared for the Highways Agency [HA] by Arup R & D and incorporated in the HA Design Manual for Roads and Bridges. The agency recognises that stainless steel reinforcement can improve the durability of the structure, reduce maintenance, and minimise the costly effect of lane closures and traffic disruption during periods of maintenance/repair. As shown in Tables 4 and 5, BA84/02 recommends the use of stainless steel reinforcement according to the location of the bridge, and indicates where stainless steel should be used, and the appropriate grade of stainless steel.

Table 1   Strength and size range of stainless steel reinforcing bar

STRENGTH GRADE	NOMINAL SIZE
500N/mm^2	3, 4, 5, 6, 7, 8, 10, 12, 14, 16, 20, 25, 35, 40, 50
650N/mm^2	3, 4, 5 ,6, 7, 8, 10, 12, 14, 16, 20, 25

Table 2 Material grades of stainless steel reinforcing bar & service conditions

GRADE	A	B	C	D
1.4301	1	1	5	3
1.4436	2	2	1	1
1.4429	2	2	1	1
1.4462	2	2	1	1
1.4529	4	4	4	4
1.4501	4	4	4	4

Key to Table 2
1 – Appropriate choice for corrosion resistance and cost
2 – Over specification of corrosion resistance for the application
3 – May be suitable in some instances: specialist advice should be obtained
4 – Grades suitable for specialist applications which should only be specified after consultation with corrosion specialists
5 – Unsuitable for the application

Key to service conditions
A – For structures or components with either a long design life, or which are inaccessible for future maintenance
B – For structures or components exposed to chloride contamination with no relaxation in durability design
C – Reinforcement bridge joints, or penetrating the concrete surface and also subject to chloride contamination (e.g. dowel bars or holding down bolts)
D – Structures subject to chloride contamination where reductions in normal durability requirements are proposed (e.g. reduced cover, concrete quality or omission of waterproofing treatment.)

The guidelines recognise the contents of BS6744: 2001, and that stainless steel reinforcing bar is issued with a CARES (a standards compliance certification body) certificate. The agency also accepts that stainless steel reinforcement can be used cost-effectively by siting the stainless steel in the high-risk elements of the structure and using carbon steel reinforcement in protected low risk areas. The stainless steel content can vary, therefore, from a small percentage to 100% depending upon the bridge location.

Due to the high corrosion resistance of stainless steel, the agency recognises that the rules developed to improve the durability of carbon steel reinforced structures can be relaxed when using stainless steel reinforcement. This relaxation, however, should not be taken to imply that the concrete quality and workmanship can be relaxed when using stainless steel, and it is important for these qualities to be maintained to realise the full durability benefits of the improved structure. The relaxation rules for using stainless steel are shown in Table 3.

Table 3 Relaxation rules for using stainless steel reinforcement

DESIGN CONDITION	RELAXATION
Cover	Cover for durability can be relaxed to 30mm where stainless steel is used irrespective of the concrete quality or exposure condition
Design crack width	Allowable crack width increased to 0.3mm
Silane treatment	Not required on elements with stainless steel

Classification of Structures

The classification of structures and elements by the Highways Agency identifies the areas of potential corrosion risk and the potential disruption of carrying out future maintenance. The first classification identifies where stainless steel may not be appropriate.

Table 4   Classification of structures and elements

CLASSIFICATION	RECOMMENDATION
U	New structures where stainless steel is not considered appropriate as they are unlikely to be exposed to high concentrations of chlorides. For structures in the UK motorway and local trunk road system this category is likely to include minor crossings such as footbridges, farm access crossings, buried structures and elements that are remote from the highway
Category C	New structures – stainless steel is appropriate for the partial replacement of reinforcement in selected substructure and superstructure elements exposed to chlorides from road de-icing salts. For example parapet edge beams, slab soffits, columns, and walls adjacent to the road.
Category B	New structures – stainless steel is appropriate for the partial replacement of reinforcement in substructure elements exposed to seawater. As for category C plus all parts of the substructure (columns, abutments, walls etc) within the tidal and splash zones.
Category A	New structures – stainless steel is appropriate for the complete replacement of reinforcement in the substructure, superstructure, and deck slab, except foundations or piles.

Selection of steel grade

With regard to BS6744:2001 and the HA BA84/02 guidelines, the advised grade of material is shown in Table 5 for a range of exposure conditions.

Table 5   Stainless steel grades and appropriate bridge applications

MATERIAL GRADE	EXPOSURE CONDITION
1.4301	Stainless steel embedded in concrete with normal exposure to chlorides in soffits, edge beams, diaphragm walls, joints & substructures
1.4301	As above but where design for durability requirements are relaxed in accordance with Table 3.7.
1.4436	As above but where additional relaxation of design for durability is required for specific reasons on a given structure or component i.e. omission of a waterproof membrane
1.4429	Direct exposure to chlorides and chloride bearing waters, for example, dowel
1.4436	bars, holding down bolts and other components protruding from the concrete
1.4462	Specific structural requirements for the use of higher strength reinforcement
1.4429	and suitable for all exposure conditions

INITIAL AND WHOLE LIFE COSTS

Initial Costs

Stainless steel reinforcement is more expensive than carbon steel therefore there will be an initial cost attached to the improved durability. This has been studied by Arup's for the British Stainless Steel Association (BSSA) and their report is included on a CD-Rom which is available free of charge from the BSSA[6]. The study was based on two actual bridges designed in accordance with BS5400 Part 4 [7] using conventional carbon steel reinforcement with a characteristic strength of $f_y = 460N/mm^2$.

Taking cognisance of the new British Standard BS6744:2001 and the highways agency document BA84/02, the carbon steel reinforcement was replaced, where appropriate, by stainless steel in the parapets, piers, abutments, and decks which were assessed individually.

In this paper, the multi-span bridge shown in Figures 1 and 2 has been taken for the evaluation of weight saving and initial cost difference using reinforcement prices current at July 2004. It should be noted that, due to the world-wide steel shortage, prices for both carbon and stainless steel are at a ten year high.

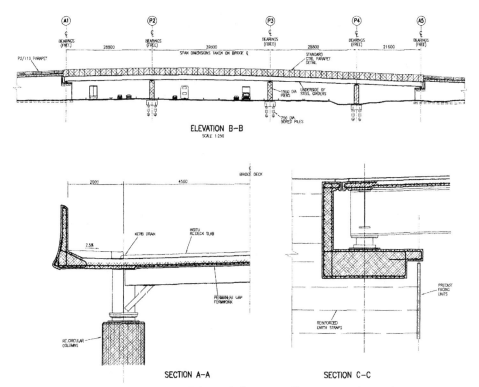

Figure 1   Location of stainless reinforcement for category A structures

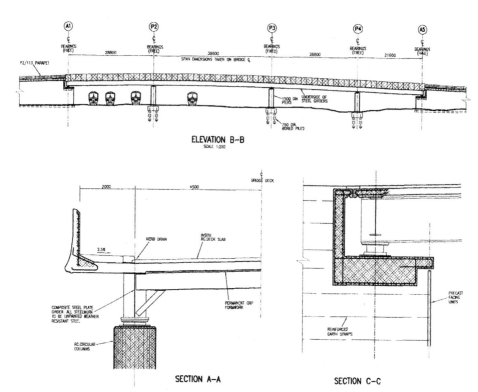

ELEVATION B-B
SCALE 1:250

SECTION A-A   SECTION C-C

Figure 2   Location of stainless reinforcement for category B and C structures

The output from the BSSA CD-Rom is tabulated in table 6. The cost effects of using stainless steel strengths of $500N/mm^2$ and $650N/mm^2$ are shown as a percentage of the overall construction costs.

Whole life costs

Corrosion of the reinforcement in highway bridges can severely impair the design life unless expensive remedial treatment is carried out. Inevitably, this means lane closures and traffic delays during periods of repair. Whole life costs include the initial construction costs and the maintenance cost over the 120 year life of the bridge. The maintenance costs include routine maintenance for surfacing and drainage, winter salting, inspections, new deck movement joints, typically, every 5 years, and new deck waterproofing membranes every 20 years. Whole life costs include the cost of repair or even replacement if required.

In comparing the cost of different strategies to achieve a 120 year life, it is necessary to take account of commuted maintenance costs. This is the sum that needs to be invested now, which, with interest added will be sufficient to pay for the work when it is required. This can be calculated using the following:

$$NPV = C/(1 + i)^n$$

Table 6. Use of stainless steel and effect on weights and initial costs

ELEMENT	STRUCTURE CLASS A				STRUCTURE CLASS B & C			
	Steel Content		O/A Weight Reduction	O/A Cost Increase	Steel Content		O/A Weight Reduction	O/A Cost Increase
Parapets	CS	SS			CS	SS		
500N/mm²	0	100%	24.42%	2.72%	37%	63%	15.94%	1.91%
650N/mm²	0	100%	37.00%	2.68%	42%	58%	37.00%	2.68%
Piers	CS	SS			CS	SS		
500N/mm²	0	100%	13.50%	0.22%	0	100%	13.5%	0.22%
650N/mm²	0	100%	35.36%	0.18%	0	100%	35.36%	0.22%
Abutments	CS	SS			CS	SS		
500N/mm²	55%	45%	4.64%	0.38%	55%	45%	4.64%	0.38%
650N/mm²	55%	45%	10.50%	0.41%	55%	45%	10.50%	0.41%
Deck	CS	SS			CS	SS		
500N/mm²	56%	44%	4.67%	3.43%	54%	46%	0.21%	4.23%
650N/mm²	62%	38%	13.00%	3.20%	54%	46%	0.14%	5.24%

CS – Carbon steel   SS – Stainless steel   O/A – Overall

Note to Table 6. Overall weight reduction relates to the tonnage of reinforcement in the element, Overall cost increase relates to the percentage on total construction cost.

Reinforcement steel costs:
Carbon steel = £500/tonne
Stainless steel $f_y = 500N/mm^2 < 20mm$ dia. = £2100/tonne
Stainless steel $f_y = 500N/mm^2 > 20mm$ dia. = £2240/tonne
Stainless steel $f_y = 650N/mm^2 < 20mm$ dia. = £2500/tonne
Stainless steel $f_y = 650N/mm^2 > 20mm$ dia. = £3000/tonne

where

NPV = net present value
C   = sum required at the end of n years
i   = discount rate which is interest rate less inflation rate
n   = number of years

The effect of using a discount rate of 3.5% (the current Highways Agency figure) is that work carried out after approximately 50 years carries comparatively little NPV.

As an illustration of the effect of using commuted costs, consider a simple replacement strategy for obtaining a 120 years bridge life. Bridge engineers would expect a 30 year life even if minimal maintenance was carried out, and so the simple strategy could be to replace the bridge every 30 years.

The following example is extracted from a report by WSP consulting engineers for the Nickel Institute. Whole life costs were calculated on a 37m single span composite bridge, designed with carbon steel reinforcement, for which contract drawings and schedules were available. Traffic delay costs of £500/hour for a low traffic location, and £10,000/hour in a high traffic location, were taken with a road closure time of 5 weeks for reconstruction.

The whole life costs are shown in table 7.

Table 7   Whole life costs

YEAR	CONSTRUCTION/ Replace £	TRAFFIC DELAY Low	TRAFFIC DELAY High
0	1,000,000	(420,000)	(8,400,000)
30	356,278	149,637	2,992,739
60	126,934	53,312	1,066,248
90	45,224	18,994	379,881
120	0	0	0
Cost	£1,528,436	£221,943	£4,438,868

Total Cost, Low traffic location, £1,750,379
Total Cost, High traffic location, £5,967,304

The reinforcement in this bridge at high risk to corrosion can be replaced by stainless steel to increase the durability of the structure. Carbon steel can be used for the balance of the reinforcement where it is at low risk of corrosion. This combination of materials, referred to as selective substitution, see figure 3, will provide increased durability at minimal initial cost.

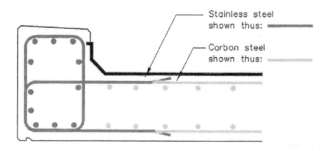

Figure 3   Selective substitution

For maximum durability, however, all reinforcement should be stainless steel.   In this particular bridge, the following elements require stainless steel reinforcement to realise increased durability. The tonnage of reinforcement is shown for each.

Parapet edge beams	3.2 tonnes
Deck top	14.0 tonnes
Deck soffit	14.5 tonnes
Ballast wall front faces	2.0 tonnes
Bearing shelves	2.3 tonnes
Abutment wall faces	13.5 tonnes
Total tonnage	49.5 tonnes

The cost of using stainless steel reinforcement would be an additional cost of about £2000/tonne = £99,000.

In this example, using stainless steel selectively, whole life costs are 28% less than the replace option if no traffic delay costs are taken into account, 37% less than the replace option in the low traffic location, and 81% less in the heavy traffic location.

Figure 4   Typical repair of a bridge pier using stainless steel reinforcement

CONCLUSIONS

This paper has shown that increased durability in highway bridges is readily achievable by the intelligent use of stainless steel reinforcement. The guidelines now available to bridge designers can eliminate corrosion in the elements of the structure that are well known to be at highest risk of corrosion. The removal of unexpected and premature repair costs is highly desirable and impacts on the environment in reducing traffic hold-ups and delays, lane closures and possible detours.

On first cost, the use of stainless steel reinforcement has been shown to add less than 3% of the capital cost to increase the durability of the parapets, piers, and abutments. Early repair of these elements is common, and in many cases has become necessary after less than twenty years of service. Stainless steel reinforcement is increasingly being used in the repair of corrosion damaged bridge structures.

The impact of using stainless steel on whole life costs can be considerable on motorways and trunk roads carrying a high traffic density where high traffic delay costs are taken into the equation. Whole life costs using the current discount rate indicate that the stainless steel option is over 80% less in a high traffic location than the option of using carbon steel with regular replacement.

The benefits of using stainless steel reinforcement in the infrastructure are clear; durable structures, less repair and disruption, and significant cost savings over the design life. But in addition, because stainless steel is recyclable, when the structure is ultimately replaced, the reinforcement will have a scrap value – and then be recycled.

REFERENCES

1. EUROPEAN FEDERATION OF CORROSION. Stainless steel in concrete, state of the art report, number 18, Institute of Materials, London. 1996.

2. CONCRETE SOCIETY. Guidance on the use of stainless steel reinforcement. Technical Report No. 51. 1998.

3. COX R.N., OLDFIELD J.W., The long term durability of austenitic stainless steel in concrete. Proceedings of the 4th International Symposium on Corrosion of Reinforcement in Concrete Construction, Society of Chemical Industry, 1996, pp 662-669.

4. COCHRANE D.J., New build or repair – The suitability of stainless steel reinforcement for durable and cost-effective structures. Proceedings of the 7th International Conference on concrete in hot & aggressive environments. Bahrain 2003,

5. BRITISH STANDARDS INSTITUTION, BS6744:2001 Stainless steel bars for the reinforcement of and use in concrete – Requirements and test methods.

6. THE HIGHWAYS AGENCY. BA84/02: Design manual for roads and bridges – Volume 1, Section 3, Part 15, Use of stainless steel reinforcement in highway structures.

7. BRITISH STAINLESS STEEL ASSOCIATION, CD-Rom, Stainless steel reinforcement for concrete.

8. BRITISH STANDARDS INSTITUTION, BS 5400: Part 4: 1990: Steel, concrete and composite bridges, Part 4. Code of practice for design of concrete bridges.

CONSTRUCTION WASTE MANAGEMENT: USING THE 7Rs GOLDEN RULE FOR INDUSTRIAL ECOLOGY

M S Al-Ansary
University of Cambridge
United Kingdom
S M El-Haggar
The American University in Cairo
Egypt

ABSTRACT. The construction industry is criticised for consuming extensive amounts of finite natural resources as well as producing substantial quantities of leftover, named as construction waste. Various laws and regulations have been issued to regulate the negative impacts experienced by the construction industry including the generated pollution; however, only partial success has been attained. This paper suggests the application of the Industrial Ecology (IE) concept as a potential method to help the construction industry to become a sustainable responsible business. The Industrial Ecology concept is defined as the study of industrial systems that operate more like natural eco-systems. This concept can be applied to the construction industry by utilising the site wastes as raw materials for other industries or applications, where materials and energy could be circulated in a complex web of interactions. A possible strategy for the implementation of the IE concept could be through the full adaptation of the 7Rs Golden Rule. The 7Rs Golden Rule is a waste management hierarchy that consists of Regulations, Reducing, Reusing, Recycling, Recovering, Rethinking and Renovation. The employment of the 7Rs Golden Rule could help in eradicating the landfilling process as well as substituting the end-of-pipe treatment techniques. Consequently, the Cradle-to-Cradle concept could be fulfilled, where considerable amount of cost savings could be achieved. This paper explains the basic elements of the 7Rs Golden Rule and the possible routes to apply the Industrial Ecology concept to the construction industry. Furthermore, the paper recommends some management techniques to mitigate wastage of various construction materials such as concrete, masonry works, tiles, steel, excavated soil, wood, paper, plasterboard, asphalt, metal, glass, paper, plastics, paints and coatings and insulation materials.

Keywords: Construction/Site waste, Industrial ecology, Waste management.

Miss M S Al-Ansary, PhD Student, Engineering Department, University of Cambridge, United Kingdom.

Professor S M El-Haggar, Professor of Energy and Environment, Engineering Department, The American University in Cairo, Egypt.

INTRODUCTION

Construction or site waste is defined as any material that enters the construction system of any civil engineering project in its form or as an auxiliary substance, with the purpose of being utilized at any phase or in any activity towards the project completion, and left unused. Construction wastes are heterogeneous, mixed wastes that normally consist of relatively clean building materials. The quantity and quality of the construction waste generated from any specific project will vary depending on the project's circumstances and types of utilised materials.

Possible sources for generating construction waste can be classified under six main categories as follows [1, 2]: (a) "Design source" that could be caused by errors in the blueprints and lack of detailing, and design change orders. (b) "Procurement source" that could be caused by the inaccurate estimate of the material quantities. (c) "Material handling source" that could be caused by improper storage and inappropriate on-site or off-site material handling. (d) "Operational source" that could be caused by human errors, inadequate equipment or Force Majeure. (e) "Residual source" that could be caused by poor site management and bad on-site housekeeping and (f) "Other sources" that could be due to various causes such as incompliance of manufactured dimensions with designed panels or materials, excessive material packaging, mismatching of the material quality with the required specifications, poor training in material handling, sorting and disposal methods, and incorrect use of material that require replacement.

INDUSTRIAL ECOLOGY AND 7Rs MANAGEMENT TOOLS

The Industrial Ecology (IE) paradigm is defined as the study of industrial systems that operate more like natural eco-systems [3]. In a natural eco-system there is a closed materials and energy loop, where animals' and plants' wastes are decomposed, by the effect of microorganisms, into useful nutrients to be consumed by plants. Afterwards, plants provide food to the animals. Finally the loop is closed when animals die and are either converted to fossil fuels or useful nutrients to be consumed by plants. Industries can follow the same system through making one industry's waste into another's raw material, where materials and energy could be circulated in a complex web of interactions. A possible tool for the implementation of the IE concept to an industry, to implement the concept of Cradle-to-Cradle, could be through the full adaptation of the 7Rs Golden Rule. The 7Rs Golden Rule is one of the established waste management hierarchies, which encompasses Regulations, Reducing, Reusing, Recycling, Recovering, Rethinking and Renovation. In the following section the elements of the 7Rs Golden Rule will be illustrated.

(1) Regulations: Regulations are one of the essential tools for enforcing laws and legislations through incentive mechanisms. There is a need for regulations that could encourage the full adaptation of the IE concept. This could be fulfilled through adding incentive articles or provisions to the existing environmental regulations.

(2) Reducing: Reduce is a precautionary technique aimed at minimising the waste generated from the source before it becomes a physical problem. Reducing is also called source reduction, waste minimization and prevention [4]. Reducing technique helps in minimising the waste quantity and toxicity by either decreasing the utilised quantity of raw materials or by replacing hazardous materials with more environmentally friendly ones.

(3) Reusing: The reuse technique is defined as re-employment of materials to be reused in the same or in lower grade applications, without carrying out any further modifications or changes [4]. Waste separation and sorting are the essential procedures for the success of this technique. Reusing endeavours could be feasible if (a) the waste materials are of proper quality and (b) the material and the waste disposal costs are greater than the reusing expenses.

(4) Recycling: The recycle technique is defined as utilising wastes as raw materials in producing other products. The recycling procedures normally depend on (i) the availability of sufficient recyclable wastes, (ii) the marketing for the recyclable products, (iii) the generation of a profit, (iv) the landfill tipping fees and (v) the governmental incentives and regulations regarding the recycling opportunities [5]. On the other hand, the most prevailing problems that could possibly hinder the recycling technique are: limited quantity of waste, space limitation, unavailability of recycling equipment, logistics and facilities, inappropriate source separation, elevated initial costs, low production rates of the recycling facility, high set-up and start-up times, inadequate staff training and equipment failure [6]. There are several methods for preparing the site waste in order to be recycled such as [4, 5]: (i) site separation and waste sorting, (ii) site separation and processing, (iii) co-mingled waste, off-site separation and processing, and (iv) transfer stations.

(5) Recovering: The recovery technique is defined as generating energy from waste materials. This stage is also called 'waste transformation' since physical, chemical or biological transformation has to be applied to the wastes [7]. There are a number of recovery techniques such as incineration, pyrolysis, gasification and biodigestion that could produce energy in the form of steam, electricity, synthetic gases, or liquid and solid fuel.

(6) Rethinking: After applying the previous management tools and there are still inevitable, unmanageable wastes, then rethinking should be employed before taking any further action for treatment or disposal.

(7) Renovation: Renovation is the last stage to close the loop of the industrial ecology cycle where alternative, innovative technologies for mitigating the inevitable, unmanageable wastes are developed. The essence of this stage is to develop renewable resources by reducing or preventing the disposal option.

It should be noted that the rethinking and the renovation techniques should be applied to the entire process starting from Regulation and Reduction to Reuse and Recycle. Rethinking and renovation techniques give room to the stakeholder to think about their waste as renewable resources throughout the entire life cycle of any construction project.

APPLYING 7Rs IN MITIGATION CONSTRUCTION MATERIAL

It is recommended that the 7Rs Golden Rule tools should be applied to the whole project life cycle; from the early design phase, the planning phase, the tender and contract formulation phase, the construction and the maintenance phases [1, 8, 9]. Furthermore, all the project designs, specifications and documents should enforce the implementation of all the waste mitigation techniques. This should be developed by the project participants such as government, local authority, owner, engineer, designer, architect, planner, quantity surveyors, contractor and sub-contractors, suppliers and HSE personnel. This section will discuss the application of the 7Rs management tool in mitigating some of the construction materials.

A) Applying Regulations Techniques

Taking Egypt as an example, the Egyptian environmental legislator has regulated the dumping and treatment of solid wastes including construction wastes. In the Egyptian Environmental law No.4/1994, the dumping and treatment of solid wastes have been defined and regulated in many provisions [10]. Moreover, the same law has established a system of incentives to encourage the implementation of environmental protection activities and projects. However, the Industrial Ecology concept has not been recommended or encouraged. Therefore, there is a vast need to add relevant provisions in the existing law to incorporate the IE paradigm into the Egyptian construction industry.

Table 1   Reducing Techniques

MATERIAL	REDUCING TECHNIQUE
Wood	• The specified wood panels dimensions should correspond to the standard wood dimensions to minimise leftovers. • Efficient framing techniques should be implemented. • Detailed formwork working/shop drawings should be available to help in ordering wood lengths according to the detailed drawings. • Wooden formwork should be kept covered, off the ground and in a dry secure area to be protected from the atmospheric conditions (such as rain, sun and humidity), physical deterioration (such as twisting), theft or loss. • Wood cutting workshops should be centralised to optimise the use of wood scraps in other applications.
Excavated Soil/Earth Works	• Balance the volume of cut and fill of excavated materials. • Subject the soil, if possible, to appropriate remediation process (e.g. solidification/ stabilization) rather than excavating and clearing soil waste outside the site. • Avoid excavating unnecessary soil; for example, in case of isolated footing just excavate the place of the footings, if possible.
Concrete	• Cement bags should be ordered on time and stored properly to avoid any deterioration.
Plasterboard	• The designed plasterboard should correspond to the standard market/factory plasterboard dimensions to minimise the cut-off wastes. • Plasterboard should be ordered in optimal dimensions to minimize leftovers. • Plasterboard panels should be stacked in such a way that the sequence of removal is the same as the order they need to be erected. Spacers or 'sleepers' should be placed in-between the panels to separate them. • Plasterboard panels should be kept covered, off the ground and in a dry secure area to be protected from atmospheric conditions, deterioration or theft. • The plasterboard should be utilised in the most efficient way and with skilled installation that could reduce leftovers.
Insulation Materials	• Use the cold method of applying bitumen rather than the hot one to avoid carbon dioxide emissions. • If hot bitumen should be used, a good combustion system with complete emission control should be applied.
Packaging and Shipment	• Avoid excessive packaged materials. • Assure packaging is adequate to prevent material damage. • Choose the supplier who is willing to recycle packaging. • Reuse shipment containers in further shipment, if possible.
Cardboard and Paper	• Avoid unnecessary reproduction and copying of drawings and sketches. • Segregate cardboard and paper from other waste streams and store them in a separate area.
Hazardous Wastes	• Employ materials that produce minimum hazardous/toxic effects. • Prepare well-documented Material Safety Data (MSD) sheets that identify all the hazardous wastes that could be generated from the project.

B) Applying Reducing Techniques

The reducing technique should be applied to the entire life cycle of the construction project. Reducing techniques can possibly be applied to various construction materials to prevent the transformation of such materials to become potential wastes. Table 1 summarises some of the suggested techniques to reduce the generation of construction wastes.

C) Applying Reuse Techniques

The reuse technique can help in maximising the reuse of the materials in the same or other applications within the same or different site. Reuse techniques can be applied during the construction and maintenance phases and be adapted by the contractor team. Table 2 summarises some of the suggested techniques to reuse the construction waste.

Table 2   Reusing Techniques

MATERIAL	REUSING TECHNIQUES
Wood	• Increase the number of times for reusing wooden formwork. • Clean wood waste could be reused in many applications such as: super structure and sub-structures formwork, blocking wall and floor cavities, surfacing material and landfill cover.
Excavated Soil/Earth Works	• Excavated soil could be used for landscaping and as noise reducing embankment. • If excavated soil is clean and complies with the quality specifications and code of practice, it could be used in backfilling between tie beams, under plain concrete and slabs, under foundations and under structural walls.
Steel	• Steel reinforcement wastes could be straightened and reused in the reinforcement of any pavement, sidewalks and curbs, concrete lintels and openings. • Steel reinforcement wastes could be used as spacers between the main reinforcement grids, mainly in stairs.
Concrete	• Wastes of green concrete could be used in manufacturing on-site curbs and concrete blocks and in architectural decoration (e.g. hard landscape) applications. • Concrete wastes could be used in non-structural works such as in windows and door openings and in road construction applications. • Spoiled cement can be used in non-structural purposes such as masonry works, cement paint, plain concrete and tile mortars.
Masonry (blocks and bricks)	• Masonry wastes can be used in non-structural applications such as capping layers for road embankment, landscape cover and sub-base fill and backfill. • Blocks and bricks waste can be used in producing lightweight concrete. • All bricks leftovers in the different floors could be collected on the top floor where it could be used as temperature insulating material instead of using foam or other chemicals.
Plasterboard	• Plasterboard wastes could be used to insulate wall cavities, doors and windows to improve both the sound and thermal proofing. • Large pieces of plasterboard waste could be used as fillers.
Insulation Materials	• Bituminous waste could be used to improve the surfaces of footpaths. • Leftovers from insulation materials could be used in filling the interior wall cavities or on the top of installed insulation to enhance thermal performance. • Large pieces of hard insulation materials could be used under concrete floors.
Carpet	• Carpets if left in good condition could be used in areas where aesthetics play fewer roles such as in basements. • Carpets leftovers could be used to make mats for hallways and entryways.
Others	• Try to reuse leftovers of materials such as paints and sealants on other areas.

D) Applying Recycle Techniques

All recycled materials should be in accordance with all quality control tests and specifications in order to verify the suitability of each material for the intended purpose. Table 3 summarises some of the suggested techniques to recycle some of the construction wastes.

Table 3 Recycle Techniques

MATERIAL	RECYCLE TECHNIQUE
Concrete	• In order to recycle concrete waste, the following procedures should be executed. Firstly, concrete waste should be crushed, and then ferrous metals (such as steel bars and bolts) should be separated from concrete either manually or by specialized scissors. Afterwards, screening for sizing should be performed to separate different sizes to meet with the quality control specifications. • Recycled concrete could be used as a fill material (e.g. backfill, general fill, and base or sub base course) and/or as secondary aggregate.
Wood	• In order to recycle wood to be used in further applications, the following procedures should be followed. Firstly, wood grinders should shred the wood wastes. Secondly, ferrous metals should be separated magnetically from the other wood particles. Then, the shredded parts should be passed through trammel or mechanical screens to separate between the different sizes. Oversized wood pieces could be reused in other applications. Normal sized wood particle could be used in landscaping or could be burnt to produce energy/fuel. Under size or fine particles could be used in landscaping or as animal bedding. • Wood waste can be reduced to fibers to be used in producing processed wood products such as composite panels (e.g. particleboard) and reconstituted boards. • Wood waste could be used in the manufacture of pulp and paper since the secondary wood fiber could provide the required longer and stronger fibers. • Clean sawdust (untreated and unpainted) could be composted to produce compost, soil amendments, mulch and soil fertilizers and conditioners. It could also be mixed with yard mulch to produce boiler fuel. • Wood waste could be mixed with cement to produce cement-wood composite and used in structural applications, in the manufacture of construction panels and the construction of highway sound barriers.
Steel	• Steel reinforcement wastes could be recycled into new steel bars.
Masonry	• Masonry wastes could be used in manufacturing new blocks and bricks.
Tiles	• Tiles wastes could be crushed and used in the manufacture of agloromated marble (i.e. tiles with polished marble/granite pieces) or the manufacture of mosaic tiles.
Asphalt	• In order to recycle asphalt waste, the following steps should be fulfilled. Firstly, asphalt waste (e.g. from pavement or roofing applications) should be crushed followed by the magnetic removal of ferrous metals. Then, the crushed asphalt should be screened and graded. Afterwards, the graded material could either be used as: (a) road base with other crushed and screened aggregates, (b) in producing new paving material and (c) in producing new asphalt products by mixing it with new asphalt binders. • Asphalt waste could be used as an additive in the production of hot asphalt mix, after being ground to 0.5 inch. • If asphalt waste is mixed with rock and gravel, it could be used as a groundcover that is used in rural roadways or temporary construction surface to control the generated dust from the construction activities. • Asphalt wastes could be utilised in the manufacture of new types of composite roofing shingles.
Packaging	• Corrugated cardboard could be recycled to be used in the manufacture of outside skin layers and internal rolled layer for new containers. • Paper packaging, in general, could be recycled into paper products.
Plastic	• Recycle plastic wastes into further plastic products.

Table 3 Recycle Techniques (Con't)

MATERIAL	RECYCLE TECHNIQUE
Glass	• Glass could be recycled to produce new containers and bottles. The presence of glass helps in reducing the required furnace temperature, thus saving energy and prolonging the furnace lifetime.
Plasterboard	• Plasterboard wastes could be shredded and pulverized to 1 to 0.25-inch size, after removing the paper faces. The powdered gypsum could be utilised as soil conditioner to improve plant growth due to its high calcium and sulphur contents. • Plasterboard wastes could be utilized as raw materials in the manufacture of new gypsum products. Generally, the new plasterboard products could utilize about 10% - 20% of the recycled gypsum content and 100% of the recycled liner paper. • Plasterboard wastes could be used as animal bedding after being crushed and mixed up with wood chips. • Plasterboard wastes could be processed and blended with gypsum rock and then utilised as a granular gypsum product. • Plasterboard wastes could be used in the manufacture of non-structural building such as lightweight interior walls, sound barrier walls, textured wall sprays, acoustical coatings, fire barriers and absorbent products.

E) Applying Recovery Techniques

Recovery techniques can be classified into two main groups; waste-to-energy (waste recovery) technique and waste-to-material technique (material recovery) technique. Organic wastes can be used to recover energy according to waste-to-energy recovery technique. For example, wood wastes that could not be reduced, reused nor recycled, could be incinerated to produce energy in the form of fuel or collected in digester (container) for anaerobic biological reaction to produce gaseous fuel (biogas). Proper segregation of construction waste will help waste-to-material recovery techniques and reuse material again as discussed in Table 2.

F) and G) Applying Rethinking and Renovation Techniques

Construction wastes that could not be reduced, reused, recycled or recovered should not be sent to disposal facilities before applying rethinking and renovation techniques. In this stage, these kinds of waste should have another thought in order to develop some innovative technologies in order to mitigate them. Table 4 shows some of the recommended innovative techniques.

Table 4 Renovation Techniques

MATERIAL	RENOVATION TECHNIQUE
Steel	• Coupling technique could be used for the reinforcement connection instead of the overlapping technique (i.e. with effective length) to provide the required strength.
Tiles	• Tile wastes (Ceramics, Mosaic, Terrazzo, Marble, Granite …etc.) could be used in architectural and decorative applications.
Glass	• Glass could be used in manufacturing glasswool, fiberglass insulation and glasphalt (paving material) and in the production of bricks, ceramics, terrazzo tiles and lightweight foamed concrete.

CONCLUSIONS

A possible means for fulfilling the 'sustainability triple bottom line' of attaining environmental benefits, economical development and social enhancement is to incorporate the Industrial Ecology paradigm. The Industrial Ecology paradigm is defined as the study of industrial systems that operate more like natural eco-systems. Similarly, industries can develop the IE concept by making one industry's waste as another's raw materials, where materials and energy could be circulated in a complex web of interactions. The adaptation of the IE concept could be possibly fulfilled by the full implementation of the 7Rs Golden Rule. The 7Rs Golden Rule is a developed waste management hierarchy that consist of Regulations, Reducing, Reusing, Recycling, Recovering, Rethinking and Renovation.

The application of the IE concept to the construction industry can help in rendering the industry to become a more sustainable business. This could lead to many benefits such as attaining resource optimisation and waste management enhancement as well as complying with the environmental protection regulations. The paper illustrated some examples of the mitigation techniques that could optimise the use of the construction materials by utilising the IE concept. It could be concluded that if the 7Rs Golden rule are well implemented throughout the whole life cycle of a construction project, ultimately zero waste could be reached and hence there would be no need for landfills. Furthermore, this approach could fulfil the Cradle-to-Cradle concept, save the finite materials and hence re-define the construction waste as renewable resources.

REFERENCES

1. AL-ANSARY M.S.M. (2001). Recommending Egyptian Guidelines for Managing Construction and Demolition (C&D) waste. MSc. Thesis, The American University in Cairo, Egypt.

2. GAVILAN, R. (1994). Source Evaluation of Solid Waste in Building Construction. Journal of Construction Engineering and Management 120.3: 536-552.

3. EL-HAGGAR, S.M. (2004). Industrial Ecology for Renewable Resources. Proceedings of International Conference for Renewable Resources and Renewable Energy: A Global Challenge, 10 – 12 June 2004, Terista, Italy.

4. EL MADANY, I.M. (1999). Integrated and Sustainable Management for Municipal Solid Wastes. Cairo: Proceedings of Arab and Middle East International Conference and Fair on Solid Waste Management, 8-6 December 1999.

5. MINKS, W. R. (1996). Reducing and Recycling Construction and Demolition Waste. Online Posting. 2nd of March 2000.

6. HECKER, T. (1993). "10 Common Problems with Recycle Operations". Pit & Quarry July: 43.

7. TCHOBANOGLOUS, G., THEISEN, H. AND VIGIL, S. (1993). Integrated Solid Waste Management: Engineering Principles and Management Issues. McGraw-Hill, Inc.

8. AL-ANSARY, M.S., EL-HAGGAR, S.M. AND TAHA, M.A. (2004). Sustainable guidelines for managing demolition waste in Egypt, Proceedings of the International RILEM Conference on the Use of Recycled Materials in Building and Structures, Barcelona, 9-11 November 2004.

9. AL-ANSARY, M.S., EL-HAGGAR, S.M. AND TAHA, M.A. (2004). Proposed Guidelines for Construction Waste Management in Egypt for Sustainability of Construction Industry, Proceedings of the International Conference on Sustainable Construction Waste Management, Singapore, 10-12 June 2004.

10. EGYPTIAN ENVIRONMENTAL LAW NO. 4 for year 1994 and its Exwcutive Regulations. Promulgating a law concering Envitoment. Egypt.

ENVIRONMENTAL ENGINEERING WITH CEMENT AND CONCRETE

IMPROVEMENT OF CONCRETE MIX DESIGN FOR HAZARDOUS MATERIAL STORAGE OR IMMOBILISATION OF WASTES BY NUMERICAL SIMULATION

D Damidot

S Kamali   **F Bernard**

Ecole des Mines de Douai

France

ABSTRACT. The performance of concrete as a barrier or a confining material, depends strongly on its mix design and its interaction with the environment. This last parameter may be difficult or impossible to reproduce in laboratory experiments and thus numerical simulation is very helpful to estimate the durability and the environmental impact of the concrete – waste system in various environments. Reactive transport codes developed in geochemistry can be used in order to relate the physico-chemical evolution of concrete with time in a given environment. However, the evolution of the microstructure of concrete is not taken into account and as a consequence, it is not possible to estimate accurately the spatial and temporal evolutions of physical properties such as the diffusion coefficient and the elastic modulus. Thus, we have developed a simulation based on a multi-scale 3D concrete microstructure and several computer codes applied to it in order to simulate the evolution of this digital microstructure in a defined environment.

Keywords: Concrete, Multi-scale modeling, Reactive transport, Mechanical strength.

Prof D Damidot, is head of the civil engineering department at Ecole des Mines de Douai, France. His main field of interest is the physico-chemistry and the numerical simulation of the hydration of hydraulic binders and their durability.

Dr S Kamali, is a civil engineering researcher at Ecole des Mines de Douai. Her main field of interest is the durability of cement based materials and structures from a physico-chemical and mechanical point of view.

Dr F Bernard, is assistant professor in the civil engineering department at Ecole des Mines de Douai; he works on numerical modelling of building materials long term mechanical behaviour.

INTRODUCTION

Concrete is a cheap and reliable material that is often a good candidate in order to confine hazardous materials or to immobilize waste. However, the performance of concrete as a barrier or a confining material, depends strongly on its mix design and its interaction with the environment. This last parameter may be difficult or impossible to reproduce accurately in laboratory experiments for very long periods and thus numerical simulation is very helpful to estimate the durability and the environmental impact of the concrete – waste system in various environments. As a consequence, numerical simulation has been more and more often used in the last decade mainly on cement paste that is the reactive part of concrete if we except pathologies such as alkali-silica reaction. First a purely thermodynamic approach has been developed in order to define the stability of the solid phases contained in cement paste and submitted to various environments. The application of this approach was initially devoted to the confinement of radioactive waste [1]. It was then extended to the knowledge of phase diagram of systems relevant to Portland cement containing various secondary cementitious materials subjected to common aggressive environment [2]. The main advantage of phase diagram relies on some simplified drawings of parts of the phase diagram that enable to understand the evolution of phase assemblages when the environment changes. However, a purely thermodynamic approach cannot give the time dependence of the process that is a very important parameter for designing reliable confining or immobilizing materials.

The time dependency can be obtained by using reactive transport codes developed in geochemistry. Here thermodynamics, kinetics of chemical reactions and transport of matter (and sometimes heat) are taken into account. Thus, it is possible to obtain the spatial and temporal evolution of the concentration of matter. However, some simplifications are often made, such as considering a unidirectional transport of matter in an isotropic porous material [3]. In the most accurate simulations, the retroaction of chemical reaction (mainly dissolution and precipitation) on the porosity and thus on the diffusion coefficient can be taken into account by using empirical laws for diffusion coefficient. Thus, from this purely physico-chemical approach, it is only possible to estimate the spatial and temporal evolutions of physical properties such as the elastic modulus by empirical laws. There does not exist laws that enable to calculate these parameters like for concentration of matter with Fick's laws. As a consequence the need for simulations based on 3D concrete microstructure becomes more and more important in order to relate the influence of the concrete mix design, on concrete microstructure and thus on the evolution of concrete properties in a given environment. This approach needs to numerically define the concrete microstructure over several orders of magnitude (from nanometre to decimetre) that is not yet possible due to computer power even with clusters working with parallel codes. Thus alternative methods can be developed such as methods based on a multi-scale simulation of the microstructure of concrete. The aim of this paper is to describe one of these alternative methods in its actual state of development. The main computer codes and the methods developed in order to exchange data from a scale to another larger one are defined.

3D SIMULATION OF THE DIFFERENT SCALES

Concrete microstructure is defined at three scales. The finest scale corresponds to the cement paste scale in which the cement phases, the hydrates and porosity (filled or not with water) are considered. A representative elementary volume of cement paste can be represented by a cube having a length between 150 and 300 μm and a resolution between 0,25 and 1μm. Then the mortar scale is made of cement paste, sand, air void and sometimes an interfacial transition zone around the larger sand grains. At this scale, the representative elementary volume of mortar can be a cube having a length between 5 and 10 mm and a resolution between 25 and 100 μm. Finally the concrete scale contains mortar, aggregate and an interfacial transition zone. Here the representative elementary volume of concrete can be a cube having a length between 100 and 150 mm or a prism of 70x70x280 mm and a resolution between 500 and 1000 μm. From this description, concrete can be accurately defined if the data obtained at the smaller scales, mortar and cement paste, are homogenised at that scale.

Several computer codes exist in order to simulate the evolution of a 3D digital microstructure during the hydration of the cement paste [4-6]. Among these codes, virtual cement and concrete testing laboratory (VCCTL) in its version 5, is the most advanced one for generating an accurate cement paste microstructure as it allows the real shape of cement grains to be used, to define the particle size distribution, to accurately distribute the cement phase and to use a complete set of chemical reactions corresponding to cement hydration. Figures 1 and 2 represent a 2D cut of a 200x200x200μm cement paste microstructure having a W/C ratio of 0.5, before hydration (unhydrated) and after 28 days hydration respectively. Moreover, VCCTL enables to calculate some physical properties from these digital microstructures such as the diffusion coefficient and the elastic modulus.

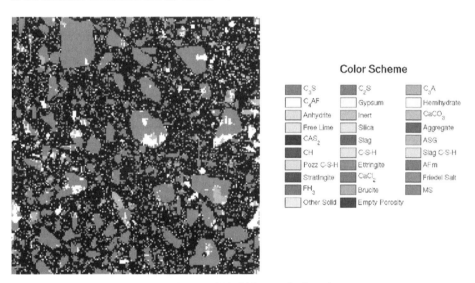

Figure 1   2D cut of a 200x200x200μm unhydrated cement paste
microstructure having a W/C ratio of 0.5

VCCTL version 5 is also used at the mortar and concrete scales as it contains a huge database of aggregates. The shape of the aggregate is very close to the real shape as it has been modeled from data collected by X-ray tomography measurements (Figure 3).

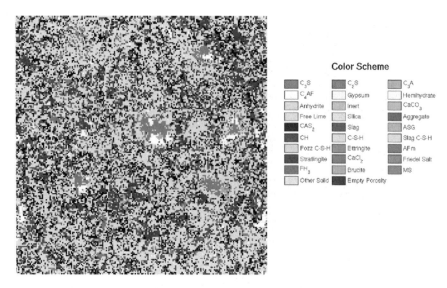

Figure 2. 2D cut of a 200x200x200μm cement paste microstructure
having a W/C ratio of 0.5 after 28 days hydration

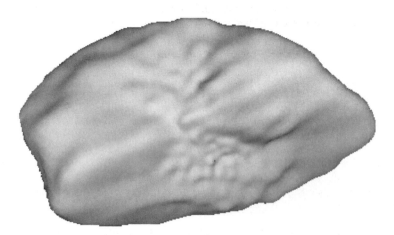

Figure 3   Example of 3D aggregate generated by VCCTL [7]

Figure 4 represents a concrete cube having a length of 150 mm. In this case only three kinds of voxels are considered : mortar, ITZ and aggregate. The mineral distribution in the aggregate is not yet taken into account. ITZ is simulated using VCCTL at the cement paste scale by adding a plate that corresponds to the surface of a coarse sand grain or an aggregate. As the resolution of a voxel at the mortar and concrete scales is greater than the length of ITZ, the properties of the ITZ voxels are homogenised relative to the amount of the different phases.

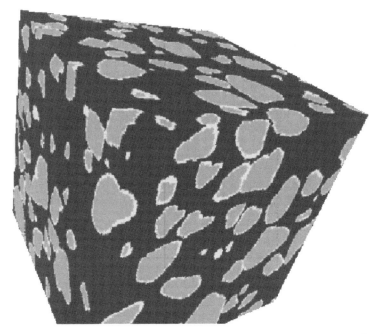

Figure 4   Concrete cube having a length of 150 mm with of voxel size of $500^3 \mu m^3$; black voxels correspond to mortar, green voxels correspond to ITZ and brown voxels correspond to aggregate [7]

APPLYING CODES TO THE 3D CONCRETE MICROSTRUCTURE

Once the multi-scale 3D microstructure of concrete is built, some codes have to be applied in order to follow its evolution when it interacts with the environment. In the present case, where the durability issue is assessed, we have considered first a code that performs reactive transport in order to estimate the impact of the transport of matter on the concrete but also the retroaction of the modification of its microstructure on the diffusion coefficient. Secondly a finite element code is used in order to calculate the spatial and temporal evolutions of the elastic modulus. This last point is a first attempt in order to link the impact of chemistry on mechanics.

The code HYTEC [8] is used for the simulation of the reactive transport. This code first calculates the transport of matter and secondly performs an equilibration between the modified aqueous phase and the solids by dissolution, precipitation and adsorption mechanisms. A 3D concrete piece having the desired size is made from the assemblage of several 3D concrete microstructures obtained by VCCTL. Then a 3D mesh is applied to the concrete piece with the addition of some cells that represent the environment. After fixing the initial conditions, HYTEC starts the calculations and at each time step, a specific application updates the 3D concrete microstructure at the different scales. In a first approximation, if we do not consider the reactivity of sand or aggregate and that the temperature remains constant (no dilation), the transport of matter only impacts the cement paste microstructure. Of course the induced changes at the paste level are also observed at the higher scales, mortar and concrete as a part of cement paste is contained at these scales. Thus in each mesh of the 3D concrete, the amount of the aqueous phase and its modified composition due to the transport, but also the quantity of mortar and thus of cement paste is known. The cement paste is considered to be homogeneous in a given mesh of concrete and is represented by a cement paste microstructure generated by VCCTL corresponding to a cube having a length of 200μm and a resolution of $1\mu m^3$ by voxels. Once the aqueous phase is equilibrated with the mineral phases present in the cement paste, it is possible to calculate the amount of precipitated and dissolved solids and thus to update the 3D microstructure. Then the effective diffusion coefficient for the cement paste can be calculated from this updated microstructure. Afterwards the diffusion coefficient at the mortar and then concrete scales are updated using the new diffusion coefficient for the paste. A similar microstructure and diffusion coefficient update is made for ITZ. The thermodynamic database of HYTEC has been modified in order to model accurately C-S-H behavior from 10 to 85°C. Moreover, this database is in the process of being validated by the Common Thermodynamic Database Project [9].

Also once the microstructure at the cement paste scale is updated, the evolution of the elastic modulus can be calculated. ABAQUS code [10] that is a general purpose FEM code, has been used to calculate the mechanical behavior and in the present case the elastic modulus. This last calculation required a specific program to convert VCCTL microstructure to ABAQUS input file considering that the size of a finite element corresponds to a voxel (1 μm³). The elastic modulus of each phase is another input that is needed for the calculation. Once elastic modulus of the cement paste is calculated, it is possible to update the elastic modulus of mortar and then of concrete.

Then the new diffusion coefficient is updated in each mesh and HYTEC continues its calculation for the next time step ; figure 5 summarizes the links between the different codes used in the simulation. The process continues up to the desired time ;

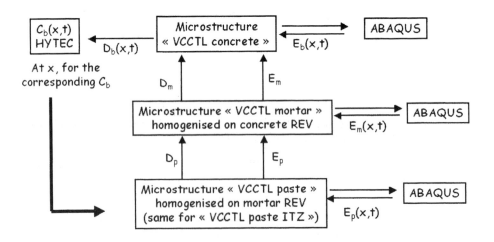

Figure 5   Flowchart summarizing the different steps and the links between
the different codes used in the simulation ;
b, m and p indexes refer to concrete, mortar and cement paste respectively for diffusion
coefficient (D), elastic modulus (E) and concentration (C).

It is thus possible to have the spatial and temporal evolutions of the elastic modulus and diffusion coefficient in addition to the ionic concentrations. Indeed, the variations of elastic modulus and diffusion coefficient are expected to be important during a degradation process. For example, if we consider a simple leaching of a cement paste by pure water, it can be observed that portlandite dissolves first and then the other hydrates inducing a zoning process [3,11]. The evolution of diffusivity and elastic modulus in areas where portlandite is completely dissolved shows a drastic modification of these parameters that is associated to an increase of the porosity ; the diffusion coefficient is 20 times higher and the elastic modulus is reduced by half [11].

CONCLUSIONS

We have reported a method that allows for the first time a 3D multi-scales simulation of a concrete subjected to a defined environment thanks the description of the microstructure at the concrete, mortar and cement paste scales. This method can be applied in order to improve concrete mix design for hazardous material storage or immobilisation of waste.

Further improvements can be added to the present simulation such as the addition of a nanometric scale to have a better representation of the assemblage of the hydrates and especially of C-S-H layers and thus the porosity related to C-S-H. Other physical phenomena such as heat transfer could be accounted for by the addition of modules. Indeed, the present model has been developed in order to build a platform called Open Digital Concrete (ODC). ODC defines a method of coding the multi-scales 3D microstructure and allows specific modules to be added that can be developed by other laboratories. ODC runs on parallel computers such as Beowulf clusters running under Linux, and manages the CPU resources and the data transfers between codes automatically.

ACKNOWLEDGEMENTS

The authors would like to acknowledge ATILH for financial support.

REFERENCES

1.   GLASSER F. P. Fundamental aspects of cement solidification and stabilisation . Journal of Hazardous Materials, Vol. 52, 1997, pp 151-170.

2.   DAMIDOT. D. BARNETT S.J. MACPHEE D. and GLASSER F.P. Investigation of the $CaO-Al_2O_3-SiO_2-CaSO_4-CaCO_3-H_2O$ system at 25°C by thermodynamic calculation. Advances in Cement Research, Vol 16, 2004, pp 69-76

3.   ADENOT F. Durabilité du béton: Caractérisation et modélisation des processus physiques et chimiques de dégradation du ciment, PhD Thesis, Université d'Orléans (France), 1992.

4.   VCCTL. Virtual Cement and Concrete Testing Laboratory. NIST's software. Web Site http://www.bfrl.nist.gov/861/vcctl/

5. YE G. VAN BREUGEL K. and FRAAIJ A. L. A. Three-dimensional microstructure analysis of numerically simulated cementitious materials, Cement and Concrete Research, Vol. 33, 2003, pp 215-222

6. NAVI P. and PIGNAT C. Three-dimensional characterization of the pore structure of a simulated cement paste Cement and Concrete Research, Vol. 29, 1999, pp 507-514

7. BULLARD J. et al. Minutes of the VCCTL bi annual meeting ,Washington, November 2004

8. http://www.cig.ensmp.fr/~vanderlee/hytec/index.html

9. http://ctdp.ensmp.fr/about.html

10. ABAQUS: User's manual, Version 6.4, Karlson & Sorensen Inc. 2004

11. KAMALI S. Comportement et simulation des matériaux cimentaires en environnements agressifs : lixiviation et température. LMT-ENS de Cachan, France, PhD Thesis, 2003

MIX PROPORTIONING AND SOME PROPERTIES OF SLAG CONCRETE

M C Nataraja

Sri Jayachamarajendra College of Engineering

B M Ramalinga Reddy

Sri Siddhartha Institute of Technology

India

ABSTRACT. There is currently no well-tried mix proportioning methods available for slag concrete. Presently this supplementary cementitious material is being used widely in India in almost all fields of construction, as it is a cheaper material. The use of such mineral admixture is also recommended in IS: 456-2000. This paper presents a simple method to produce concrete having a 28-day compressive strength varying from 30 MPa to 60 MPa at different percentages of slag namely 30%, 60% and 90% as a replacement to cement. Though the strength of slag concrete decreases as the slag content increases, it is found that the slag concrete possessed significant compressive strength even at higher replacement levels. Concrete mixes are designed for three exposure conditions as per IS: 456-2000 satisfying the durability requirements. Their compressive strength characteristics are reported in this paper. Mix design procedure suggested in the draft IS: 10262(2003) is adopted.

Keywords: Ground granulated blast furnace slag (GGBFS); Concrete Mix Proportioning; Compressive strength; Workability.

Dr M C Nataraja, Assistant Professor, obtained his B.E.(Civil) in 1983 with First Class Distinction, securing a Rank from Mysore University and M.Tech in Industrial Structures from Karnataka Regional Engineering College, in 1985 with First Class Distinction. Later, he obtained his Ph.D from IIT, Kharagpur, India. He has 19 years of Post Graduate teaching and research experience. He has published 70 technical papers in leading International and National Journals and Conferences. His areas of interest include concrete materials, HPC and marginal materials in concrete. He has attended many International conferences and chaired few sessions. He is a member to many professional organizations. He is a recipient of best paper award from Indian Concrete Institute.

B M Ramalinga Reddy, Assistant Professor, obtained his B.E.(Civil) in 1983 with First Class Distintiction from Mysore University and M.E(PSC-Structures) from UVCE, Bangalore in 1991. He has 19 years of teaching experience and has published a few papers in National and International Conferences. Presently he is working for his Ph.D in the area of High performance concrete.

INTRODUCTION

Ground granulated blast furnace slag (GGBFS) is now recognised as a desirable cementitious ingredient of concrete, and as a valuable cement-replacing material that imparts some specific qualities to the composite-cement concrete superior to those of concrete made from ordinary Portland cement (OPC) alone. The practice of grinding slag and Portland cement separately, and combining them at the mixer, has been widely accepted because of the practical advantages of matching the cement and slag to optimise the properties of the composite cement, and of adjusting their proportions to suit the needs of a particular situation and the properties required of the resulting concrete [1,2].

At the moment there is no specific mix proportioning methods designed for slag concrete. The material is used as a direct replacement of cement by weight, at proportions of 25 to 70 percent by mass of the total cementitious contents and then current proportioning techniques for concrete made with Portland cement is followed. Generally when slag fineness is of the same order of magnitude as Portland cement, the compressive strength of concrete containing slag is lower than that of a control concrete without slag, particularly at early ages and at replacement levels of 50 percent and above. At lower water-binder ratio, 28-day cube compressive strengths of 30 to 50 MPa have been obtained for such cements without much difficulty. Much higher 7 and 28-day cube compressive strengths of more than 50 MPa can be obtained by grinding slag to much higher fineness than OPC or by using reproportioning techniques to enhance the strength [1, 3].

Because the hydration of slag in combination with Portland cement is generally a two-stage process and because slag hydration tends to lag behind that of the Portland cement component's hydration, slag concretes are likely to be more susceptible to poor or inadequate curing conditions than with concrete containing only Portland cement. Thus an adequate curing is needed for slag concrete [1].

RESEARCH SIGNIFICANCE

The main focus of this paper is to emphasize and identify some of the engineering implications when high volume of slag is used in concrete as a replacement to cement. A simple method of mix proportioning suggested in the draft IS 10262 code is employed and presented. High volume of slag as replacement to cement decreases the compressive strength substantially and the decrease is less significant for lower percentages of replacement. The paper also emphasises the fact that by using 350 kg of cementitious materials (10% cement and 90 % slag), concrete of strength equivalent to or higher than M25 grade of concrete can be produced at a w/c ratio of 0.5.

EXPERIMENTAL PROGRAMME

The primary objectives of the study are;

1. To develop mix proportions for concrete containing varying percentages of slag as a replacement to cement for a 28 day cube compressive strength in the range of 30 MPa to 60 MPa.
2. To study the strength development characteristics of slag concrete
3. To produce concrete of reasonable good workability
4. To demonstrate the applicability of the mix design procedure proposed in the draft IS: 10262 code (Private circulation).

Materials

The Portland cement used was OPC 43 grade satisfying the requirements of IS: 8112-1987[4] with a specific surface of 310 m^2 / kg. The 28-day compressive strength of cement is 59 MPa. The slag supplied by Andhra cements[5] had a specific surface of 320 m^2/kg. The fine aggregate is a washed natural river sand with a specific gravity of 2.65. The coarse aggregate is crushed granite with a specific gravity of 2.65 with 20 mm as maximum size. Aggregates conform to IS standard[6]. To achieve mixtures of good workability at low water binder ratio, all mixtures contained superplasticiser in varying percentages by weight of cement and slag (cementitious materials).

Mix Proportioning

The mix proportioning method adopted in this study is based on the criteria mentioned in the draft code. In addition few guidelines mentioned elsewhere [7,8] for mineral admixtures are followed. Some of them are:

1. The water reduction associated with fine mineral admixtures is not a mechanical effect, but the result of dispersion and deflocculation similar to the effect of inorganic admixtures.

2. Setting time is a better criterion of mix design, particularly for early strength development. a low water-binder ratio, irrespective of water demand, will avoid the increased setting times associated with admixtures

3. The water-binder ratio is far more sensitive to changes in water content of composite-cement concrete than in OPC concrete. To maximise the contribution to strength of pozzolanic/cementitious admixtures, the water-binder ratio needs to be as low as practical.

4. With siliceous admixtures that contribute strength through pozzolanic/cementitious reactivity, early strength development is mostly governed by the reactivity of the cement, whereas long-term strength depends more on the reactivity of the mineral admixtures. A low but adequate water-binder ratio and an organic dispersing admixture (superplasticiser) are therefore necessary to achieve high early strength development.

5. Durability considerations also requires a low water-binder ratio in addition to excellent cohesion and flow properties for the fresh concrete.

6. The simplest method of incorporating a mineral admixture in concrete is a straightforward replacement of cement, weight for weight or by considering the specific gravity of mineral admixture in the design. In the present work, specific gravity of slag is considered in the mix design.

Based on the above guidelines, trial mixes are designed for three arbitrary water-binder ratios of 0.40, 0.45 and 0.51 which are the maximum values for extreme, severe and moderate exposure as per IS: 456-2000[9]. A cement content of 350 kg/m^3 is selected, which satisfies the minimum requirement for the three exposures considered. For these values, water content is calculated for the above cement content. Fine and coarse aggregates are then found as per the guidelines of IS:10262 draft code[10,11,12,13,14]. To compensate for the workability, superplasticiser is used at 1.5% by weight of cementing materials in case of slag concrete. For OPC concrete 0.5%, 1% and 1.5 % SP is used for 0.51, 045 and 0.4 water cement ratio respectively. Details of mix proportioning are presented in Table 1.

Table 1 Mix proportion for concretes for different w/b ratio

TYPE OF EXPOSURE	MODERATE	SEVERE	EXTREME
0% Slag			
Water/binder ratio	0.51	0.45	0.40
Water content (l)	180	160	140
Superplasticiser (l)	1.5	3	4.5
Cement content (kg / m^3)	350	350	350
Slag content (kg / m^3)	0	0	0
Weight of C.A. (kg / m^3)	1177	1207	1239
Weight of F.A (kg / m^3)	663	679	696
30% Slag			
Water/binder ratio	0.51	0.45	0.40
Water content (l)	180	160	140
Superplasticiser (l)	4.5	4.5	4.5
Cement content (kg / m^3)	245	245	245
Slag content (kg / m^3)	105	105	105
Weight of C.A. (kg / m^3)	1165	1198	1232
Weight of F.A (kg / m^3)	656	674	693
60% Slag			
Water/binder ratio	0.51	0.45	0.40
Water content (l)	180	160	140
Superplasticiser (l)	4.5	4.5	4.5
Cement content (kg / m^3)	140	140	140
Slag content (kg / m^3)	210	210	210
Weight of C.A. (kg / m^3)	1159	1192	1225
Weight of F.A (kg / m^3)	652	671	689
90% Slag			
Water/binder ratio	0.51	0.45	0.40
Water content (l)	180	160	140
Superplasticiser (l)	4.5	4.5	4.5
Cement content (kg / m^3)	35	35	35
Slag content (kg / m^3)	315	315	315
Weight of C.A. (kg / m^3)	1170	1187	1220
Weight of F.A (kg / m^3)	658	668	687

Details of Mix Design

The mix design of concrete with admixtures is designed as per draft code.Step by step procedure for mix design as given in IS: 10262-draft code is as follows. Water cement ratio curves of IS: 10262-1982 are removed in the draft code. Provisions are made for the use of different admixtures in the draft code. Absolute volume method followed in other methods of mix design is used to calculate the quantities of various ingredients.

Step 1 - Calculation of the target mean strength

The target mean compressive strength should be calculated as

$$f_{target} = f_{ck} + 1.65\ \sigma$$

where σ is the standard deviation. This can be established by testing minimum of 30 concrete cubes or the values given in IS: 456-2000 can be used. These values are reproduced here.

Table 2   Assumed standard deviation as per IS: 456-2000

GRADE OF CONCRETE (f_{ck})	ASSUMED STANDARD DEVIATION (σ), N/mm^2
M10 and M15	3.5
M20 and M25	4.0
M30, M35, M40, M45 and M50	5.0

Note: The Above values correspond to the site control having proper storage of cement, weigh batching of all materials, controlled addition of water; Regular checking of all materials, aggregate grading and moisture content; and periodical checking of workability and strength. When there is deviation from the above, the values given in Table 1 shall be increased by 1 N/mm^2

Step 2 - Selection of water-cementitious ratio or fluid binder ratio

Approximate water content in kg/m^3 for nominal maximum size of aggregate has to be selected based on the following table. In case of high strength concrete this water content should be further reduced and the workability should be adjusted by using plasticisers or superplasticisers in suitable dosage. However w/c ratio should be equal to or less than the maximum specified value based on durability requirements as per IS: 456-2000 (Table 3). W/c ratio can also be fixed based on the practical experience for the required target compressive strength without violating the durability requirements. If possible it is advisable to establish to w/c relation for the given set of ingredients.

Table 3   Maximum water content as per IS: 456-2000

MAXIMUM SIZE OF AGGREGATE, mm	MAXIMUM WATER CONTENT, kg/m^3
10	208
20	186
40	165

Step 3 - Determination of cementitious content

Based on the cater content and w/c ratio fixed, determine the cementitious contents. The maximum cementitious content should not be more then 450 kg/m^3. In addition to cement, various mineral admixtures can be used.

Step 4 - Estimation of coarse aggregate content

Approximate volume of oven dry coarse aggregate should be fixed based on the maximum size of aggregate and the Zone of sand as per the following table

Table 4   Estimation of coarse aggregate content

NOMINAL MAXIMUM SIZE OF AGGREGATE, mm	VOLUME OF OVEN DRY COARSE AGGREGATE PER UNIT VOLUME OF CONCRETE FOR DIFFERENT ZONES OF FINE AGGREGATE			
	Zone VI	Zone III	Zone II	Zone I
10	0.50	0.48	0.46	0.44
20	0.66	0.64	0.62	0.60
40	0.75	0.73	0.71	0.69

Step 5 - Determination of fine aggregate content

The fine aggregate content can be finally obtained by either of the two procedures namely the mass method or absolute volume method as explained in the code. Necessary corrections should be made for change in aggregate type and field conditions. Final mix should be selected after designing and testing trial mixes. These steps are followed for the design of mixes in the present work and is presented below.

Illustrative example of concrete mix proportion

Case 1 For 100% cement and moderate exposure
- Type of mix: design mix
- Type of cement: OPC, 43 grade
- Maximum size of aggregates: 20 mm
- Cement content: 350 kg/m^3
- Water content: 180 l
- Water/binder ratio: 0.51
- Admixture type: Superplasticiser –Conplast 430.

Mix Calculations

- Sand = zone III
- Volume of coarse aggregates = 0.64
  (Volume of dry rodded coarse aggregates per unit volume of concrete)
- Volume of fine aggregates = 0.36
- Volume of concrete = 1 m^3
- Volume of cement = 350/ (3.1* 1000) = 0.112 m^3
- Volume of water = 180/ (1*1000) = 0.18 m^3

- Volume of admixture (0.5%) = $1.75/(1.2* 1000) = 0.0015 \text{ m}^3$
- Volume of all in aggregates = $1- (0.112 + 0.18 + 0.0015) = 0.707 \text{ m}^3$
- Weight of all in aggregates = $0.707 * 2.6 * 1000 = 1839 \text{ kg}$
- Weight of coarse aggregates = $1839 * 0.64 = 1178 \text{ kg}$
- Weight of fine aggregates = $1839 * 0.36 = 663 \text{ kg}$

Mix proportion 1: 1.89: 3.36

Case 2   For 30% replacement of cement by GGBS and moderate exposure

- Type of mix: design mix
- Type of cement: OPC, 43 grade
- Maximum size of aggregates: 20 mm
- Cement content: 350 kg/m^3
- Water content: 180 l
- Water/binder ratio: 0.51
- Admixture type: superplasticiser –Conplast 430.

Mix Calculations

- Sand = zone III
- Volume of coarse aggregates = 0.64
- Volume of fine aggregates = 0.36
- Volume of concrete = 1 m^3
- Volume of cement = $(0.7* 350) / (3.1* 1000) = 0.079 \text{ m}^3$
- Volume of GGBS = $(0.3* 350) / (2.82* 1000) = 0.037 \text{ m}^3$
- Volume of admixture (1.5%) = $5.25 /(1.2 * 1000) = 0.0044 \text{ m}^3$
- Volume of water = $180 / (1 * 1000) = 0.18 \text{ m}^3$
- Volume of all in aggregates = $1 - (0.079 + 0.037 + 0.0044 + 0.18) = 0.7 \text{ m}^3$
- Weight of all in aggregates = $0.7 * 2.6 * 1000 = 1820 \text{ kg}$
- Weight of coarse aggregates = $1820 * 0.64 = 1165 \text{ kg}$
- Weight of fine aggregates = $1820 * 0.36 = 656 \text{ kg}$

Mix proportion 1: 1.87 : 3.33

Similarly other mixes are designed and the details are presented in Table 1.

TEST DETAILS

The workability of fresh concrete is determined by the conventional slump and compacting factor tests. Compressive strength of concrete is based on the 150 mm cubes[15]. Mixing, casting and curing are done as per relevant IS standards. For slag concrete, cement and slag are mixed in dry condition before mixing with other ingredients. After mixing, a needle vibrator is used to ensure good compaction.

The surface of concrete is then smoothed and wet gunny cloth is used to cover the concrete until the specimens are de-moulded one day after casting. The specimens are air dried for 3 hours prior to testing for every mix at the required age and the average strength of three specimens was used as an index.

TEST RESULTS AND DISCUSSIONS

Workability

The control mixture had a workability of 10 mm slump and 0.87 compacting factor even with superplasticiser. All the other mixtures are found to be cohesive and dense with workability more or less same as that of control concrete. The water binder ratio of concrete ranged from 0.4 to 0.51. Though the workability is less, concrete was quite workable under vibration. In case higher workability is required, superplasticiser type and dosage can be varied.

Compressive strength

The variation of compressive strength of concrete for different percentages of slag, for different water/binder ratio and for different ages are presented in Figures 1-3, 4-6 and 7-9. The results are presented in Table 5.

Table 5 Variation of Compressive Strength at Various Ages for Different water/binder Ratio and Percentages of GGBS

% OF GGBS	WATER/BINDER RATIO	COMPRESSIVE STRENGTH (MPa)*		
		7 DAYS	28 DAYS	2 MONTHS
0	0.40	47.03	56.80	66.97
	0.45	41.11	48.83	61.00
	0.51	37.20	47.01	58.01
30	0.40	43.14	52.03	62.02
	0.45	37.10	45.10	56.60
	0.51	33.32	43.15	53.19
60	0.40	37.29	48.03	57.70
	0.45	32.03	42.20	52.10
	0.51	28.50	40.21	49.03
90	0.40	20.79	38.35	39.01
	0.45	16.94	32.70	34.30
	0.51	15.01	31.01	31.70

* Average of 3 samples.

The results show that all the mixtures were able to develop substantial compressive strength at all replacement levels. Strength of slag concrete is less compared to OPC concrete. Concrete with lower percentage of slag developed greater strength than concrete with higher percentage of slag replacement. This is true as per the literature [1, 2].

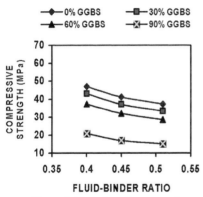

Figure 1 Variation of compressive
strength with fluid-binder ratio at 7 days

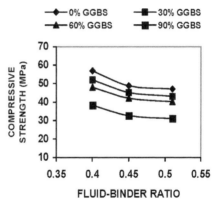

Figure 2 Variation of compressive
strength with fluid-binder ratio at 28 days

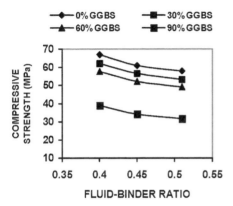

Figure 3 Variation of compressive
strength with fluid-binder ratio at 60 days

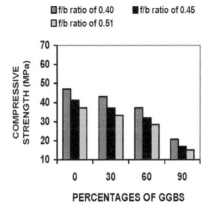

Figure 4 Variation of compressive strength
with percentages of GGBS at 7 days

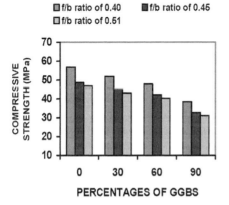

Figure 5 Variation of compressive strength
with percentages of GGBS at 28 days

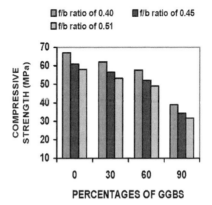

Figure 6 Variation of compressive strength
with percentages of GGBS at 60 days

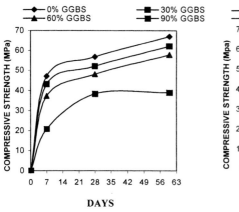

Figure 7 Variation of compressive
strength with days at 0.40 f/b ratio

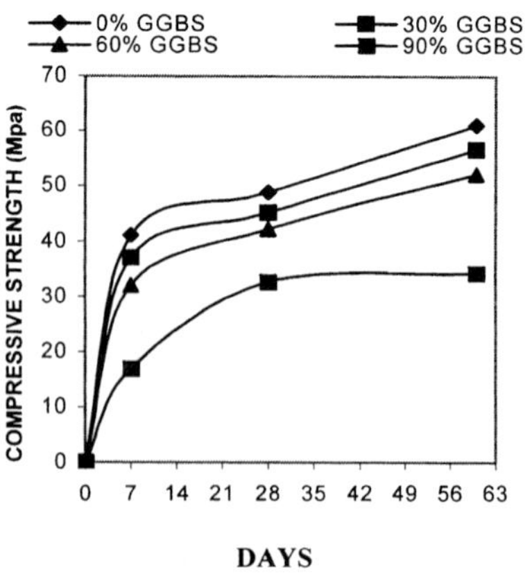

Figure 8 Variation of compressive
strength with days at 0.45 f/b ratio

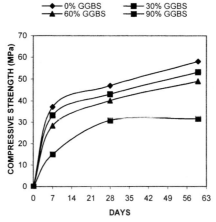

Figure 9 Variation of compressive
strength with days at 0.51 f/b ratio

Control concrete at w/b ratio of 0.4 has produced compressive strength of 57 MPa at 28 days.
For 30%, 60%, and 90% of slag addition, the decrease in strength is about 8%, 15% and 35%
respectively based on the 28 day strength. This is more or less same for all w/b ratios. The
corresponding decrease based on the 7 day strength of control concrete is 10%, 25% and 60%
respectively. It means the rate of hydration of slag concrete is rather slow at early ages (7
days) and for higher percentage of slag the hydration process further decreases. The rate of
hydration of slag increases beyond 7 days due to delayed hydration of slag. However the rate
of increase is not sufficient to attain the strength of control concrete. These differences will
vanish with time (6 months to 1 year) as reported in the literature. At lower percentage of
slag, concrete attains strength very close to that of control concrete, but for higher percentage
of slag, the strength is significantly lower. The same order of difference is noticed even at 60
days and the strength of slag concrete is always less compared to that of control concrete.

What is important from the point of mix design is that the strength of control concrete is equivalent to about M40 grade as its characteristic strength is about 50 MPa. With the addition of slag at 30, 60 and 90 percent as replacement to cement, the low concrete strength attained is equivalent to about M35, M30 and M25 grade concrete. Thus with the addition of slag at the above percentages, the concrete mix will result in lower strength which can be used as equivalent to the next lower grade concrete. Another point to be noted here is that by using 350kg (per cubic meter of concrete) of cementing material with varying slag content, concrete having wider range of compressive strength can be designed satisfying the durability requirements of IS: 456-2000. However, a compatible superplasticiser in required dosage has to be used to get the desired workability.

It is also possible to modify the proportion (i.e. re-proportioning) of slag concrete which will result in the same strength as that of control concrete by suitable methods suggested by the earlier researchers [16,17]. This is achieved by increasing the total cementitious contents or cement content of the original mix with an appropriate reduction in w/b ratio. Other mixture adjustments, if required can also be made to achieve a cohesive, dense and homogeneous mixture [18,19]. However this is not tried in the present investigation.

CONCLUSIONS

Mix design proportioning method used here is as per the guidelines of draft IS: 10262 code, specially applied for slag concrete. Earlier IS: 10262 is salient about the use of mineral admixtures. The method is for OPC concrete, which is based on the use of w/c curves developed long back, which results in very high cement content. Draft code insists on the development of w/c curves for a given set of ingredients. Thus the properties of the ingredients are reflected in these curves, which form the reference for the subsequent mix designs. In addition the guidelines given in the code can be used for the production of any type of concrete using various mineral admixtures and superplasticisers. Since superplasticiser is used the water content can be reduced suitably to produce concretes of higher strengths as w/b ratio reduces significantly.

The method presented here is only to demonstrate the mix design method for slag concrete for low, medium and high percentage of replacements. Though the strength of concrete decreases for higher percentages, the concrete still possess significantly higher strength at par with M20 or M25 concrete. It is also possible to design mixtures that have the same or similar strength development as OPC concrete, particularly at lower ages, without using excessive amounts of cement at all replacement levels. However this methodology is not presented in the paper and further refinements to the mix proportioning methodology is in progress as a part of the research activities of the authors.

REFERENCES

1. SWAMY.R.N, AMMAR BOUIKINI., "Some engineering properties of slag as influenced by mix proportioning and curing", ACI Materials Journal, 87, 1990.

2. GENGYING LI and XIAHOUA ZHAO., "Properties of concrete incorporating fly ash and ground granulated blast furnace slag", Cement & Concrete Composites, 25, 2003, 293-299.

3. MALHOTRA V.M, ZING M.H, READ P.H, RYELL J., "Long term mechanical properties an durability characteristics of high strength/high performance concrete incorporating supplementary cementing materials under outdoor exposure conditions", ACI Materials Journal, 2000, 97, 518-25

4. IS: 8112-1989: Specifications for 43-grade ordinary Portland cement, Bureau of Indian Standards, New Delhi.

5. Product brochure-Andra cements, Bangalore, India

6. IS: 383-1970: Specifications for coarse and fine aggregates from natural sources for concrete, Bureau of Indian Standards, New Delhi, 1993

7. SWAMY R.N, ALI A.R.S, THEODORAKOPOULOS D.D., "Early strength fly ash concrete for structural applications", ACI Journal, 1983, 414-23.

8. ALASALI M.M, MALHOTRA V.M., "Role of structural concrete incorporating high volume of fly ash in controlling expansion due to alkali aggregate reaction", ACI Materials Journal, 1991, 88, 159-63.

9. IS: 456-2000, "Plain and Reinforced Concrete – Code of practice", (Fourth Edition), BIS, New Delhi.

10. IS: 10262-1982, "Recommended Guidelines for Concrete Mix Design", BIS, New Delhi.

11. IS: 10262, "Draft code on private circulation", BIS, New Delhi.

12. NATARAJA M.C., NIRJHAR DHANG and GUPTA A.P., "A Simple Equation for Concrete Mix Design Curves of IS: 10262- 1982, The Indian Concrete Journal, 1999, Vol. 73, No 2, pp111-115.

13. NATARAJA M.C., NIRJHAR DHANG and A.P.GUPTA A.P., "Computer Aided Concrete Mix Proportioning", The Indian Concrete Journal, Vol. 71, No 9, pp487-492.

14. NATARAJA M.C., NIRJHAR DHANG and GUPTA A.P., "Computerised Concrete Mixture Proportioning Based on BIS method- A Critical Review", Fifth International Conference on Concrete Technology for Developing Countries, Proceedings volume 2, conducted by National Council for Cement and Building Materials on 17 to 19 November 1999, New Delhi, India, V1-96 – V1-105.

15. IS: 516-1959: Methods of tests for strength of concrete (Eleventh reprint), Bureau of Indian Standards, New Delhi, (April.1985).

16. NAGARAJ.T.S and ZAHIDA BANU, A. F., (1999). Relative efficacies of different concrete mix proportioning methods, Journal of Structural Engineering, SERC Chennai, India, Vol.26, No.2, pp 107-112.

17. NAGARAJ.T.S and ZAHIDA BANU, A.F., (1996). Generalization of Abram's law. Cement and Concrete Research J, 26(6): pp 933- 942.

18. UMESH CHANDRA, "Water-cement ratio curves and mix design for metakaoline concrete", M. Tech. thesis submitted to VTU, Belgaum, S.J. College of Engineering, Mysore, 2003.

19. M.C. NATARAJA, B. M. RAMALINGA REDDY and D. UMESH CHANDRA, "Mix design for concrete containing metakaoline", Submitted to Indian Concrete Institute Journal for publication.

TECHNOLOGIES FOR THE USE OF HAZARDOUS WASTES FLY ASH AND CONDENSED SILICA FUMES FOR ECO-FRIENDLY BUILDING MATERIALS

R Chowdhury

Independent Consultant

India

ABSTRACT. Huge quantities of fly ash and condensed silica fumes are generated by the Indian steel Industries as wastes and which are environmentally detrimental. They can be profitably used for cement and concrete making. Fly ash generated from the power plant of the steel plant creates disposal problems. They can be used in Fly ash pozzolana cement. The inclusion of fly ash in concrete affects all aspects of concrete properties. As a part of the composite that forms the concrete mass fly ash influences the rheological properties of fresh concrete, strength, finish, porosity, durability and the cost and energy consumed in manufacture. Condensed silica fume is a by-product of the smelting process of alloys. Other forms of condensed silica fumes are micro silica, Ferro silicon dust, arc furnace silica, blue dust, amorphous silica and volatilized silicon. It imparts more cohesiveness than normal concrete. Several Investigators have explored the possibility of using condensed silica Fume in combination with fly ash. This purpose has generally been to use the highly reactive condensed Silica Fume to compensate for slow strength development in association with fly ash in concrete. By the addition of condensed silica fume, durability is remarkably improved. The coal ash is used for developing self-compacting concrete and high volume ash concrete.

Keywords: Waste, Fly ash, Silica fume,

Dr Ranajit Chowdhury did M. Tech in Chemical Engineering from I. I. T. Kharagpur, PhD (Tech) in Chemical Engineering from Bombay University. He is a life fellow of Institution of Engineers, life fellow of Institute of Chemical society, life fellow of public health Engineering. He took six month training in coke oven U.S.S.R., (1982). He was Ex-senior Manager of Steel Authority of India Ltd.

INTRODUCTION

Huge amounts of fly ash are generated in power plants of facilities such as integrated steel plants. The fly ash in dust form or slurry form is environmentally detrimental; disposal in profitable manner is necessary. Technologies developed for the utilization of fly ash in several industries including lightweight aggregate, cellular concrete etc.

FLY ASH UTILISATION

The inclusion of fly ash in concrete affects all aspects of concrete properties. As a part of the composite that forms the concrete mass, fly ash acts in part as fine aggregate and in part as a cementitious component. It influences the rheological properties of fresh concrete, the strength, finish, porosity, durability, cost, and the energy consumed in manufacturing the final product.

Proportioning Concrete Containing Fly Ash

In most applications, the objective for using fly ash in concrete is to achieve one or more of the following benefits:

(i)   reducing cement content to reduce cost;
(ii)   obtaining reduced heat of hydration;
(iii) improving workability;
(iv)   attaining the required level of strength in concrete at ages beyond 90 days

In practice, fly ash can be introduced into concrete in one of the two ways:

(a)   A blended cement containing fly ash may be used in place of port land cement.
(b)   Fly ash may be introduced as an additional component at the concrete mixing plant.

Two common assumptions are made in selecting an approach to mix proportioning concrete:

(i)   Fly ash usually reduces the strength of concrete at early ages.
(ii)   For equal workability, concrete incorporating fly ash usually requires less water than concrete containing only Portland cement.

AARDELITE PROCESS OF MAKING PELLETS OF LIGHT WEIGHT AGGREGATE

The AARDELITE process is an energy efficient cold bonded process which utilizes the pozzolanic properties of fly ash in combination with the binder. The green pellets produced after pelletisation of the AARDELITE mix are steam cured and hardened. They are screened, separated and sorted out by size ranging from 2 mm to 32 mm. The Tata electric company is setting up an AARDELITE plant at the Trombay power station. The plant will utilize 18,000 tons of fly ash per annum.

The aggregate produced will be light in weight and will substitute natural aggregate in a variety of applications including sub-base for concrete roads and concrete pavements, prefabricated concrete elements, such as floor slab, concrete, pillars, shore protection building blocks, for protection walls and other non load bearing civil engineering structures. The artificial aggregate will be lighter than natural aggregate.

The possible presence of heavy metals hindered the application of coal residues. The AARDELITE process encapsulates these components in the cementitious matrix in such a way that leaching into the environment is drastically restricted.

Pellet Making

For making pellets a disc pelletiser was used. The pelletiser was fixed at an inclination of 45° and rotated at a speed of 10 R.P.M. However, the speed was increased or decreased as required. Initially 100 ml of tap water used to mix the mixture of composite fly ash, sand, lime or cement and activator. The pelletiser was run for about 10 minutes with this feed charge during which time nuclei started forming. Remaining water was sprayed slowly on the charge in running pelletiser. This helped in the growth of pellets and ultimately gave 11-22 mm size Cement – Fly ash pellets (B S L Fly ash)

In the process of making of cement fly ash pellets only two variations in ash content were made: i.e. 5 % and 10 %. Both 5 % and 10 % pellets were measured using the green strength test and the dry strength test. The pellets were watered twice a day up to 28 days. After 3, 7 and finally 28 days, their strengths were measured. The data of the 5 % and 10 % fly ash pellets is given on Table 1 and is shown in Figure 1.

Table 1 Green and dry strength test

TEST DATE	5 % CEMENT		10 % CEMENT	
	A. V. strength kg/pellet	A. V. diameter in mm	A. V. strength kg/pellet	A. V. diameter in mm
3 days	5.19	22.22	4.49	15.50
7 days	8.31	22.13	6.42	15.49
28 days	16.73	22.13	17.39	17.08

Lime – Fly ash pellet

Lime – fly ash pellets have been made by varying parameters such as drying time, autoclaving lime, percentage of lime, percentage of sand and tem/pressure. In this process, hydrated lime available on the local market was used. The diameter of pellets was also measured in mm with the help of a screw gauge.

In this method, pellets were autoclaved after air drying in a 5 litre capacity autoclave.

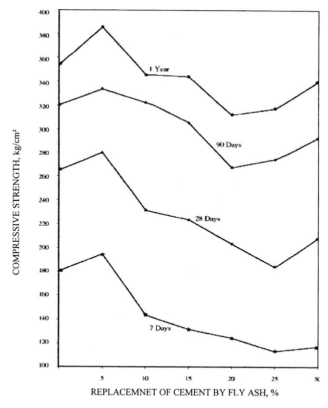

Figure 1 Relationship between compressive strength and replacement of cement by fly ash

Table 2 Optimisation data for lime fly ash pellets

PARAMETERS	VARIATION	OPTIMUM
1) Air drying time (h)	18 – 96	24
2) Lime (%)	5 – 13	11
3) Sand % (< 1 mm)	8 –16	14
4) Autoclaving time (h)	3 – 8	7
5) Autoclaving temperature (°C)	150 – 300	200

The activator used in a limited number of experiments indicated that it helped in improving the crushing strength of the pellets. The crushing strength increased from 4 kg / pellet to 22 kg / pellet with increase of diameter 14 mm to 16.5 mm diameter pellets.

Fly ash cellular (Gas) concrete

Cellular blocks posses an aerated cementitious matrix made by introducing gas/air through suitable foaming agents and admixtures. The resulting slurry is poured into mould and allowed to set. The blocks are removed after initial set and steam cured. The compressive strength depends upon the cement content and fineness of the fly ash. The average compressive strength is 40-85 kg/cm^2. The average bulk density is 600 kg/m^3. It is used in Poland, U.S.S.R, China etc.

Utilisation of fly ash in the production of cellular concrete instead of sand solves the problem of ash disposal. Fly ash cellular concrete technology is suitable for the utilization of dry fly ash transported directly from the power plant, as well as for wet fly ash delivered from existing lagoons.

Advantages

i)   Reduced mortar consumption
ii) Reduced construction time
iii) Reduced demand for steel reinforcement
iv) Low building mass
v)   Very good fire resistance and sound absorption.

Fabrication method

i)   Raw material (fly ash ash + quick time + gypsum + aluminium powder + cement – only for large size panel)
ii) Raw materials milled to obtain binder
iii) Mixing of ingredients
iv) Filling of moulds and transfer of cellular concrete to curing chamber
v)   Demoulding
vi) Cutting of concrete
vii) Curing of cellular concrete in autoclaves, using 12 atm saturated steam for 12 h

Light weight concrete foamed polystyrene

- Polystyrene fly ash concrete

This building material is produced using Portland cement (approx. 50 %), foamed polystyrene and fly ash (approx. 50 %).

Fly Ash use by the cement Industry

Fly ash can be used for a number of purposes by the cement Industry

i)   To replace the clay fraction of the raw materials of cement manufacture
ii) To improve the particle size distribution of cement to obtain a dense product
iii) As an addition in or after the clinker mills of the cement plant

The most important quality of parameter (i.e. physical + chemical properties) that determines the possible destination of the fly ash is:

(i)   Residual care on cement
(ii) Granulation and fineness
(iii) Chemical and alkaline composition

Latest development

Pilot plant investigations are under way to produce fly ash bricks consisting of 99% fly ash and 1% binder composition.

Portland and pozzolana cement (PPC)

PPC is made by blending 15 – 20 % fly ash with Portland cement or grinding the ash with cement clinker. The PPC shows good resistance to sulphate attack, leads to energy saving, reduced permeability of concrete and ultimate strength approximately equal to that of Portland cement, which is achieved by (i) proportioning of the fly ash concrete mix (ii) and Batching and mixing of different ingredients. The fly ash concrete is so proportioned as to attain its 28 days compressive strength equal to that of a corresponding plain concrete. The batching and mixing of different ingredients is generally done at a central batching and mixing plant.

Precast Fly ash concrete building units

Fly ash concrete has been used in the production of various types of precast building units such as concrete building blocks (solid or hollow), columns, beams, hollow core slabs, door and window frames, etc.

Sintered Fly ash Lightweight Aggregate

Conversion of fly ash into lightweight aggregate is one of the potential ways to bulk disposal of fly ash in an economic manner. Sintered fly ash light weight aggregate (SELA) with a bulk density of 640 – 750 kg/m^3 has been produced at a pilot plant at CBRI.

USE OF CONDENSED SILICA FUMES

Condensed silica fume (CSF) is a by-product of smelting process of alloys. Other forms of condensed silica fumes are microsilica, ferro silicon dust, arc turnace silica, silica blue dust, amorphous silica and volatilised silicon.

Condensed silica fume has the following characteristics:

(i)   The SiO$_2$ content is between 85 and 98 %
(ii)   The mean particle size is in the range 0.1 to 0.2 cm
(iii)   The particles are spherical in shape with a number of primary agglomerates.

CSF for use in concrete is either densified or in slurry form mixed with 50 % water by weight. In Norway cement contains 7.5 % condensed silica fume (CSF).

CSF

CSF is used in two different ways:

(i)   As a cement replacement: To obtain reduction in cement content usually for economic reasons.
(ii)   To enhance concrete properties – both in the finished and hardened state.

For low grade structural concrete the required strength can be obtained with an extremely low cement content where CSF is used. Here the CSF content is 10% or less, unless otherwise specified. High condensed silica fumes with superplasticizers and extremely low ratios of water to cement content (W/C) are the basis for a new type of concrete.

The w/c is calculated on the basis of cement content only. The water to cement + silica fume (w/c + s) is used for CSF mixes where 'S' stands for the amount of CSF. The CSF content given as a percentage of the weight of cement.

POZZOLANIC AND FILLER EFFECTS

CSF acts as both a reactive pozzolan and a very effective filler.

Reactivity and reaction product

The pozzolanic reactivity of CSF in cement paste has been demonstrated by the amount of $Ca(OH)_2$ at different times in the paste with varying content of CSF. Thermal gravimetry DTG and X-ray diffraction methods were used. The results generally show high pozzolanic reactivity.

Hardened concrete – strength development

The main contribution of CSF to concrete strength development at 20°C takes place from about 3 to 28 days. For condensed silica fume and control concrete of equal 28 days strength, the strength of condensed silica fume control will be lower over the entire time period with 20°C curing.

Fly ash – CSF combination

Several investigators have explored the possibility of using CSF in combination with fly ash. The approach has generally been to use the highly reactive CSF to compensate for the slow strength development associated with fly ash in concrete.

Abrasion – Erosion – Resistance

Mortars having very low water to cement ratios (<0.25) and high CSF levels (>20%) are known to be highly resistant to abrasion and wear. This property is of use in a variety of applications.

Frost resistance

The need for more durable concrete with improved resistance to freeze – thaw exposure in the presence of salts has been the motivation for a number of investigations involving CSF concrete. These investigations include studies of air – void system characteristics, basic studies of ice formation and pore structure as well as freeze – thaw testing.

Sulphate Resistance

Based on measurements carried out on concrete stored in 10% Na_2SO_4 solution, Bernhardt concluded that sulphate resistance was improved when 10% to 15% cement was replaced by CSF.

Alkali – aggregate reaction

It is well known that reactive pozzolans can be used to control the expansion associated with alkali – aggregate reaction. Pore water analysis of CSF – Cement paste demonstrated the utility of the CSF to reduce alkali concentrations in the pore water quite rapidly, thus making it unavailable for reaction with reactive silica in the aggregates.

Other chemicals

Improved resistance of concrete to a large number of chemically aggressive agents including NO_3^-, Cl^-, $SO_4^=$ and acids has been reported by using a proprietary product which contains 80 % CSF.

CONCLUSIONS

Fly ash and condensed silica fume which present environmental problems can very well be used as Portland cement substitutes for making concrete of improved quality, manufacture of bricks, light weight aggregates and so forth with lesser price and simultaneously reduce the environmental pollution.

The present evidence suggests that the use of CSF as an addition to improve concrete durability will also improve its ability to protect embedded steel from corrosion.

REFERENCES

1. GUPTA, S.C., MEHROTRA, V.K., JAIN, S.C., Use of fly ash to produce Strong and Durable Concrete – National Workshop on utilization of Fly ash, May 14-20, 1988 Roorkee.

2. DASS, K., RAJ, T., Corrosion Protection Aspects of Fly ash cement concrete of Fly ash, May 89-90, 1988 Roorkee.

3. MULLIK, A.K., Quality considerations for use of Fly ash in cement and concrete.

4. RISHI, S.S., GARG, S.K., Long Term Study on Stability of Hugh Magnesia Cement containing Fly ash.

5. RISHI, S.S., KISHANLAL, GARG, S.K., Cement from non-conventional material.

6. RISHI, S.S., GARG S.K., Heat resistance of Portland Fly ash cement.

7. TANEJA, C.A., TEHRI, S.P., A masonary cement based on slag and fly ash.

8. GARG, S.K., CHAKRAVORTY, S.C., Use of fly ash in present concrete.

9. RISHI, S.S., GARG S.K., Production of cement clinker using fly ash.

10. RISHI, S.S., Ready mixed Fly ash concrete.

11. JINDAL, B.K., SHISE, N.N., LAL, K., SHARMA, K.N., Use of Light weight concrete in Precast construction.

12. RISHI, S.S., GARG S.K., Factor affecting strength of structural Fly ash concrete.

13. DUTTA, R.K., RISHI, S.S., GARG, S.K., Studies on the incidence of corrosion of steel Reinforcement in concrete Casting Fly ash.

MECHANICAL PROPERTIES AND REACTION PRODUCTS IN THE TERNARY HYDRAULIC SYSTEM ANHYDRITE - BLAST FURNACE SLAG – METAKAOLIN

P E Fraire-Luna

J I Escalante-Garcia

A Gorokhovsky

Mexico

ABSTRACT. The report presents results of an investigation on composite cements made of industrial wastes such as Fluorgypsum (50-75%) and blast furnace slag (15-25%). Metakaolin (0-10%) was used to replace the slag as a secondary cementitious material. Pastes were prepared and cured under water after setting. The results indicated that the Fluorgpypsum set and in the presence of the slag the pastes developed strength after curing under water, the addition of metakaolin enhanced the strength development. Compressive strength at 28 days for blends of fluorgypsum-slag was of 8-9 MPa; the trends indicated that higher proportions of fluorgypsum showed better strength, and after 120 days the system 75%fluorgypsum-25%slag reached 10.5 MPa. The incorporation of 10% metakaolin to the latter, resulted in enhanced strengths with values up to 14.7 MPa after 120 days. Results of DTA, XRD and SEM indicated the presence gypsum, C-S-H and ettringite as the main hydration products.

Keywords: Gypsum, Fluorgypsum, Blastfurnace slag, Metakaolin

P E Fraire-Luna, BSc. Chemical Engineering, MSc. Ceramic Engineering, Currently PhD. Student

J I Escalante-Garcia, BSc. Chemical Engineering, MSc. Metallurgical Engineering, PhD. Materials Engineering

A Gorokhovsky, Ph.D., D.Sc., BSc. Physics, PhD. Physical Chemistry, DSc. Glass

INTRODUCTION

The worldwide interest in reducing the amounts of industrial wastes represents an interesting opportunity for the construction industry to make use of raw materials of low price to generate products of low environmental impact. In Mexico there exist large amounts of industrial by products, many with potential as cementitious materials. Such is the case of Fluorgypsum, a by product from hydrofluoric acid production, just one of the plants generates about 100,000 tons yearly [1], and about 700,000 tons arise yearly in México.

There are a number of studies in the literature oriented towards the generation of cementitious materials out of flyorgypsum waste. Investigations on blends of the latter with Portland cement and pozzolans such as fly ash [2,3], reported compressive strengths of about 50MPa after 28 days; the hydration products were gypsum, C-S-H and ettringite after characterisation by X ray diffraction and thermal analysis. Other papers [4] investigated blends of fluorgypsum (50-70) and blastfurnace slag, they obtained compressive strengths up to 35MPa, after 28 days, for systems 50%-50%, similar hydration products were reported as mentioned previously.

On the other hand, the use of clay minerals such as montmorillonite, muscovite, talc, kaolinite, metakaolinite, etc, has been reported with favourable effects when used as admixtures in gypsum systems [5]. The use of metakaolin is reported to increase the compressive strength of gypsum by up to 13 and 15 MPa after 180 days of curing under water and under 90%RH, respectively.

From the above, this work considered a systematic study on systems that combined fluorgypsum and blast furnace slag to obtain a hydraulic cementitious material; the addition of metakaolin was considered to improve the compressive strength.

EXPERIMENTAL DATA

Materials

Fluorgypsum (F) was obtained from the company "Industrial Química de México" (San Luis Potosi, SLP, Mexico), it was pulverized to a Blaine of $4000 cm^2/g$. It had small amounts (1%) of sulfuric acid, and hydrofluoric acid. Thus neutralization with $Ca(OH)_2$ was found to be necessary. The XRD pattern is shown in figure 1.

Blastfurnace slag (BFS) was obtained from the company "Altos Hornos de México" (AHMSA, Monclova, Coahuila, Mexico), the granulated slag was conditioned by means of a ball mill to a Blaine fineness of $4000 cm^2/g$. Its XRD pattern is also shown in Figure 1; the slag is predominantly amorphous with a small crystalline fraction of gehlenite ($Ca_2Al_2SiO_7$) and akermanite ($Ca_2MgSi_2O_7$).

Metakaolin (MK) was selected for its well known pozzolanic character and potential contribution to mechanical strength; it was kindly provided from the manufacturer Advanced Cement Technologies (Blaine WA, USA) as the commercial product PowerPoozz[TM]. It was reported to have a Blaine fineness of $9000 cm^2/g$ and its chemical analysis (as Table 1).

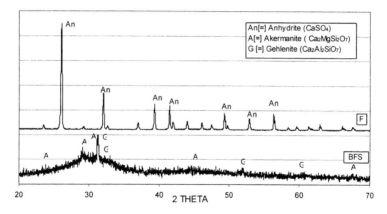

An[=] Anhydrite (CaSO₄)
A[=] Akermanite (Ca₂MgSi₂O₇)
G [=] Gehlenite (Ca₂Al₂SiO₇)

Figure 1   X-ray diffraction patterns of the fluorgypsum and blastfurnace slag

Table 1   Chemical composition of metakaolin PowerPoozz [TM] reported by the manufacturer

MATERIAL	Kaolin	Quartz	TiO₂	Fe₂O₃	K₂O	MgO	CaO	Na₂O	Loss on ignition
% WEIGHT	96-98	Max =2	1.90	1.40	0.18	0.15	0.11	0.05	13

Preparation of Samples

Cubic samples of 5 cm were prepared of composite pastes of F-BFS-MK, the proportions were those shown in table 2. All systems were mixed with $Ca(OH)_2$, Na_2SO_4 and $CaSO_4·\frac{1}{2}$ H_2O (commercial hemihydrate) as neutralizing and activating agents. The water/solids ratio was kept at 0.5 for all systems. Setting was carried out at 25°C for 24 hours. After demoulding the cubes were further cured at 20°C under water in an environmental chamber. The setting time was measured by the Vicat procedure ASTM C472-93 [6]. Cubic samples were used for compressive strength characterisation, and the crushed samples were dried for further characterisation. X-Ray Diffraction was performed with Cu-Kα radiation (X'Pert 3040, Phillips) with a step of 0.05° 2θ at 2 s/step; differential thermal and thermogravimetric analysis (Pyris Diamond TGA/DTA, Perkin Elmer) was carried out with a heating rate of 5°C/min. Fractured samples cured for 120 days were carbon coated for analysis under the Scanning Electronic Microscopy (SEM) (XL 30 ESM, Philips).

RESULTS AND DISCUSSION

Setting Times

The results of setting times are shown in table 2. It was noted that as the amount of F was increased, the initial and final setting times were reduced. The addition of MK in substitution of BFS resulted in a reduced initial and final setting time. Only the system with 65%F-25%BFS-10%MK showed increased initial and final setting time with respect to the MK free systems. We do not have an explanation for such a result

Table 2 Composition of pastes prepared (% weight) and setting time (min)

MIXES			SETTING TIME, (MIN)	
F	BFS	MK	Initial	Final
	50		270	400
50	45	5	240	360
	40	10	180	300
	35		200	290
65	30	5	180	270
	25	10	260	310
	25		180	300
75	20	5	170	270
	15	10	180	250

Compressive Strength

Compressive strength results are shown in Table 3. All systems presented strengths above those obtained with commercial gypsum. The latter presented initially 3.1 MPa after curing for one day and the strength reduced gradually by about 30% down to 2.2 MPa after 120 days, this is due to lixiviation in water. After one day of curing most of the composite systems showed strengths below those of commercial gypsum. Only the systems 75%F-20%MFS-5%MK and 75%F-15%MFS-10%MK presented higher strength than commercial gypsum from day one. However, at later ages all composite pastes outperformed commercial gypsum.

Data from Table 3 shows that the compressive strength increased with the F content at all curing ages, indicating that the initial strength is mainly due to the transformation of anhydrite to gypsum. Moreover, the strengths improved in the presence of MK. The systems of the highest strengths were those containing 75%F, after 120 days. They registered 10.6, 13.9 and 14.7, for 0, 5 and 10% MK respectively. The presence of 10%MK increased the strength by approximately 38%. The further increase in compressive strength in composite blends with respect to gypsum is due to the formation of additional C-S-H; all of the blended systems studied did not show lixiviation after curing under water and developed strength.

Table 3 Compressive strengths of the systems F-BFS-MK and commercial gypsum

COMPOSITION (%wt)			COMPRESSIVE STRENGTH (MPa)			
F	BFS	MK	1	7	28	120 Days
	50		1.2	4.9	7.6	8.8
50	45	5	1.7	4.9	7.7	9.8
	40	10	1.7	4.6	7.6	10.4
	35		2.0	4.4	8.3	9.7
65	30	5	2.9	5.8	8.8	10.6
	25	10	2.5	5.0	9.2	14.2
	25		2.6	4.4	9.4	10.6
75	20	5	3.8	6.5	9.3	13.9
	15	10	3.4	6.0	8.6	14.7
100	Commercial		3.1	2.7	2.4	2.2

XRD Results

The results of the systems with 75%F - 0%MK for curing times from 1 to 120 days are shown in Figure. 2. The presence of gypsum (d=8.530 Å), increased as a function of age, along with a reduction of the intensities of the main F peak (d=6.953 Å). The latter was noted even after 120 days, indicating that the conversion to gypsum was still in progress. The presence of ettringite (d=11.173 Å) was noted from 7 days of curing. Ettringite results from the participation of the slag in the reactions after at least 1 day.

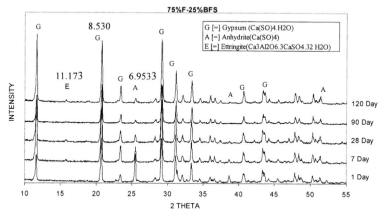

Figure 2   XRD pattern of the system 75%F-25%BFS

The systems with 5 & 10% MK displayed similar behaviour. Figure 3 shows the XRD pattern for 10%MK. The most abundant phase was gypsum (d=8.530Å); the intensity of the peaks increased as a function of curing time. The presence of F (d=6.953 Å) was also noted even after 120 days of curing; however it seems faster in the presence of MK. Ettringite peaks (d=11.173 Å) were observed weakly after 1 day and was clear after 7 days of curing. It appears that ettringite seems to be more abundant in relation to the system with 0% MK.

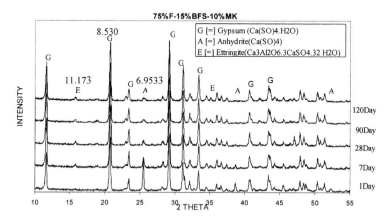

Figure 3   XRD pattern of the system 75%F-15%BFS-10%MK

DTA/TGA Results

The DTA analysis for the systems with 75% F, showed the presence of two endothermic peaks, one at approximately 95°C that is commonly attributed to ettringite and C-S-H [7,8] and one at ≈130°C, that can be attributed to gypsum [7,8,9]. The intensities of such peaks increased with age as expected from the progress of the hydration reactions, similar to the XRD results. The presence of ettringite and the C-S-H was more marked in the systems with MK,. this is shown in the figure. 4.

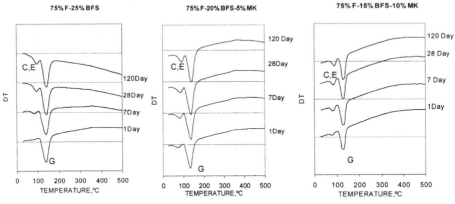

Figure 4   DTA of the various systems with 75% F

The results of TGA shown in figure 5 correspond well to those obtained by DTA. Two important steps of weight loss were noted, one at ≈95°C corresponding to the decomposition of C-S-H and ettringite, which increased as the curing time increased. The other important feature is the weight loss at 130 °C, that corresponds to gypsum. Such weight loss increased as the curing time increased.

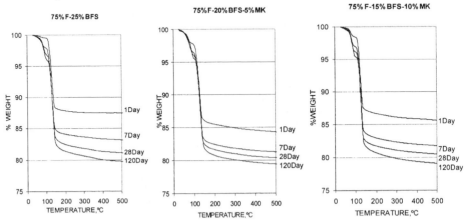

Figure 5   TGA of the systems with 75% F

C-S-H phase formation resulted from the reaction of the BFS. It was noted by thermal analysis but not by XRD. The poorly crystalline character of the C-S-H commonly results in broad XRD peaks of low intensities. The fact that gypsum was the main hydration product made it more difficult to identify C-S-H by XRD. However, thermal analysis showed the presence of C-S-H.

Scanning Electronic Microscopy SEM

Figure 6 shows micrographs of fracture surfaces from systems contains 75%F cured for 120 days. The system with 0%MK showed the presence of abundant grains of diverse morphology (rods, hexagonal crystals, irregular shapes, plates) which corresponded to gypsum. The presence of C-S-H was also observed distributed in the microstructure as a more compact phase. The C-S-H is considered to provide a waterproof character to the hardened paste and to promote strength development.

The systems 5%MK (B) and 10%MK (C) also showed the presence of abundant gypsum, and it seems that the quantity of C-S-H is increased, possibly from the reaction of the BFS and MK with the Ca(OH)$_2$ present. These micrographs show ettringite with the characteristic morphology of needles [10], in the form of localized clusters distributed throughout the microstructure. It is known that the formation of ettringite in early stages promotes the development of mechanical strength forming [11]. The ettringite needles content and length increase as a function of MK content. The presence of ettringite corresponds to the reactions of the Al$_2$O$_3$ supplied by the BFS and MK, which were the only aluminium sources.

(A) 75%F-25%BFS

(B) 75%F-20%BFS-5%MK

(C) 75%F-15%BFS-10%MK

Figure 6   Micrographs of the systems with 75% F cured at 20°C per 120 day

The systems displayed early strength from the conversion of anhydrite (fluorgypsum) to gypsum. Later ettringite & C-S-H form as reaction products of BFS and MK, contributing to enhance mechanical strength and water resistance. The formation of the hydration products increased as a function of time; compressive strength increases. The resulting material is composite of by-products, with good mechanical properties, good finish and water resistance.

CONCLUSIONS

In general, all the blended cements studied display good mechanical development and water resistance. However, the systems that displayed the best compressive strength were those that incorporated 75% F. The replacement of BFS by MK improved the strength and the highest strengths were those of the systems with 75% F-15%BFS-10%MK. The main hydration products were gypsum, C-S-H and ettringite. Fluorgypsum was noted even after 120 days of curing. It was noted that for higher MK contents, more C-S-H and ettringite were found as indicated by XRD, DTA and SEM. The amounts of ettringite and C-S-H increased as a function of curing time. These reaction products provide water resistance and mechanical strength development.

ACKNOWLEDGEMENTS

The authors are grateful to "Industrial Química de México", and "Advanced Cement Technologies, LLC" for supplying the fluorgypsum and MK, respectively. The grant from CONACyT and the financial support of CONACyT project 39902 is also appreciated.

REFERENCES

1. H SORIA-MAGAÑA, Personal communication, "Industrial Química de México".

2. P.YAN, Y.YOU, "Studies on the binder of fly ash-fluorgypsum-cement" Cement and Concrete Research Vol. 28,No 1 pp. 135-140, 1998

3. PEIYU Y.XIAN L., WENYAN Y., YI Y. "Investigation of the microstructure of fly ash-Fluorgypsum binder" Material Science concrete: Sidney Diamond Symposium, Aug 1998

4. M.SINGH, M.GARG, "Activation of gypsum anhydrite-slag mixtures", Cement and Concrete Research Vol. 25,No 2 pp. 332-338,1995

5. M.MURAT AND A. ATTARI "Modification of some physical properties of gypsum plaster by addition of clay minerals" Cement and concrete research, 21, pp.378-387,1991

6. ASTM C472-93, Annual ASTM standards, cement; lime; gypsum 04.01, 126-143, 1995

7. SINGH M., GARG M. "Phosphogypsum – fly ash cementitious binder-its hydration and strength development" Cement and concrete research, vol. 25 no.4, pp.752-758,1995

8. PEIYU Y.XIAN L., WENYAN Y., YI Y. "Investigation of microstructure of fly ash-Fluorgypsum binder" Material science of concrete: Sidney Diamond Symp., August 1998

9. SINGH M. "Treating waste phosphogypsum for cement and plaster manufacture" Cement and concrete research vol. 32, pp 1033-1038, 2002

10. R.TALERO. "Kinetochemical and morphological differentiation of ettringites by the Le Chatelier-Anstett test" Cement and concrete research, vol. 32, pp.707-717, 2002

11. P.BARNES, "Structure and performance of cement" pp.308-309,1983

CLOSING PAPER

MEDIA - A DECISION SUPPORT TOOL
FOR SUSTAINABLE URBAN DEVELOPMENT

C F Hendriks
G M T Janssen
A L A Fraaij
Delft University of Technology

The Netherlands

ABSTRACT. In this contribution attention is paid to three items: (i) Sustainable Decision Making;. This tool connects decisions concerning sustainability, content and links of decisions, options, variables, indicators and actors, to the moment in the process where the decision has to be taken. The effects of the decisions can be calculated by using several tools. The model has to be used for the whole life cycle and concerns all scale levels involved, from materials and constructions to the build environment. (ii)Sustainable urban planning; In this part the presentation of a new model inclosing important indicators are dealt with in relation with spatial scale levels. (iii) Quality assessment; using computer applications for realising sustainable use of materials in constructions.

Keywords: Sustainability, Decision making, Quality assurance, Sustainable materials and constructions, Modelling

Professor Dr Ir C F Hendriks, was, until he passed away on November 13th 2004, Professor in Materials Science and Sustainable Construction at Delft University of Technology, Faculty of Civil Engineering and Geosciences. His research interests included durability and sustainability of construction materials.

Ir G M T Janssen, is a commercial Engineer and Master of Total Quality Management. She specialises in waste management and recycling. She is Assistant Professor of the Delft University of Technology, Director of the consultancy Enviro Challenge and Director of The Dutch and Flemish associations of mobile recycling companies.

Dr Ir A L A Fraaij, is an Associated Professor at the Materials Science Group of the Faculty of Civil Engineering and Geosciences. His research interests include the use of fly ash and recycled aggregates in concrete. He is involved with the micro lab at the faculty where about 15 PHD students perform their research, often in combination with the development of mathematical tools.

INTRODUCTION

The days that designers could focus on finding technical solutions for technical design challenges are past. Policy and design complexity is further enhanced by concept of Sustainable Development (SD). Urban (re)development, and large building projects in general, are design challenges where this trend is clearly visible. Decision makers in such projects, with ambitions regarding sustainable development, cannot ignore this increasing complexity. In order to realize technically and functionally sound projects that are acceptable to the actors involved and also comply with the principles of sustainable development, it is necessary that policy-makers, designers and SD-experts work together. Co-operation between these actors can be enhanced by tools that make design processes and decision making processes more transparent and offer insight in the impacts of decision options [1]. So far, most tools focus on one of these three complexities described above, resulting in process-oriented tools, design tools, and analytical/technical tools, respectively. Although such tools generate useful information, integration is a challenge. The DuBes Sustainable Decision Making Project makes an attempt to provide an integrated approach for making sustainable decisions for an urban (re)development project. This project has so far resulted in two major products: the computer based conceptual modelling tool MEDIA (Modelling Environment for Design Impact Assessment) and a gaming exercise for sustainable decision making.

COMPLEXITY OF SUSTAINABLE URBAN (RE)DEVELOPMENT

An integrated approach to sustainable urban (re)development must do justice to these three types of complexity.

Content-oriented complexity, or the equivocal character of SD

Sustainable development and its derivative for the built environment, sustainable building, are concepts that are difficult to define. The most cited definition of sustainable development is the one found in the Brundtland report: 'a development that meets the needs of the present without compromising the ability of future generations to meet their own needs [2]. This definition was succeeded by principle 3 of the Rio Declaration on Environment and Development [3]: 'to equitably meet developmental and environmental needs of present and future generation'. Both definitions leave room for interpretation. When are present needs met? What are the needs of future generations? Are we compromising their abilities? These questions cannot be answered univocally and give SD its equivocal character. This equivocality manifests itself on philosophical, political and operational levels. All in all, it is fair to say that SD is a concept to be further defined and negotiated in a political context.

Policy complexity

Large design challenges typically involve a complex network of actors striving to realise ambitions and protect interests [4], determined not by objective rationality, but by perceptions and value systems, causing competing assumptions about problems and solutions, means and ends, cause and effect. From the actor network perspective, there is no single correct view to policy problems and their solutions. Decisions are made in a complex process of negotiation between actors. The interaction process is often sub-optimal in terms of both substantive quality and (public) support. The solution to this problem is sought in process management [5]. This however does not make urban redevelopment projects themselves simpler.

Design complexity

Large building projects typically require of hundreds of major design decisions and many thousands of more detailed choices to be made. Decisions are often interdependent, and decision options may influence the values of multiple variables relevant for SD. Reductions approaches to deal with this complexity tend towards mono-disciplinary models which, whilst scientifically valid, lack an overall picture of the problem in a societal context. More pragmatic and holistic approaches, on the other hand, are often not transparent and for that reason not acceptable for generic use. Others have described this as the dilemma of the choice between scientific disciplinary rigor and practical relevance. What needed is a balance between integration on one hand and scientific validity and precision on the other.

Together, the equivocality of SD, policy complexity and design complexity constitute a formidable barrier for an integrated and effective approach aimed at sustainable urban (re)development. The procedural rationality, typical of most model-based approaches, alone does not suffice for complex problems such as urban (re)development. An integrated approach must address each of the three complexities described earlier. MEDIA is proposed as a suitable underlying model for such an approach.

Conceptual foundations of media

From the analyses that have been briefly described in section 2, these requirements can be drawn up: the structure of the model should be based on the physical decision making process; the model should adequately represent the policy complexity that is typical of urban (re)development; relationships between decisions should be included in the model; content related knowledge and insights regarding the consequences of decisions, e.g. costs, environmental impacts, should be linked to the relevant decisions; the model should not be normative with regard to SD-issues; the model should help the users focus on the most relevant issues only; the model should be able to function in a dynamic environment, e.g. a gaming exercise.

Addressing design complexity

To represent decisions and their interdependencies, Analysis of Interconnected Decision Areas (AIDA) has been used as a basis [6]. AIDA is a technique that makes large design challenges transparent and manageable. Design challenges are described as a set of decision areas, each with at least two mutually excluding decision options. Relations between decision areas and decision options can be formalized, creating transparency and insight in the direct and indirect consequences of one specific decision or a comprehensive set of decisions (scenario). In figure 1, three typical decision areas and their options are presented as an example. Two types of relations exist: exclusions between options of different decision areas and preclusions between a decision option and other decision area(s). These relationships can be definite, probable or possible.

The AIDA concept has been implemented in MEDIA, without any major conceptual changes. Based on several case studies, currently over 200 major decision areas have been distinguished for a typical urban redevelopment case. To maintain overview, the concept 'Theme' has been implemented in MEDIA. Themes are categories of decision areas related to the same topic.

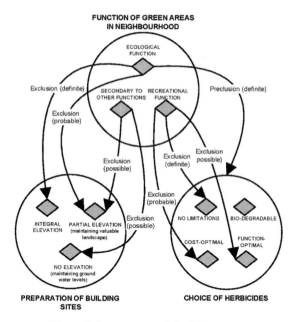

Figure 1: Interconnected decision areas

Addressing the equivocality of SD

The problem of the equivocality of SD is hard to solve conceptually. Any attempt to develop a generic set of indicators and calculation models can be regarded as 'yet another interpretation' of this concept. To deal with this problem, MEDIA has been designed as flexible as possible. MEDIA is not conceptually based on any single definition or model of sustainability, but structured in such a way that different definitions can be facilitated. The model can accommodate different types of data and models. In figure 2, this flexibility is presented with an example from practice. Direct impacts of options are expressed as descriptive variables, which can directly or through the use (complex) functions be used for defining SD-indicators.

This flexible structure may seem indecisive, but the benefits, no approaches are precluded or forced upon actors, and dialogue between actors is facilitated, outweigh the drawbacks, no 'easy' answers are provided.

Addressing policy complexity

Differences in actor perceptions are a main driver of policy complexity, and actor network analysis provides concepts to deal with this complexity: Actor perceptions are made explicit in a conceptual language, making different types of comparative analysis possible. By doing this, the analyst sharpens her insight not only in the policy situation at hand, but also in her own reasoning. In figure 3, the implementation of actor network concepts in MEDIA in presented, with an example.

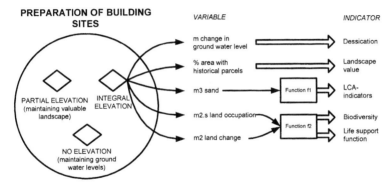

Figure 2   Flexible multimodel structure

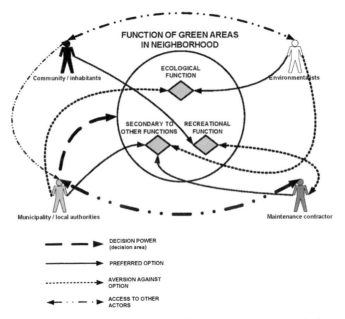

Figure 3   Conceptual foundations: dynamic actor analysis

Several functions in MEDIA represent the concept actor network analysis. First, each separate actor can be described. Second, the decision power of each actor can be formalized, as well as their preferences and aversions for separate options. Finally, the relations between actors can be formalized.

In figure 4 a screenshot of the most relevant menus of MEDIA is presented. Top left is the menu in which decision areas, decision options and variables (impacts) are defined. Top right shows the menu in which preclusions and exclusions are specified. The relations between decisions, options and actors are defined in the down left menu. In the final menu, the agenda is presented. The agenda is used in the gaming exercise, were it helps the users to focus on the most relevant decisions.

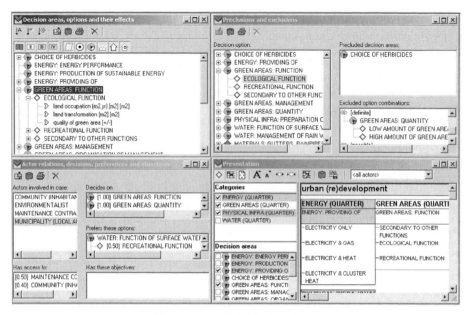

Figure 4   MEDIA interface

Results and outcomes

The results of working with MEDIA can be divided in two. First, the model has been used stand-alone for the modelling of a number of cases. Second, MEDIA has been used as a supportive tool in the gaming exercise. The results of these exercises are briefly discussed below.

MEDIA has been used for the modelling of a urban redevelopment project of a typical Dutch post-war residential area. The definition of decision areas, decision options, variables, preclusions and exclusion has turned out to be both possible and useful, since it forced the user to look at the project in a systematic and consistent manner. Altogether, about 200 decision areas have been distinguished. Most decision areas relate to the quarter and building-level. The decision areas are categorized into 30 themes. In addition to defining themes, decision areas and options, variables have been attributed to some of the decision options. Almost no values of variables (data) have been added to the database, as this proved to be very time consuming, since a lot of data is unavailable, or not available on the level of aggregation needed for MEDIA.

The implication of the limited amount of data is that MEDIA at the moment is not able to assess the SD-impacts of scenarios. More consistency and completeness was achieved in the definition of preclusions and exclusions. A total of 438 of preclusions and exclusions have been distinguished, mostly exclusions. As far as actors and their relationships concerned: because of the very case-specific character, this feature has not been the focal point of our efforts. The aggregation of decision areas and options to the level of a coherent design turned out to be possible, but still needs further development in order to be useful.

MEDIA has also been used as a supportive tool in gaming exercises. For this purpose, a paper version of the model, containing only themes, decision areas and options was used, as well as the agenda-function of the computer version. In this form, MEDIA served mainly a communicative, educative and informative purpose. No assessments could be performed. Since the gaming exercises were developed for the early stages of a project, in which there is less need for evaluation, this posed no real problems. The lack of assessment functionality was perceived by the players as a shortcoming of MEDIA, especially if the model is to be used in design stages.

The DuBes simulation game

Simulation games

A simulation game provides a safe environment, based on reality, in which the participants can experiment with decisions and negotiations [7]. The participants in a simulation game play various 'roles' that are derived from existing organisations and individuals. As in reality, the participants make decisions, form coalitions and make compromises based on their own goals and interests. A simulation is not primarily intended as a 'game' but rather as a serious, policy-oriented study, and is therefore also designated as a policy exercise [8]. Apart from being instructive, participation in a simulation game is simply enjoyable.

Urban restructuring in the game

DuBes is a method which can be applied to all types of construction projects. In the pilot project, MEDIA and DuBes have been developed for the restructuring of a post 1945 residential area with multi-storey apartment buildings. In Europe, there are some 80000 flats which need to be improved. The neighbourhoods in which these flats are located are characterised by problems of a physical and social-economic nature [9]. The DuBes simulation game can be played for neighbourhoods like these, real or fictitious ones.

In preparation for the game, the participants are sent the DuBes file, as it is known, containing the scenario. The scenario describes the urban planning history and characteristics of the municipality and the neighbourhood concerned, the housing stock and demographics of the neighbourhood and its problems with respect, for example, to housing, public space, water, energy, safety, traffic and transport. The DuBes file also contains an overview of the roles, the tasks and competencies of each role, and the role distribution. From 20 to 40 people can take part in each simulation. Prior to the simulation game, each participant is allotted a certain role, such as councillor, director of the housing association, director of the 'Welfare' foundation or project leader of the municipal departments for Planning and Housing, and Energy, Water and the Environment. In addition, the participants are asked to make use of their own knowledge and experience when playing the simulation [10].

The assignment for all participants in the simulation game is to jointly draw up a programme of requirements for the sustainable redevelopment of the neighbourhood concerned. The principal stakeholders, the municipality and the housing association owning the apartments in the neighbourhood, have decided to consult the principal interested parties in drawing up the programme of requirements. With a view to obtaining a subsidy, sustainability has to form an explicit component of the programme of requirements.

The DuBes method and tools

The project managers have a number of tools to help them achieve this difficult task in a short space of time. The most important tool is the DuBes Table. This table, in A0 poster format, schematically shows all the themes, decision fields and options from the MEDIA computer programme.

DuBes provides the participants with an overview of the decisions that can be made in the course of a restructuring assignment. This overview is not exhaustive. The aim of the table is to offer the participants a guide for determining the agenda of the programme of requirements. They can therefore use the agenda when the discussion falters or runs short of inspiration or expertise, but they must draw up their own agenda.

In two sessions, one in the morning and the other in the afternoon, the participants are split up into groups to draw up their own DuBes table for the fictional neighbourhood or for a real neighbourhood. The discussions and negotiations in the groups are guided by process managers. Process managers are game participants who are instructed by the game leaders, prior to the simulation game, on the best way that they can chair and guide the process within the working groups and planning studios.

In the morning, the participants are divided into three working groups. These work from various angles of approach, such as public space, housing and welfare, to list the themes and decision fields and decide which ones must be included in the programme of requirements. As they do this, they can refer to the existing DuBes table and the DuBes advisors for guidance. The DuBes advisors are members of the Sustainable Decision Making project team (DuBes team) and they assist and advise the working groups and register their results as well as possible in the MEDIA computer program. Before the working groups commence, the participants individually and anonymously prioritise all the decision fields in the DuBes table based on how important they think a theme or decision field is for the programme of requirements. An electronic conferencing system is used for this, so that the results of the vote are available to the working groups within a quarter of an hour. The result of the first session, just before lunch, is a DuBes table for each of the three working groups to serve as the basis for the programme of requirements.

During the lunch break, the DuBes team takes the three DuBes tables produced by the working groups in the morning session and, aided by MEDIA, processes and combines them into a single, integrated DuBes table. All the themes, decisive points and options that the participants consider important for the neighbourhood's programme of requirements are now put into order and placed on the agenda in this integrated DuBes table. However, no choices have been made or strategies decided upon as yet.

In session 2, in the afternoon, the participants are asked to work out strategies for the programme of requirements in three planning studios oriented toward different aspects of sustainability, such as the environment, quality of life and feasibility. This is done by choosing, in discussions, from the options that the participants drew up in the morning. The participants are asked to reason out their choices by devoting attention to the various effects of options, including those on sustainability, and the connection with other decisions, for example decisions relating to other themes and at other levels of scale. The chosen options are marked in the table and selected in MEDIA.

The planning studios are also guided by the process managers and the DuBes advisors. The MEDIA computer model registers the choices and gives extra information on the consequences of decisions, effects, consistency etc.

At the end of the day, the presentation, analysis and comparison of the three DuBes tables shows on which points the participants agree and on which points the participants have differences of opinion. On some points, the participants will have come to the conclusion that further research is necessary in order to arrive at the right choice. The DuBes tables, together with the arguments for the choices during the group discussions and evaluations, form the basis upon which a sustainable programme of requirements is drawn up for the fictional or real neighbourhood.

At the end of Session 1 and Session 2, the participants answer some questions about the process in their working group or planning studio, and about the result. This is done using an electronic conferencing system. At the end of the simulation game, the process, the result and the simulation game itself are thoroughly evaluated in a plenary discussion. Afterwards, the participants are asked to fill in an evaluation form about the simulation game.
The advantage of the game is from the point of view of the University two-fold: professionals can come together and learn in a quick way more about the possible interests, ideas and blockings of the different stakeholders, their positions, goals - open and hidden -, but as a bonus they exercise themselves in meeting techniques, disputing, and reasoning, selling and convincing. Because their performance is monitored by special assigned experts, they can be guided in their learning process.

Computer application

The computer application 'Crushing with a big Q' is a brand new digital system to support crushing plants to guarantee the quality of their products. The application is an intelligent system with 'thinking forms' and is installed on a central server. This means that central support, benchmarking and virtual auditing is possible. Meanwhile there have been requests from several countries to adapt the Dutch version of the computer application to the specific requirements of these countries. This application is not only a first for the recycling sector, not even for the building sector, but according to the suppliers of DSL and computer programs it is a world first in the field of guaranteeing product quality.

CONCLUSIONS

If the results of working with MEDIA are analysed, it can be concluded that the model with its current functionality facilitates the early stages in a project (initiative and programming) much better then later stages (design and realization). For these early stages the model can, in combination with the gaming exercise, play an important communicative, educative and informative role.

For the later stages of urban redevelopment projects the model currently lacks the finesse to model and visualize actual designs. Also, the lack of data and the subsequent inadequate assessment functions of the current model, limit its application to the early stages of projects. Future research is therefore expected to focus on these two main shortcomings

REFERENCES

1. GEURTS, J., et al. (1998). Gaming/simulation for policy development and organizational change. TUP, Tilburg.

2. WCED, World commission on Environment and Development (1987), Our common future, Oxford: Oxford University Press.

3. UNITES NATIONS (1992), Report of the United Nations Conference on Environment and Development, UNCED report A/Conf.151/5rev.1.

4. MARIN, B. et al. (1991) Policy Networks: Empirical Evidence and Theoretical Considerations. Boulder, Colorado: Westview Press.

5. BRUIJN DE, H. et al. (2001) Networks and Decision Making, Utrecht: Lemma publishers.

6. MORGAN, J.R. (1971) AIDA - A Technique for the Management of Design, Coventry: Tavistock Institute of Human Relations, Institute of Operational Research.

7. DUKE, R.D. (1980). A paradigm for game design. in: Simulation and Games, 11, pp. 364-377.

8. TOTH, F., (1988) Policy exercises, in Simulation and Gaming (19), pp. 235-276

9. EUROPEAN COMMISSION. (1996). European Sustainable Cities. Report of the Expert Group on the Urban Environment, Luxembourg: Office for Official EC-Publications.

10. VAN BUEREN, E., et al. (2002). Duurzaam Beslissen Spelsimulatie Alphen aan den Rijn. Verslag van de spelsimulatie op dinsdag 11 december 2001 te Delft. TBM, TU Delft, Delft.

LATE PAPER

THE ROLE OF CONCRETE POST KYOTO:
THE UNRECOGNIZED SEQUESTRATION SOLUTION?

J Harrison

TecEco Pty

Australia

ABSTRACT: The Kyoto treaty came in to force on the 16th February, 2005 and member nations are wondering how they can meet their objectives. This paper demonstrates the potential of the concrete industry to deliver a large proportion of the emissions reduction required. With production at over 14 billion tonnes concrete is the largest material flow on the planet. It is also already a very sustainable material with relatively low embodied energies. The challenge is to develop manufacturing technologies to capture chemically released CO_2 during manufacture and usage technologies that encouraged carbonation as a strength giving process. It points out that doing so is easier with TecEco kiln and binder technologies, and profitable, particularly under Kyoto.

Keywords: Built environment, Carbon credits, Economic, Emissions, Trading, Sequestration, mitigation, Abatement, Sustainability, CO_2

John Harrison has degrees in science and economics, is the managing director and chairman of TecEco Pty. Ltd. and is known around the world for the invention of tec, eco and enviro-cements. He is an authority on sustainable materials for the built environment, has been a speaker at many conferences and is committed to finding ways of "materially" improving the sustainability of the built environment.

INTRODUCTION

We have a major problem to address. Life on this planet may not be very tenable in a few hundred years if global warming occurs as predicted.

The first to seriously point out the ramifications of continued high use of Portland cement as it is formulated today and global warming was probably Prof. Joseph Davidovits in his paper published in World Resources Review in 1994 (Davidovits, J., 1994). His writings attracted the attention of Fred Pearce, a renowned writer with New Scientist Magazine who wrote an article in New Scientist with the attention getting title of "The Concrete Jungle Overheats"(Pearce, F., 1997). Fred's article caught the attention of the world and was one of the influences that inspired me to research more sustainable cement technologies. Later Fred was to publish the article "Green Foundations" about the ramifications of our eco-cement concretes for global warming (Pearce, F., 2002).

THE KYOTO TREATY

On the 16th February 2005 the 1997 Kyoto Protocol, drawn up in Kyoto, Japan in 1997 to implement the United Nations Framework Convention for Climate Change, finally became international law.

Signatory countries are legally bound to reduce worldwide emissions of six greenhouse gases (collectively) by an average of 5.2% below their 1990 levels by the period 2008-2012.

For the protocol to become law it needed to be ratified by countries accounting for at least 55% of 1990 carbon dioxide emissions. The key to ratification came when Russia, which accounted for 17% of 1990 emissions, signed up to the agreement on 5th November 2004. Ratification of the agreement means Kyoto will receive support from participating countries that emit 61.6% of carbon dioxide emissions.

Member countries have developed their own methods to meet targets. The EU for example has established quotas and a market to buy and sell credits. Unfortunately however some major emitters have not joined making it difficult for resident companies to trade their credits. The official view in the US and Australia is that it would ruin their economies. The Australian government has developed its own scheme called "The National Greenhouse Strategy" that will attempt to reduce emissions by only 10.1% by 2012, which is an 8% increase on 1990 levels. Only a week ago (April, 2005) the combined Australian states defied the Federal government and announced they were developing a scheme.

It will be a difficult task for most of the member countries to meet their Kyoto targets and already nations are falling behind. Spain and Portugal in the EU were 40.5% above 1990 levels in 2002. Canada, one of the first countries to sign, has increased emissions by 20% since 1990, and they have no clear plan to reach their target. The Japanese are also uncertain about how they will reach their 6% target by 2012.

CONCRETE - POTENTIALLY A VERY SUSTAINABLE MATERIAL

Contrary to lay understanding Portland cement concretes have low embodied energies and relatively high thermal capacity compared to other building materials such as aluminium and steel and are therefore relatively environmentally friendly.

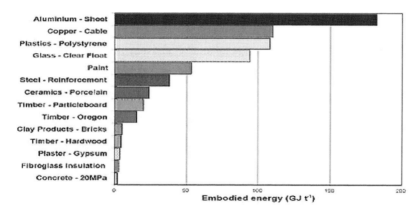

Figure 1 - Embodied Energy of Building Materials (Tucker, S., 2000)

WHAT MATTERS IS IMPACT

However concrete, based mainly on Portland cement clinker, is the most widely used material on Earth.

As of 2005 some 2.00 billion tonnes of Portland Cement (OPC) were produced globally (USGS, 2004) (see Figure 2), enough to produce over 7 cubic km of concrete per year or over two tonnes or one cubic metre per person on the planet.

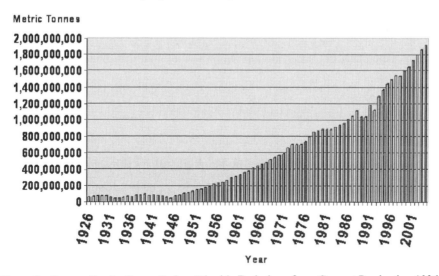

Figure 2 - Cement Production = Carbon Dioxide Emissions from Cement Production 1926-2002 (Van Oss, H., G., Hendriks, K. et al., 2003)

As a consequence of the huge volume of Portland cement manufactured, considerable energy is consumed (see Figure 3 - Embodied Energy in Buildings (Tucker, S., 2000)) resulting in CO_2 emissions. CO_2 is also released chemically from the calcination of limestone used in the manufacturing process.

Various figures are given in the literature for the intensity of carbon emission with Portland cement production and these range from 0.74 tonnes CO_2 / tonne cement (Hendriks, C. A., Worrell, E. et al., 2002) to as high as 1.24 tonne determined by researchers at the Oak Ridge National Laboratories (Wilson, A., 1993) and 1.30 tonne (Tucker, S., 2002). The figure of one tonne of carbon dioxide for every tonne of Portland cement manufactured (Pearce, F., 1997) given by New Scientist Magazine is generally accepted accept in the concrete industry.

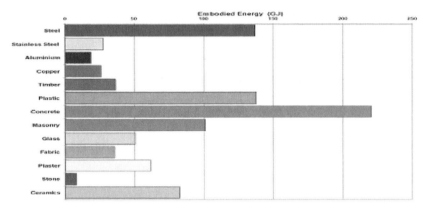

Figure 3 - Embodied Energy in Buildings (Tucker, S., 2000)

Because of the huge volume used, Portland cement concrete is the biggest single contributor to embodied energy in most buildings. As a consequence Portland cement concretes account for more embodied energy than any other material in the construction sector (Tucker, S., 2000).

The manufacture of Portland cement is one the biggest single contributors to the greenhouse effect after the burning of fossil fuels, accounting for between 5% (Hendriks, C. A., Worrell, E. et al., 2002) and 10% (Pearce, F., 1997) of global anthropogenic CO_2 emissions.

WHY CONCRETE IN THE BUILT ENVIRONMENT IS SO IMPORTANT?

When trying to solve a problem of immense proportions a good strategy is to first assess the solution, the means of applying it and the outcome from the effort of applying it. Put simply the input/outcome ratio or "bang for buck" is important for success. The obvious place that seems to have been missed by just about everybody to focus sustainability efforts is the built environment. It is our footprint on the globe. Given the size of the built environment there are huge opportunities for improving the techno-process and whilst doing so solving pollution and climate change problems.

"In 1999, construction activities contributed over 35% of total global CO_2 emissions - more than any other industrial activity. Mitigating and reducing the impacts contributed by these activities is a significant challenge for urban planners, designers, architects, engineers and the construction industry, especially in the context of population and urban growth, and the associated requirement for houses, offices, shops, factories and roads (UNEP, 2001)."

According to the Human Settlements Theme Report, State of the Environment Australia 2001(CSIRO, 2001), "Carbon dioxide (CO_2) emissions are highly correlated with the energy consumed in manufacturing building materials."

There is huge scope for emissions reduction and conversion of waste to resource given the massive size of the materials flows involved in the built environment of which concrete is around 30%. With the right materials and manufacturing technologies, because of its sheer size, concrete could be a big part of the solution to global warming and other environmental problems.

GREENING CONCRETES

The challenge is to reduce the impact of concrete production and usage and there are several ways this could be done.

1. Reduce chemical emissions during the manufacture of cements

2. Reduce the process emissions (resulting in embodied energy) during the manufacture of cements.

3. Change building technology and practice towards carbonating concretes.

4. Use carbon in concrete.

5. Introduce other properties to concretes that would result in reductions in the lifetime energies and emissions of structures.

6. Change concrete placement practices.

Reduce chemical emissions during the manufacture of cements

Alkali metal oxides are the basis of Portland, TecEco and most other hydraulic cements. Alkali metals like calcium and magnesium release a large amount of chemically bound CO_2 when their oxides are made from their carbonates. If this chemically released CO_2 could be captured during manufacture there would be significant net reductions in emissions.

The capture of CO_2 at source during the manufacturing process is easier and more efficient for the calcination of magnesium carbonates than any other carbonates mainly because the process occurs at relatively low temperatures.

TecEco Pty. Ltd. own intellectual property in relation to a new tec-kiln in which grinding and calcining[1] can occur at the same time in the same vessel for higher efficiencies (grinding releases heat) and CO_2 is easily captured. Geopolymers will be used to so it is easy to mass produce the new TecEco kiln and it will at first be used to calcine magnesite and make reactive magnesia. When materials improve sufficiently to withstand 1500 deg C the technology will be able to be used for the manufacture of Portland cement. Unfortunately considerable funds will be required to develop the technology beyond the resources of existing shareholders.

---

[1] Calcining in the context of this document refers to the heating of limestone or magnesite to drive off CO_2 and produce the oxide.

Reduce process emissions (resulting in embodied energy) during manufacture of cements

It is important to not just improve the efficiency of producing cement as is being done. Complete changes in the basis of energising the process are essential. A paradigm shift in technology is required.

As can be seen from Table 1, fossil fuel energy is only a very small fraction of total energy on the planet.

Table 1   Different Energy Fluxes on The Planet[2]

10^{13} WATTS	TOTAL	HEAT	WIND	EVAPORATION	PHOTOSYNTHESIS
Solar	12100	8000	37	4000	4
Earth Heat	3.2	-	-	-	-
Tidal	0.3	-	-	-	-
World Techno-Process Energy Demand	1.5	-	-	-	-

Cement could be made using non fossil fuel energy. There are several enticing possibilities – tapping deep geothermal energy or using direct or indirect energy from the sun.

TecEco's new kiln is shaped like a proboscis and designed to utilise direct solar energy in a solar concentrator, or waste heat. It can also be powered using electricity from intermittent sources such as wind or wave.

Change building technology and practice towards carbonating concretes

Carbonation is the basis of strength gain in carbonating lime concretes. Unfortunately insufficient strength gain occurs for use other than in mortars and renders. It is also well known that carbonation increases the strength of Portland cement concrete (PCC). Disadvantages of deliberately carbonating PCC however outweigh the advantages. The pH drops to below about 8.9 and steel reinforcing rusts. Besides, why go to all the trouble and cost of producing strong silicates only to carbonate them?

The alternative that works most efficiently is that proposed by TecEco. Reactive magnesia incorporated in hydraulic cement in porous materials will first hydrate and then carbonate producing large quantities of strength giving minerals. The main reason for the incredible efficiency of this process is the huge volume of binding material produced essentially out of water and CO_2.

---

[2] Estimated from various sources

Figure 4   The World's First Eco-Cement Porous Pavement,
Windsor Park, Glenorchy, Tasmania

When magnesia in eco-cements re-carbonates, more CO_2 is captured that in calcium systems because magnesium has a low molecular weight. There is more CO_2 content per tonne as in the calculations below.

$$\frac{CO_2}{MgCO_3} = \frac{44}{84} = 52\,\%$$

$$\frac{CO_2}{CaCO_3} = \frac{44}{101} = 43\,\%$$

Consider the volume changes that occur when magnesia hydrates to Brucite and then carbonates to nesquehonite:

$$MgO + H_2O \rightarrow Mg(OH)_2 + CO_2 + 2H_2O \rightarrow MgCO_3.3H_2O$$

$11.2 + (l) \rightarrow 24.3 + (g) + (l) \rightarrow 74.77$ molar volumes.

Overall, the molar volume expansion (11.2 to 74.77 molar volumes or 568%) is significant. Considerable quantities of binding material are produced! Absorbing CO_2 out of the air attracts carbon credits. The amount of water used will hardly impact on fresh water shortages. Kinetically the carbonation of eco-cements in the built environment will only proceed rapidly in porous materials such as bricks, block, pavers, mortars, porous pavement etc., which, fortunately, make up a large proportion of construction materials. All these materials, including mortars, require the use of appropriately course aggregates for carbonation to occur efficiently.

The effect of the substitution of Portland cement with MgO in a simple concrete brick formulation containing 15% cement with and without capture of CO_2 during manufacture of magnesia is depicted in Figure 5 below.

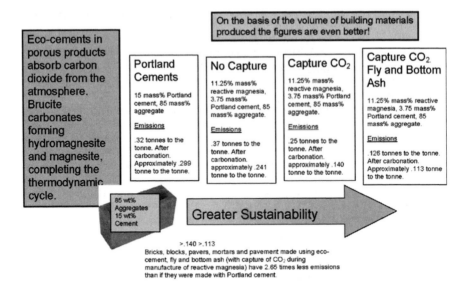

Figure 5 - Abatement in a Concrete Brick Containing 15% Eco-Cement

From Figure 5 it can be deduced that with the formulation specified almost 2/3 of the CO_2 that would have been produced is potentially abated if substituted by eco-cement in porous products such as bricks, blocks, pavers, mortars etc. With CO_2 capture or the inclusion of organic fibre materials and fillers for strength and insulation or both, cementitious building materials that act as net carbon sinks are feasible.

Use carbon in concrete

Not only could carbon be used as part of the binder system as in eco-cements discussed above, it could be used as a reinforcing element or aggregate.

As a reinforcing element

Carbon containing fibres including wood fibre, straw, sugarcane, bagasse, kenaf, hemp, guayule, shredded plastic, carpet clippings etc. have tensile strength and often go to waste. The use of fibre in concrete, and especially waste fibre is one way to add strength without using steel and other reinforcing elements that have high embodied energies and emissions. Apart from adding strength carbon is permanently encapsulated.

As an aggregate or filler

There are many carbon containing materials that go to waste that can add value to concretes as a composite because they have desirable physical properties. For example plastics are generally light in weight and have tensile strength.

With the introduction of robotics to construction all sorts of material composites that have the flow characteristics of concretes will be required. Incorporating carbon containing materials as aggregates would convert wastes to resources and also permanently encapsulate carbon.

Introduce other properties to concretes that would result in reductions in the lifetime energies and emissions of structures.

Concrete is a self setting solid. The self setting solids industry needs to evolve by vertical integration – to not just supply concrete but complete walling and flooring solutions[3]. To vertically integrate cost effectively the industry need to undertake proper market research and by doing so will discover that strength is not all that is demanded. Other properties in a material that transforms from easily placable to solid will be required including insulation, sound proofing and weight reductions. The introduction of robotics to construction will help drive this.

Architects are pushing the boundaries of design to reduce the lifetime energies of structures. As the physical properties of the materials they use have a strong influence on lifetime energies further improvements will mainly occur as a result of paradigm shifts in the materials technologies they use and if the concrete industry wish to remain in the mainstream, they will have to evolve to supply the properties required in settable solids. Such evolution also implies vertical integration and product differentiation with an increasing range and value of properties.

Change concrete placement practices

One of the major problems in the industry is the fact that finishers are getting lazier and want to do as little work as possible to place concrete. Because there is a direct relationship between the water/binder ratio and strength and they virtually always add water on site, a correspondingly higher percentage of cement is used than would be required if less water were added. All this extra cement used produces emissions.

I have been experimenting with Roman methods of placing concrete. The Romans tamped a dry concrete mix into place and the result was not only strength but extreme durability (Herring, B., 2002). The Pantheon for example was built in 118-35 AD, and still stands today as testament to the quality of their work. The nearest equivalent today is roller compacted concrete which is laid relatively dry.

My method is as follows and it works. Deliver concrete as a very dry mix similar to what is produced for making concrete blocks, move it about with bobcats, mini bob cats or "Kangas" as they are called in Australia and finally rakes. For a slab shave the surface level using long sharp cutting edges and then tamp down with a vibrating compactor. Finish by using a finishing slurry over the top to any colour and you have it – approximately 20 – 30% less cement and potentially lower cost. It takes not longer and costs less so why are we not doing it? Worse still why aren't people throwing the research dollar at the method?

Thinking Big in the Industry – Sequestration on a Massive Scale

TecEco are thinking big so why aren't the industry? We are seeking to mimic nature looking right up and down the supply chain to integrate the manufacture of self setting solids with the waste from other processes and utilise the huge flux of free energy on the planet.

---

[3] This would help get over a lot of other problems in the industry

Mineral Sequestration

The deposition of carbonate sediments is a slow process and involves long periods of time. Ways of accelerating sequestration using carbonates include geological sequestration and mineral carbonation. Although promoted by the petroleum industry as a means of extracting remaining reserves of oil "there are significant fundamental research needs that must be addressed before geologic formations can be widely used for carbon sequestration.(NETL, 2004)" Mineral carbonation, the reaction of CO2 with magnesium silicates as inputs such as peridotites, forsterite or serpentine, to form magnesite has been identified as a possible safe, long-term option for storing carbon dioxide by many authors (NETL, 2004), (Seifritz, W., 1990), (Lackner, K., Wendt, C. et al., 1995), (Morgantown, W. V.), (Dahlin, D. C., O'Connor, W. K. et al., 2000), (O'Connor, W. K., Dahlin, D. C., Nilsen, D. N., Walters, R. P., and Turner, P. C, 2000) (Fauth, D. J., Baltrus, J. P. et al., 2001).

Mineral sequestration is a process and number of universities and research organizations around the world are working on. Although there still some kinetic issues it is workable and potentially profitable under Kyoto. TecEco want to make it more profitable so it happens and ask - why waste the magnesite produced?

Making Mineral Sequestration More Feasible by Interfacing with Other Processes.

The tec-kiln technology previously mentioned provides a method of calcining the magnesium carbonate produced during mineral sequestration processes using solar derived intermittent energy or waste energy from other sources.

The magnesium oxide (MgO) produced can be used to directly sequester more CO_2 in a scrubbing process or to sequester carbon as hydrated magnesium carbonates in the built environment. The idea of capturing CO_2 as carbonate in the built environment mimics what has in fact naturally been occurring for millions of years[4]. Carbonates formed in seawater are the natural, large scale, long term sink for carbon dioxide, however the process takes over 1000 years to equilibrate. Good evidence of the enormous volumes of CO_2 that have been released from the interior of the earth during many volcanic episodes over the last few billion years is the high percentage (7%) of the earths surface covered in rocks such as limestone, dolomite and magnesite.

If carbon dioxide is captured during the calcining process for the manufacture of reactive magnesia then it can either be geologically sequestered or recaptured as eco-cements in porous cementitious materials for the built environment. Eco-cements gain strength with the formation of magnesium carbonates including lansfordite, nesquehonite and an amorphous phase mineral all of which because of their generally acicular shape add microstructural as well as innate strength as binders.

Oil has remained trapped in strata for millions of years and on this basis it is argued that carbon dioxide pumped down to push it up would also remain trapped. Given the fact that on average the pH of the earth is the same as that of seawater (8.2) then some neutralisation with the formation of immobile carbonates is also expected. A process diagram showing combined mineral and eco-cement sequestration is included as shown in Figure 6 below.

---

[4] There have been at least seven (7) other epochs of global warming easily discernable from the geological evidence.

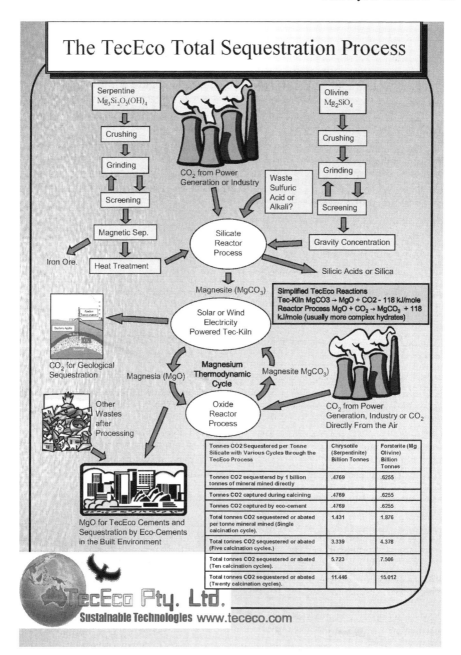

Figure 6 - The TecEco Process for Sequestration on a Massive Scale

The above explanation is simplistic. As the energy considerations are complex readers are directed to the web site of TecEco at www.tececo.com where further papers are available.

The only component of the entire process that is not proven is the TecEco kiln technology. It will take about half a million Australian dollars to do so but this is small change for saving the planet.

The Politics of a Massive Sequestration Process

To achieve the massive sequestration possible it will need the co-operation of governments, the power industry, the cement industry and the construction industry. The incentive is that it could be profitable. Besides, sequestration on a massive scale is far more politically acceptable than energy rationing, the only other alternative. Practically we should adopt a holistically integrated approach and also work on reducing emissions. The pain will however be a lot less if we can also sequester massive amounts of carbon.

The Economics of a Massive Sequestration Process

The cost of mineral sequestration even without the advantages added by TecEco technology is predicted to be quite reasonable.

Assuming thermodynamically efficient processes a cost including rock mining, crushing and milling, of around $ US 20/tonne of CO_2 are suggested. For a 66% efficient power plant this would add less than 1 cent to the cost of a kilowatt hour of electricity (Yegulalp, T. M., Lackner, K. S. et al., 2001).

Should a price of carbon under the now legally binding Kyoto protocol be anything over $ US 20/tonne of CO_2, the process would be economic once kinetic issues for chemically processing magnesium silicates are overcome.

A combined process involving TecEco tec-kiln technology would sequester several times more carbon and involve the eventual production of eco-cement concrete components that also sequester carbon to create the built environment and is therefore potentially very profitable even if an even lower price for carbon of say less than $ 10.

The Kyoto protocol will also encourage the development of other technologies whereby CO_2 becomes a resource and as a result the process will eventually be supported by an economically driven price for CO_2 The use of TecEco eco-cements concretes would also be favoured as magnesite, the raw material, would not have to be mined. An added advantage would be the permanent disposal of carbon dioxide with no possibility for leakage.

REFERENCES

CSIRO (2001). Human Settlements Theme Report, State of the Environment Australia 2001, Australian Government Department of Environment and Heritage,

DAHLIN, D. C., O'CONNOR, W. K., ET AL. (2000). A method for permanent CO2 sequestration: Supercritical CO2 mineral carbonation, 17th Annual International Pittsburgh Coal Conference.

DAVIDOVITS, J. (1994). "Global Warming Impacts on the Cement and Aggregates Industries." World Resources Review 6(2): 263-278.

FAUTH, D. J., BALTRUS, J. P., ET AL. (2001). "Carbon Storage and Sequestration as Mineral

CARBONATES." Prepr. Symp. Am. Chem. Soc., Div. Fuel Chem 46(1): 278.

HENDRIKS, C. A., WORRELL, E., ET AL. (2002). Emission Reduction of Greenhouse Gases from the Cement Industry, International Energy Agency (IEA),

HERRING, B. (2002). The Secrets of Roman Concrete, Constructor. Virginia, Associated General Conractors of America (AGC),

KIMBALL, J. W. (2004). Kimball's Biology Pages, Secondary Kimball's Biology Pages. Secondary Kimball, J. W., Place Published, John W Kimball, http://users.rcn.com/jkimball.ma.ultranet/BiologyPages/C/CarbonCycle.html,

LACKNER, K., WENDT, C., et al. (1995). "Energy." 20: 1153 - 1170.

MORGANTOWN, W. V., Secondary. Secondary Morgantown, W. V., Place Published, http://www.fetc.doe.gov/products/ggc,

NETL (2004). Chemical and Geologic Sequestration of Carbon Dioxide, Secondary Chemical and Geologic Sequestration of Carbon Dioxide. Secondary NETL, Place Published, National Energy Technology Laboratory (NETL), http://www.netl.doe.gov/products/r&d/annual_reports/2001/cgscdfy01.pdf,

O'CONNOR, W. K., DAHLIN, D. C., NILSEN, D. N., WALTERS, R. P., AND TURNER, P. C (2000). "Carbon Dioxide Sequestration by Direct Mineral Carbonation:

Results from Recent Studies and Current Status." Proceedings of the 25th International Technical Conference on Coal Utilization & Fuel Systems: 153 - 164.

PEARCE, F. (1997). "The Concrete Jungle Overheats." New Scientist(2097): 14.

PEARCE, F. (2002). "Green Foundations." New Scientist 175(2351): 39-40.

SEIFRITZ, W. (1990). Nature 345, 486.

TUCKER, S. (2000). CSIRO on line brochure, Secondary CSIRO on line brochure. Secondary Tucker, S., Place Published, http://www.dbce.csiro.au/ind-serv/brochures/embodied/embodied.htm,

TUCKER, S. (2002). CSIRO Department of Building Construction and Engineering. J. Harrison, pers comm.

UNEP (2001). Energy and Cities: Sustainable Building and Construction Summary of Main Issues, IETC Side Event at UNEP Governing Council, Nairobi, Kenya, UNEP.

USGS (2004). "Mineral Commodity Summary - Cement." (2004).

VAN OSS, H., G., HENDRIKS, K., ET AL. (2003). USGS cement XLS file. Data,

WILSON, A. (1993). "Cement and Concrete: Environmental Considerations." Environmental Building News 2(2).

YEGULALP, T. M., LACKNER, K. S., ET AL. (2001). "A Review of Emerging Technologies for Sustainable Use of Coal for Power Generation." The International Journal of Surface Mining. Reclamation and Environment 15(52 - 68): 58.

INDEX OF AUTHORS

SUBJECT INDEX

This index has been compiled from the keywords assigned to the papers, edited and extended as appropriate. The page references are to the first page of the relevant paper.